THE OXFORD
BOOK OF
OXFORD

THE OXFORD
BOOK OF
OXFORD

CHOSEN AND EDITED BY

JAN MORRIS

Oxford   New York

OXFORD UNIVERSITY PRESS

Oxford University Press, Walton Street, Oxford OX2 6DP

Oxford New York Toronto
Delhi Bombay Calcutta Madras Karachi
Kuala Lumpur Singapore Hong Kong Tokyo
Nairobi Dar es Salaam Cape Town
Melbourne Auckland

and associated companies in
Beirut Berlin Ibadan Mexico City Nicosia

Oxford is a trade mark of Oxford University Press

ISBN 0-19-281424-9

First published 1978
First issued as an Oxford University Press paperback 1984
Reprinted 1985

British Library Cataloguing in Publication Data

The Oxford book of Oxford.
1. University of Oxford – History – Addresses,
essays, lectures
I. Morris, Jan
378.425'74   LF509   77-80477
ISBN 0-19-214104-X

Printed in Great Britain by
J. W. Arrowsmith Ltd., Bristol

DEDICATED GRATEFULLY TO
THE WARDEN AND FELLOWS OF
ST. ANTONY'S COLLEGE
OXFORD

EXCEPT ONE

Contents

Illustrations

All the illustrations, excluding the map,
originally appeared on Oxford Almanacks

[viii]

Explanations

Never explain (counselled Dr. Jowett, page 275, who had himself inherited the advice, I believe, from some yet earlier sage)—'get it done and let them howl'. Still, having got this anthology of Oxford University life done, I must allow myself a few lines of halting explanation.

Like all Oxford anthologies, it does not do the place justice. This is because the most interesting things that happen in Oxford are not specifically Oxford matters, and so disqualify themselves from such a collection: they are the universal events of learning and education which provide the true theme of Oxford academic history, but which express themselves in scholarly books, in careers, and in the affairs of the wider world. The Oxford that we read about in letters, diaries, novels, poems, and memoirs, the Oxford that is the stuff of this volume, is really only the setting: the reader must anthologize for himself, in his mind's collection, all the scholarship and creative labour that is the real focus of the scene.

Then again, an anthology is necessarily a personal affair. In the general interest I have restrained my preferences for swagger, sentimental Victorian verse, and *lists*: on the other hand I have indulged to the full a taste for the odd character and the oft-told tale—if there is any dear old Oxford chestnut that is missing from these pages, it is only through inadvertence (or occasionally disbelief). I have tried to take my magnificent subject with proper seriousness, but as one Oxonian said to another long ago, cheerfulness keeps breaking in.

I expect the book contains many mistakes. If so, it is probably because, generally in the amusement of the moment, I have failed to heed the injunction of another Oxford oracle (Dr. Routh, page 236): *Verify your references.*

❧ UNIVERSITAS OXONIENSIS

There is nothing like Oxford.
Francis Kilvert, 1876.

Rare, precarious, eccentric, and darkling.
Ralph Waldo Emerson, 1847.

Virtue's awful throne!
Wisdom's immortal source!
Thomas Warton, 1751.

That fair city wherein make abode
So many learned imps that shoot abroad
And with their branches spread all Britainy. . . .
Edmund Spenser, 1596.

To the university of Oxford I acknowledge no obligation, and she will as cheerfully renounce me for a son, as I am willing to disclaim her for a mother.
Edward Gibbon, 1792.

Of all the months in my life those which were passed at Oxford were the most unprofitable.
Robert Southey, 1849.

I have been at Oxford; how could you possibly leave it? After seeing that charming place, I can hardly ask you to come to Cambridge.
Horace Walpole, 1736.

The world, surely, has not another place like Oxford: it is a despair to see such a place and ever to leave it.
Nathaniel Hawthorne, 1856.

[x]

DOMINUS ILLUMINATIO MEA

I tried to work myself up to a little enthusiasm, and took a draught of the water of Isis so much celebrated in poetry, but all in vain.

James Boswell, 1763.

I was a modest, good-humoured boy. It is Oxford that has made me insufferable.

Max Beerbohm, 1899.

The finest thing in England.

Henry James, 1909.

The finest City in the world.

John Keats, 1817.

Oxford, ancient Mother! I owe thee nothing!

Thomas de Quincey, 1803.

Le rheumatisme vert.

Alphonse Daudet, 1895.

You think they would advertise this place, to let people know it was on the map.

Mr. Laurel to Mr. Hardy, *A Chump at Oxford*, 1940.

Suffice it to say of Oxford what Pomponius Mela says of Athens, 'It is too well known to be pointed out.'

William Camden, 1600.

Mighty fine place.

Samuel Pepys, 1668.

Poor dear old Oxford!

Thomas Arnold, 1819.

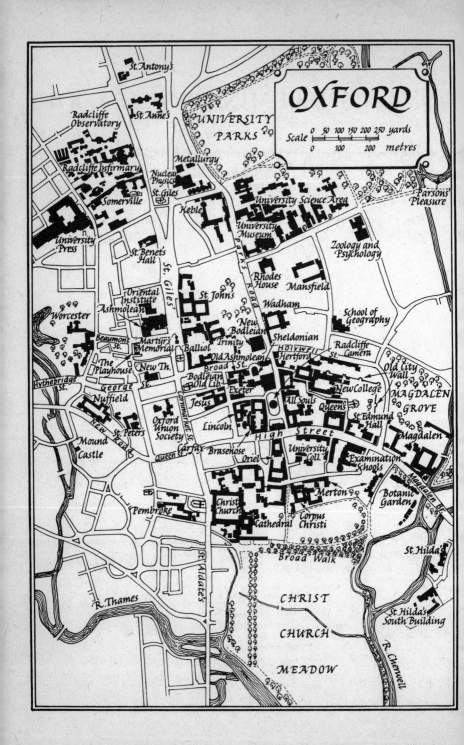

The Learned Imps
1200 – 1558

The 'learned imps of Oxford', as Edmund Spenser called them in The Faerie Queene, *spent their first centuries in tumult and uncertainty, as their University evolved a structure and established a place in English society. From the Middle Ages to the Reformation, intellectual life was dominated by often conflicting religious dogma, and the University was repeatedly at odds with the power of the State, and almost always at loggerheads with the Oxford townspeople.*

FOUNDING FATHERS

🎋 *Physical discomfort, the ancients fondly reasoned, was good for the mind, which is perhaps why the University of Oxford came into being at that catarrhal point of the English Midlands where the rivers Thames and Cherwell soggily conjoin. Nobody really knows, though, when it occurred. There had been a Saxon settlement around the river ford; the Normans built a castle there; two monasteries flourished, and a Jewry; medieval Oxford was a prosperous enough market town; but the University, which was to make it celebrated throughout the world, seems to have started fitfully, almost imperceptibly. It was left to enthusiastic antiquarians of later times to give it heroic or even Biblical origins, and here are some of their more ebullient inventions:*

About this time Samuel the servant of God was Judge in Judea, and King Magdan had two sons, that is to say Mempricius and Malun. The younger of the two having been treacherously slain by the elder, the fratricide inherited the kingdom. In the twentieth year of his reign, he was surrounded by a large pack of very savage wolves, and being torn and devoured by them, ended his existence in a horrible manner. Nothing good is related of King Mempricius except that he begat an honest son and heir by name Ebrancus, and built one noble city which he called from his own name *Caer-Memre*, but which afterwards, in course of time, was called *Bellisitum*, then *Caerbossa*, at length *Ridohen*, and last of all *Oxonia*, or by the Saxons *Oxenfordia*, from a certain egress out of a neighbouring ford; which name it bears to the present day. There arose here in after years an universal and noble seat of learning.

John Rous, *Historia Regum Angliae, c.* 1490.
(from the Latin)

The University of Oxford is found to be earlier as to foundation, more general in the number of sciences taught, firmer in the possession of Catholic truth, and more distinguished for the multitude of its privileges, than all other *Studia* now existing among the Latins. Very ancient British Histories imply the priority of its foundation, for it is related that amongst the warlike Trojans, when with their leader Brutus they triumphantly seized upon the island, then called Albion, next Britain, and lastly England, certain Philosophers came and chose a suitable place of habitation in this island. . . .

Not far from this it is known that the town of Oxford is situated, which because of the pleasantness of the rivers, meadows and woods adjoining it, antiquity formerly named Bellesitum; afterwards the Saxon people named it Oxford from a certain neighbouring ford so called, and selected it as a place of study.

From the Latin of the Oxford *Historiola, c.* 1375.

Others affirm the University of *Oxford* to have been first founded by *Arviragus*, a *British* King, according to *Juvenal*, under the reign of *Domition*, about 70 years after Our Lord's Incarnation; and that it was afterwards reduced to a Form of Government, by the Care and Policy of St. *Germain*, Bishop of *Auxerre* in *France*, who with *Lapus* Bishop of *Troyes* in that Realm, came into *Britain* to the Assistance of the Christians, in order to compose the new Divisions in the Church, arising by the means of *Agricola*, a Disciple to *Pelagius*, the Monk of *Bangor* in *Flintshire*, who had propagated his Heresy here, to the great Disturbance of the weaker Christians, not able to withstand his Errors.

John Ayliffe, *The Antient and Present State of the University of Oxford*, 1714.

In the year of our Lord 886, the second year of the arrival of St. Grimbald in England, the University of Oxford was begun; the first who presided and read divinity lectures in it being St. Neot, an abbot and able divine, and St. Grimbald, a most eminent professor of the incomparable sweetness of the sacred pages; Asser, the monk, an excellent scholar, professing grammar and rhetoric; John, monk of the church of St. David, giving lectures in logic, music and arithmetic; and John, the monk, colleague of St. Grimbald, a man of great parts and a universal scholar, teaching geometry and astronomy before the most glorious and invincible King Alfred, whose memory will dwell like honey in the mouths of all both clerics and people of his whole Kingdom.

William Camden, *Britannia*, 1586 (from the Latin).

At the first foundation of this University this noble King Alfred established at his own expense within the city of Oxford three Doctors, namely in Grammar, in the Arts, and in Theology, in three different places in the name of the Holy Trinity. In one of these which was situated in High Street towards the East gate he endowed the hall with all that was necessary for twenty-six grammarians. . . . Towards the northern walls of the city in what is now called Vicus Scholarum he founded another Hall with abundance of means necessary for twenty-six Logicians or Philosophers. . . . The third Hall which he founded in High Street, near East gate, and arranged for twenty-six Theologians who should promote the study of Holy Scripture. . . .

Besides these there grew up in short time many other Halls of the different faculties, established by the burgesses of the city and of the neighbourhood and then by those from a distance; yet not at the King's expense, but through the King's gracious example.

John Rous, *Historia Regum Angliae, c.* 1490 (from the Latin).

King Alfred sent for one Grymbald out of France, that by his great learning and method in reading, his scholars of Oxford might to the best advantage be taught. . . . Several others he sent for afterwards ('tis said) from other countries, but what their names were I am not yet certain. These came from Ireland an. 891 in a little boat made of two hides and a half of leather, but without guide or sail, yet they had provision with them sufficient for seven days. At length they set foot on English ground in Cornwall, and were known by the names of Dubslaw, Maecbeth and Maeline. . . .

Anthony Wood, *History and Antiquities of the University of Oxford*, 1674.

'But for my part,' adds Anthony Wood, 'I scarce believe it.'

THE FIRST REPORTER

However it happened, by the end of the twelfth century Oxford was a well-known place of scholarship—we hear of a scholar from Normandy in 1117, a lawyer from Lombardy in the middle of the century. Senatus, prior of Worcester, described Oxford in about 1190 as

abounding in men skilled in mystic eloquence, weighing the words of the law, bringing forth from their treasures things new and old.

The earliest extant eye-witness report of an Oxford academic event was written in Latin by Giraldus Cambrensis, the Norman-Welsh scholar and divine. Though he wrote it fastidiously in the third person (in his auto-biographical De Rebus A Se Gestis, *1204), the central figure in the event was Giraldus himself, never one for false modesty. The year was 1200, and he had just completed his book* The Topography of Ireland:

In course of time, when the work was completed and corrected, desiring not to hide his candle under a bushel, but to place it on a candlestick so that it might give light, he resolved to read his work at Oxford, where the clergy in England flourished and excelled in clerkship, before that great audience. And as there were three divisions in his work, and each division occupied a day, the reading lasted three successive days. And on the first day he received at his lodgings all the poor scholars of the whole town; on the second all the Doctors of different Faculties, and such of their pupils as were of greater fame and note; on the third the rest of the scholars with many knights, townsfolk, and burghers. It was a costly and noble act, for the authentic and ancient times of the poets were thus renewed, nor does the present or any past age recall anything like it in England.

STUDIUM GENERALE

🦃 *Boosted perhaps by a migration of scholars from Paris, presently there existed in Oxford a* studium generale, *a fully fledged medieval university. Its head was a Chancellor, its headquarters the church of St. Mary in High Street, its lingua franca Latin, and its purpose essentially the education of clerics. This was its course of studies:*

The Seven Liberal Arts
Grammar; Logic; Rhetoric; Music; Arithmetic; Geometry;
Astronomy
The Three Philosophies
Moral; Metaphysical; Natural
The Two Tongues
Greek; Hebrew

This seven-year course led to the degree of Master of Arts. The student could then proceed to the higher faculties of Law, Medicine, or the supreme study of all, Theology, often spending thirteen years along the way. Here are some examples of medieval academic usages, of one sort and another:

HEADINGS OF A TREATISE ON NATURAL MOTIONS:

Definition of Motion and Time
Unaccustomed, though not original, discussion of generation in simple elements and mixtures.
Discussion of Alteration.
Discussion of Augmentation and Diminution.
Discussion of Locomotion.
On the Possible Proportions of Velocities in Moving Bodies.
On Distinguishing Maximum from Minimum.

R. Swyneshed, *Descriptiones Motuum, c.* 1350.

A LIST OF ORNAMENTS THAT MAY PROPERLY BE
EMPLOYED IN SERMONS:

Invention of theme, winning over the audience, prayer, introduction, division, statement of parts, proof of parts, amplification, digression or transition, correspondence, agreement of correspondence, circuitous development, convolution, unification, conclusion, coloration, modulation of voice and weighing of subject matter.

Robert of Basevon, *Forma Praedicandi,* 1322 (from the Latin).

STUDIUM GENERALE

Socrates makes the following statement and no other: 'Socrates is telling a lie.'

A SUBJECT FOR DEBATE:

Whether more than one angel can occupy one place at the same time.

A MULTIPLICATION PROBLEM:

Multiply CCCXLVIII × VI
Solution: CCC × VI = MDCCC
XL × VI =   CCXL
VIII × VI =   XLVIII
_____
MMLXXXVIII

A DISPUTATION:

Opponent: What think you of this question, whether universal ideas are formed by abstraction?

Respondent: I affirm it.

Opponent: Universal ideas are not formed by abstraction, therefore you are deceived.

Respondent: I deny the antecedent.

Opponent: I prove the antecedent. Whatever is formed by sensation alone, is not formed by abstraction. But universal ideas are not formed by sensation alone: therefore universal ideas are formed by abstraction.

Respondent: I deny the minor.

Opponent: I prove the minor. The idea of solidity is an universal idea; but the idea of solidity is formed by sensation alone: therefore universal ideas are formed by sensation alone.

Respondent: I deny the major.

Opponent: I prove the major. The idea of solidity is an universal idea; but the idea of solidity arises from the collision of two solid bodies: therefore the idea of solidity is formed by sensation alone.

A CHOPPING OF LOGIC:

Three Oxford academics were deputed to wait upon Henry III in 1266 to ask permission for a postern gate through the city wall at Oxford. The King (in Latin) asked them what they wanted:

First scholar: We ask a licence for the making of a gate through the city wall.

Second scholar: No, we do not want the *making of a gate*, for that would mean the gate was always in the making, and never made. What we want is a *gate made*.

Third scholar: No, we do not want a *gate made*, for a gate made must already be in existence somewhere else, and so we should be taking somebody else's gate.

🦎 *The King told them to go away and make up their minds. When they returned in three days' time they had agreed upon a formula:*

We ask permission that the making of a gate might be made. [*Ostium fieri in facto esse*].

🦎 *Permission was granted.*

ADVICE FROM A TUTOR:

Study as if you were to live for ever; live as if you were to die to-morrow.

> Precept given to his pupils by the first recorded M.A. of Oxford University, Edmund of Abingdon (d. 1240).

THE GRETE CLERK

🦎 *The University's first Chancellor was Robert Grosseteste (1175–1253), whose name was variously spelt:*

Grosseteste; Grostest; Grostet; Grosthead; Grouthead; Grostede; Greatheade; Grostheved; Greatheved; Grosehede; Grokede; Groschede.

🦎 *Educated himself in Paris, he was a man of immense intellect and wide interests. Here are three of his theorems:*

Form is the essence of a thing or that which makes it to be what it is.

All bodies originate from light.

The functions of the angel are *knowing* and *willing*.

Matthew Paris (d. 1259) described him [in Latin] as

🦎 . . . the blamer of prelates, the corrector of monks, the director of priests, the instructor of clerks, the support of scholars, the preacher to the people, the persecutor of the incontinent, the sedulous student of all scripture. . . .

THE SCHOLAR'S LIFE

A chambre hadde he in that hostelrye
Allone, withouten any compaignye,
Ful fetisly ydight with herbes swoote;
And he hymself as sweete as is the roote

Of lycorys, or any cetewale.
His Almageste, and bookes grete and smale,
His astrelabie, longynge for his art,
His augrym stones layen faire apart,
On shelves couched at his beddes heed;
His presse ycovered with a faldyng reed;
And al above ther lay a gay sautrie,
On which he made a-nyghtes melodie
So swetely that all the chambre rong;
And *Angelus ad virginem* he song;
And after that he song the Kynges Noote.
Ful often blessed was his myrie throte.
And thus this sweete clerk his tyme spente
After his freendes fyndyng and his rente.

Geoffrey Chaucer's description of the Oxford scholar in The Canterbury Tales (c. 1390) *contains the fullest surviving inventory of a medieval student's room at Oxford, with its calculating devices (augrym stones), its deodorizing herbs, its musical instrument, and its bookshelf. From medieval records a reconstruction has also been made of the rooms of an Oxford Master, William Sprig, principal of an academic lodging-house called Broadgate's Hall:*

It is a small, sparsely furnished room, with no floor covering. Hanging on the wall is a bow, with twenty arrows hanging nearby; also a sword, two daggers, an axe, an old guitar and a lute. On the floor there is a locked chest, and Master Sprig has a chair for himself, and a bench for his pupils. There is a three-legged table, a reading desk, and on a shelf there are some wooden dishes, a pestle and mortar and a glass pitcher—for the Principal does some of his own cooking. By the fireplace there stands a pair of bellows: on the table there are perhaps a dozen manuscript books. Next door is Master Sprig's bed-room, and there his clothes are hanging on the wall—a doublet of white cloth, a white cloak, a red fur-trimmed gown, a red hood, a tabard. On his bed are four heavy blankets, very expensive—20 pence each—and his bedcover is gaily coloured, green, blue, red, or possibly 'flowryd', for two characteristics of the medieval Oxford academic are a taste for music and a love of colour.

At first most of the students lived where and how they pleased; but in 1231 Henry III ordered that every student must have his name on the roll of a master, and in 1410 it was decreed that all students must live in recognized halls of residence. These were scattered all over Oxford, and were often agreeably named:

Worm Hall	Woodcock Hall	Beefe Hall
Cat Hall	Nightingale Hall	Perilous Hall
Bull Hall	Eagle Hall	Kepeharm Hall
Hare Hall	Sparrow Hall	Aristotle Hall
Unicorn Hall	Hawk Hall	Nun's Hall
	Ape Hall	Pie Hall

✺ *Many students, some masters too, were impecunious. A sample letter drawn up by an Oxford writing-master, Thomas Sampson, in about 1380, and translated here from the Latin, is an early example of a familiar* genre:

I know not what to offer you, my sweet father, since I am your son, and after God, entirely your own creature—so completely yours that I can give you nothing. But if I can remember what the child's instinct prompts it to say, I might sing, as the cuckoo incessantly sings, 'Da, da, da, da': and this little song I am compelled to sing at this moment, for the money which you gave me so liberally for my studies last time is now all spent, and I am in debt to the tune of more than five shillings. . . .

✺ *To help such poor students university chests were established—strongboxes, generally financed by benefactors, from which loans could be drawn in return for securities deposited. The antiquarian Henry Anstey, in his* Munimenta Academica, 1868, *painstakingly reconstructed the events of a loan day in St. Mary's Church in about 1400, and from his researches this picture is drawn:*

Master T. Parys, Master of St. Mary's Hall, and his fellow-keepers of the chest, their heads tonsured, are wearing belted tunics, long over-gowns, black sleeveless capes and hoods lined with white fur. They lead the way into the Congregation House. The room is lined with some twenty heavy iron-bound chests, eight or ten feet long, perhaps five feet wide. This morning Selton's Chest is to be opened. Each keeper draws from beneath his cape a huge key, and in turn they are applied to the several locks of the chest. The lid opens. Inside there is a register of contents, and a variety of treasures: on the one side, securities—manuscript volumes, elaborately worked daggers, silver cups, hoods lined with miniver: on the other, the stock of money of the benefaction.

A small crowd has gathered round the chest. There are several masters and bachelors, but most of the onlookers are boys or very young men, dressed not in academic clothes, but in bright blues and reds and yellows—some smart, some in tatters. Their eyes are riveted on the chest. Some want to raise a loan, some are in funds now, and want to redeem their deposits—a cup pledged to cover the rent of a garret, a manuscript to pay for food bills.

Among them we see one man of evidently grander bearing, Master Henry Sever, warden of Merton Hall. The students eye him with resentment, for they think it unfair that he should borrow money from the chest, and he looks worried too. To cover the cost of some building repairs, he pledged a valuable illuminated missal. Since he failed to redeem it at the proper time, it has been taken away for approval by a possible purchaser, who in his turn has deposited a silver cup. Master Sever, who does not want the cup but badly wants his missal back, is persuaded, as he repays his loan, that he can exchange the cup for the missal: and so he walks off with it, while the less important clients cluster around the open chest.

Some redeem their pledges; some borrow more cash; some are new applicants, and have come to deposit their own books, trinkets or weapons. Presently all are dealt with. The great chest is locked again. The keepers put their keys away. 'Ye shall pray', says Master Parys, addressing the borrowers, 'for the soul of W. de Seltone and all the faithful departed.'

Business is concluded.

Despite such help the Oxford student-beggar, extemporizing poems at the doors of houses in return for food or money, was a familiar figure of the Middle Ages. When Sir Thomas More fell from power in 1534, he told his children that if the worst came to the worst they could always go to Oxford:

Then may we yet like poor scholars of Oxford go a begging with our bags and wallets and sing Salve Regina at rich mens dores.

WAR AMONG THE NATIONS

Oxford students were frequently riotous, and in particular there were bloody rivalries between the so-called Nations—Northern English, Southern English, Scots, Irish, and Welsh. For example:

[About 1258] fell out a sad dissension between the Scholars of divers Nations studying in the University, namely between the Scotch, Welsh, Northern and Southern English. The Northern and Welsh joined together against the Southern, and had banners and flags among them to distinguish each division. They also pitched their field near

Oxford . . . where each party trying their valour fell together in such a confusion with their warlike array, that in conclusion divers on both sides were slain and pitifully wounded.

This bloody Conflict during among them for some time, the event thereof was this, that the Northern Scholars with the Welsh had with much ado the victory, and were saluted by the names of Conquerors, while the other party withdrew themselves and comforted each other in their losses.

After this combustion was past and had done what it could, the Victors considered with themselves what they had done, and how in the meantime their actions would be relished by the King, and fearing also that severe punishment would fall upon them, especially forasmuch as the brother of Leolin, Prince of Wales, son of Griffyn [Llewellyn the Last] was newly deceased in prison, they summoned their council and assistance together, and upon advice had among them offer to the King 4000 marks, to Edward his son 300 marks, and to the Queen 200, to be released of their trespasses and faults committed; but the King at that time gave answer to them, that he set more value on the life of one trusty subject than on the money which they had offered. Upon which answer the Scholars without hope of reconciliation retired home with little satisfaction, notwithstanding the King was then involved in great affairs and wars, partly with Leoline and his Welsh men, and partly at home with his Nobles. Which broils diverting the King, could not (as 'twas thought) be at leisure to attend the punishment of the scholars.

🐾 *But the Welsh did not always get away with it:*

On Thursday in the fourth week of Lent [1389] Thomas Speese Chaplain and John Kirkby with a multitude of other malefactors, appointing Captains among them, rose up against the peace of the King and fought after all the Welshmen abiding and studying in Oxford, shooting arrows before them in divers streets and lanes as they went, crying out 'War, war, war, sle, sle, sle the Walsh doggys and her whelyps, and ho so loketh out of his howese, he shall in good soute be dead &c.' and certain persons they slew and others they grievously wounded, and some of the Welshmen who bowed their knees to abjure the town, they the Northern Scholars led to the gates, causing them first to piss on them, and then to kiss the place on which they had pissed. But being not content with that, they, while the said Welshmen stooped to kiss it, would knock their heads against the gates in such an inhuman manner, that they would force blood out of the noses of some, and tears from the eyes of others.

Anthony Wood, History and Antiquities of the University of Oxford, 1674.

THE FRIARS

❧ *Under the patronage of Robert Grosseteste, the first Chancellor, hardly had the University been established when the mendicant friars, Dominican, Franciscan, Carmelite, Augustinian, set up houses in Oxford and became a powerful presence there. They came from many parts of Europe to teach and to study, as these recorded names of Dominicans in Oxford show:*

Hugh of Asio, 1273, province of Provence.
Franciscus de Canalibus, 1330, province of Aragon.
Arnold of Perugia, 1339, Roman province.
Jan Moravecz, before 1348, Bohemian province.
Vincent of Lisbon, 1376, Spanish province.
William of Reymuda, 1398, German province.
Henry Gerborare, 1421, Polish province.
John Bernardi of Lubeck, 1477, province of Saxony.

❧ *They also included some of the greatest medieval scholars, whose European reputations were sustained by respectful nicknames:*

Adam Marsh: The Illustrious Doctor.
John Duns Scotus: The Subtle Doctor.
William of Ockham: The Invincible Doctor.
Roger Bacon: The Marvellous Doctor.

❧ *So extraordinary was the most remarkable of these men, Roger Bacon, who foresaw gunpowder and flying machines, and pointed the way to Copernican astronomy, that he was popularly suspected of necromancy and association with the devil. As the seventeenth-century antiquarian Anthony Wood wrote*:

For twenty years space, in which he had laboured in the study of wisdom (not without the vile censure of the vulgar) he had spent more than two thousand pounds in obtaining books of Secrets, making various Experiments, and in instructing of his helpers and such like, mostly done at Oxford to the wonder of all. He was also of so public a spirit that he did not only impart many excellent things to his auditors, but would take it for a very great kindness if any would vouchsafe to be his Scholars. . . . Admirable were those ways, which our never too much admired Frier followed in all Arts and Sciences, and from them making wonderful discoveries, followed with strange events, he passed not only with the vulgar people, but such that were accounted able Scholars, for a Necromancer, or one that busied himself in diabolical Magic . . . Doubtless though people were possest that our Frier conversed with the Devil, yet certainly what he did was purely by his learning.

So that our famous Frier being singular in his generous studies, and not understood but by few here at home, was the reason I conceive why his works were censured, and why they refused a place in their libraries to receive them.

🐾 *In later years the University itself became intensely hostile to the mendicant Orders, and in 1311 the friars formally appealed for intervention by the Pope. When a Dominican spokesman tried to deliver notice of this appeal to the Congregation of the University, meeting in St. Mary's Church, he was hustled out of the building, and had to shout the terms of the process through the window, standing on a tombstone and concluding:*

Reverend Masters, I would have notified this process to you inside, if I had been allowed to do so: but since I have been violently expelled, I have read it here before Notary and witnesses, and I call upon all you who stand around to witness the terms, and further I leave you, fastened to the door of this church, a copy of the aforesaid process.

🐾 *This was not altogether a success, for 'those who were standing around' replied, the friar reported to his superiors, with maledictions:*

It would be a sin to give alms to you friars! It would be a pious deed to shut up your doors and burn you, you arrogant fellows! You who dare, you wretches and beggars, to promote an Appeal against so great a congregation of such reverend and excellent persons!

From a Latin Roll, 1311, in the Bodleian Library.

🐾 *Twenty years later it was the University's turn to appeal to the Papacy:*

Enticed by the wiles of the Friars and by little presents . . . boys, for the friars cannot circumvent men of mature age, enter the Orders, nor are they afterwards allowed, according to report, to get their liberty by leaving the Order, but they are kept with them against their will until they make profession; further, they are not permitted, as it is said, to speak with their father or mother, except under the supervision and fear of a friar. An instance came to my knowledge this very day. As I came out of my inn, an honest man from England, who has come to this court to obtain a remedy, told me that immediately after last Easter, the friars at the University of Oxford abducted in this manner his son who was not yet thirteen years old, and when he went there he could not speak with him except under the supervision of a friar.

Richard Fitzralph, Chancellor of the University, in a speech before the Pope, 1333 (from the Latin).

THE FRIARS

🙵 *Finally the University clamped down on the friars:*

It is generally reported and proved by experience that the nobles of
this realm, those of good birth and very many of the common people,
are afraid, and therefore cease to send their sons or relatives or others
dear to them, in tender youth, when they would make most advance in
primitive sciences, to the University to be instructed, lest any friars
of the Order of Mendicants should entice or induce such children,
before they have reached years of discretion, to enter the Order of the
same Mendicants . . . therefore the said University has ordained and
decreed that if any of the Order of Mendicants shall receive to their
habit in this University, or induce, or cause to be received or induced,
any such youth, before the completion of his eighteenth year at least . . .
then *eo ipso* no one of the cloister or community of such a friar, being
a graduate, shall during the year immediately following read or
attend lectures in this University. . . .

University Statute, 1358 (from the Latin).

🙵 *The friars continued to thrive in Oxford, nevertheless, until the Reformation.*

RIOTOUS ASSEMBLIES

🙵 *Besides fighting among themselves, the scholars of Oxford constantly
fought the townspeople, who resented the University's growing authority
and arrogance. High Street, Oxford, is said to have seen more blood spilt
in anger than many a celebrated battlefield, and the first centuries of
University history are studded with skirmish. Here are three examples.*

THE WHORES, THE COOK, AND THE PAPAL LEGATE

About [1231] King Henry III, by a Precept transmitted hither, at the
Instance and Complaint of the Chancellor and Masters, order'd the
Mayor and Bailiffs to release all lewd Women then in Goal, on condition they straitways left the Town, finding Sureties not to come
again to Scholars Chambers. On the Publication of this Writ many
loose Women were expell'd from hence, maugre the Tumults then
made by some of the *French* students, whose infamous Lust had
engag'd them in their Quarrels, and by haunting Stews and Brothel-
Houses, had contracted the foul Disease almost in an Epidemical
Manner; which deprav'd Course of Life, some Years after, brought
over the Pope's Legate into *England*, sent hither to reform the Cor-
ruptions of the Place, and residing at *Osney* [Abbey] was at first
respectfully treated by all the Scholars, liberally entertaining him with

all Marks of Honour; and thereby thinking to have pleased him in the best manner.

One Day, after Dinner many went to welcome him, hereby presuming a good Reception of themselves; but as soon as they came to the Abbey-Gate, they were rudely saluted by the Porter, in his loud *Italian* Voice, demanding their Business; who reply'd, they came in Duty to attend the Legate; but he, in contumelious Language, refus'd them Admittance at the Door, which they forcing open, went in, the Legate's Retinue at the same time repelling them with Staves; but at length falling together by the Ears, many Blows ensued on both sides: Whilst some of the Scholars ran home for Arms, there happen'd a poor *Irish* scholar to wait at the Gate for Alms, on whom the chief Cook (being the Legate's Brother) threw scalding Water; which a *Welchman* perceiving, shot the Cook dead through the body, which caused an Uproar throughout the House.

The Legate hearing this Tumult, ran into the Belfry, and locking the Door, stay'd there till Midnight, but the Scholars, in no wise pacified, fought after him in every Corner, exclaiming against him as Usurer, Simoniack, and one guilty of Extortion &c. The Cardinal, in the Silence of the Night, coming out of his Fort, convey'd himself over the River to the King, then at *Wallingford*, who, on this Outrage, sent Troops to protect the Legate's Attendants, through Fear hiding themselves. Hereupon was one *Odo*, a Lawyer, with 38 other Students, seiz'd and brought before the King, and, from *Wallingford* in Carts sent to *London*, where, after much Entreaty of the Bishops, on doing of Penance, the Legate absolv'd them from their Offence.

John Ayliffe, *The Antient and Present State of the University of Oxford*, 1714.

THE MACE, THE BURGESS, AND FULK DE NEYREMIT

On Friday before the feast of St. Mathias [21 February 1298] when the bailiff was at Carfax carrying his mace, as is due, there came some clerks of the University to fight and disturb the peace, and laid hands on the bailiff and trampled on him and took away his mace; whereupon hue and cry was raised; by fresh aid one was attached and led towards the prison. There came a multitude of clerks with their followers with force and arms, and by means of an agreement made beforehand between them, rescued him who had been attached. After which rescue they went to the house of the head bailiff, entered his house with drawn swords, and would have slain him, had not force and flight saved him.

Further, on Saturday next following, came the clerks to St. Mary's church and took all the lay-folk they could find, beat them and wickedly trampled on them, and they killed one Thomas att Chirch of Iffley, coming there to sell his goods, and they took John Dorre, burgess, dragged him into the church, and beat him before the high altar, wounded and evil entreated him, and threatened all the burgesses with robbery and murder. . . .

The following Monday there came full three thousand clerks or more with bows and arrows, swords and bucklers, and divers arms, together with their manciples and force, and while folk were at their meal, by planned assault and compassed felony, assailed the town in four places at one time and broke into the spicery, cutlery, cook row, butchery, and robbed and stole the foods there. Wherefore hue and cry was raised, and people came out to keep the peace, and straightway the clerks shot and injured more than fifty, and many were wounded to death. The battle lasted from noon to compline until a clerk, Fulk de Neyremit, the strongest fighter of all, bearing a targe, looked over his targe, and an arrow came by chance and hit him in the eye, reaching to the middle of his head, so that he fell and his targe upon him; and when he was fallen, the others were dismayed and retreated to their lodgings, and so the battle ceased for this time.

After this uproar the bailiff would have pursued and taken the king's felons, but the Chancellor [of the University] would not suffer them to be imprisoned, and threatened that the town would be burnt and the burgesses slain, if he arrested them. . . . Further, when hue and cry was raised against the clerks when they made the robbery aforesaid, the Chancellor came and held an inquest by means of his riotous clerks who did the felony, concerning who raised hue and cry; and commands to arrest and imprison those who raised it and those who came to the cry; and those he cannot take he excommunicates.

Further, sire, we will show you evidence by which it appears that the Chancellor and the proctors were well warned of the coming conflicts. For the Saturday before the deed, where the rioters were assembled, there came the common bedeman of the town who prays for the souls of the dead; and they took him by force and made him pray for the soul of Walter Bost, 'who is not yet dead but he will be dead betimes'.

> Draft of an address by the Oxford City Burgesses to Edward I, 1298 (from the Norman French), in H. E. Salter, Mediaeval Archives of the University of Oxford, 1920.

ST. SCHOLASTICA'S DAY

On Tuesday 10 Feb. 1354 (being the feast of S. Scholastica the Virgin) came Walter de Springheuse, Roger de Chesterfield, and other Clerks, to the Tavern called Swyndlestock . . . and there calling for wine, John de Croydon the Vintner brought them some, but they disliking it, as it should seem, and he avouching it to be good, several snappish words passed between them. At length the Vintner giving them stubborn and saucy language, they threw the wine and vessel at his head. The Vintner therefore receding with great passion, and aggravating the abuse to those of his family and neighbourhood, several came in, encouraged him not to put up the abuse, and withal told him they would faithfully stand by him. Among these were John de Bereford (owner of the said Tavern by a lease from the Town) Richard Forester and Robert Lardiner, who out of propensed malice seeking all occasions of conflict with the Scholars, and taking this abuse for a ground to proceed upon, caused the Town Bell at St. Marton's to be rung, that the Commonalty might be summoned together into a body. Which being begun they in an instant were in arms, some with bows and arrows, others with divers sorts of weapons. And then they without any more ado did in a furious and hostile manner, suddenly set upon divers Scholars, who at that time had not any offensive arms, no not so much as any thing to defend themselves. . . . The Chancellor [of the University] perceiving what great danger they were in, caused the University Bell at St. Mary's to be rung out, whereupon the Scholars got bows and arrows, and maintained the fight with the Townsmen till dark night, at which time the fray ceased, no one Scholar or Townsman being killed or mortally wounded or maimed.

On the next day being Wednesday . . . the Townsmen came with their bows and arrows, and drave away a certain Master in Divinity and his auditors, who was then determining in the Augustan Schools. The Ballives of the Town also had given particular warning to every Townsman at his respective house in the morning that they should make themselves ready to fight with the Scholars against the time when the Town bell should ring out, and also given notice before to the country round about and had hired people to come in and assist the Townsmen in their intended conflict with the Scholars. In dinner time the Townsmen subtilly and secretly sent about fourscore men armed with bows and arrows, and other manner of weapons into the parish of St. Giles in the north suburb; who, after a little expectation, having discovered certain Scholars walking after dinner in Beaumont (being the same place we now call St. Giles's field) issued out of St. Giles's church, shooting at the said Scholars for the space of three furlongs: some of them they drove into the Augustine Priory, and others into the Town. One Scholar they killed without the walls, some they wounded mortally, others grievously and used the rest basely. All which being done without any mercy, caused an horrible outcry in the Town: whereupon the Town bell being rung out first and after that the University bell, divers Scholars issued out armed with bows and arrows in their own defence and of their companions, and having first shut and blocked up some of the Gates of the Town (least the country people who were then gathered together in innumerable multitudes might suddenly break in upon their rear in an hostile manner and assist the Townsmen who were now ready prepared in battle array, and armed with their targets also) they fought with them and defended themselves till after Vesper tide; a little after which time, entered into the Town by the west gate about two thousand countrymen with a black dismal flag, erect and displayed. Of which the Scholars having notice, and being unable to resist so great and fierce a company, they withdrew themselves to their lodgings.

The countrymen advanced crying Slea, Slea . . . Havock, Havock . . . Smyt fast, give gode knocks. . . . Finding no Scholars in the streets to make any opposition, [they] pursued them, and that day they broke open five Inns, or Hostles of Scholars with fire and sword. . . . Such Scholars as they found in the said Halls or Inns they killed or maimed, or grievously wounded. Their books and all their goods which they could find, they spoiled, plundered and carried away. All their victuals, wine, and other drink they poured out; their bread, fish &c. they trod under foot. After this the night came on and the conflict ceased for that day. . . .

The next day being Thursday . . . no one Scholar or Scholar's servant so much as appearing out of their Houses with any intention to harm the Townsmen, or offer any injury to them (as they themselves confessed) yet the said Townsmen about sun rising, having rung out their bell, assembled themselves together in a numberless multitude, desiring to heap mischief upon mischief, and to perfect by a more terrible conclusion that wicked enterprise which they had began. This being done they with hideous noises and clamours came and invaded the Scholars' Houses in a wretchless sort, which they forced open with iron bars and other engines; and entring into them, those that resisted and stood upon their defence (particularly some Chaplains) they killed or else in a grievous sort maimed. Some innocent wretches, after they had killed, they scornfully cast into houses of easment, others they buried in dunghills, and some they let lie above ground. The crowns of some Chaplains, viz. all the skin so far as the tonsure went, these diabolical imps flayed off in scorn of their Clergy. Divers others whom they had mortally wounded, they haled to prison, carrying their entrails in their hands in a most lamentable manner. They plundred and carried away all the goods out of fourteen Inns or Halls, which they spoiled that Thursday. They broke open and dashed to pieces the Scholars' Chests and left not any moveable thing which might stand them in any stead; and which was

[19]

yet more horrid, some poor innocents that were flying with all speed to the Body of CHRIST for succour (then honourably carried in procession by the Brethren through the Town for the appeasing of this slaughter) and striving to embrace and come as near as they could to the repository wherein the glorious Body was with great devotion put, these confounded sons of Satan knocked them down, beat and most cruelly wounded. . . .

This wickedness and outrage continuing the same day from the rising of the sun till noon tide and a little after without any ceasing, and thereupon all the Scholars (besides those of the Colleges) being fled divers ways, our mother the University of Oxon, which had but two days before many sons is now almost forsaken and left forlorn.

Anthony Wood, *History and Antiquities of the University of Oxford*, 1674.

🦂 *The St. Scholastica Day battle was the most notorious of all: the city was heavily punished for it, and for five hundred years the mayor and burgesses were obliged to attend a penitential service in St. Mary's Church.*

COLLEGE SYSTEM

🦂 *The first of the Oxford colleges were born in the thirteenth century. Originally self-governing religious foundations, mostly of graduates, they later became undergraduate societies and so influenced the growth of the University that in subsequent centuries it became little more than a federation of colleges. The true prototype was Merton College, founded in 1264. Walter de Merton, 'clerk, and formerly Chancellor of the illustrious Lord the King of England', established it on freehold land on the southern edge of the town, and endowed it with his manors of Maldon and Farlegh in Surrey, together with 'any other manors which I have acquired or may acquire'. Here are some extracts from its original Statutes, translated from the Latin:*

The house shall be called the House of the Scholars of Merton, and it shall be the residence of the Scholars for ever.

There shall be a constant succession of scholars devoted to the study of letters, who shall be bound to employ themselves in the study of Arts or Philosophy, the Canons or Theology. Let there also be one member of the collegiate body, who shall be a grammarian, and must entirely devote himself to the study of grammar; let him have the care of the students in grammar, and to him also let the more advanced have recourse without a blush, when doubts arise in their faculty.

The number of the Scholars is to be dependent on the means of the House itself; and each individual is to receive fifty shillings, and no more, annually, the payments to be at fitting seasons, yet so that they shall receive every week a certain proportion for their commons.

The House is to have a Superior, who is always to be denominated the Warden, and who must be a man of circumspection in spiritual and temporal affairs; and all persons, as well Scholars as ministers of the altar, brethren, managers, and bailiffs, are to obey and look up to him as their Superior.

Some of the discreetest of the Scholars to be selected; and they, in subordination to the Warden, and in the character of his coadjutors, must undertake the care of the younger sort, and see to their proficiency in study and good behaviour. There is to be one person in every chamber, where Scholars are resident, of more mature age than the others, who is to make his report of their morals and advancement in learning to the Warden.

The Scholars who are appointed to the duty of studying in the House are to have a common table, and a dress as nearly alike as possible.

The members of the College must all be present together, as far as their leisure serves, at the canonical hours and celebration of masses on holy and other days. Four ministers of the altar, or three at fewest, who are to be in priest's orders, and who must wear a respectable and suitable attire, shall be in constant residence within the House.

The Scholars are to have a reader at meals, and in eating together they are to observe silence, and to listen to what is read.

In their chambers, they must abstain from noise and interruption of their fellows; and when they speak they must use the Latin language.

Also the Scholars are carefully to observe the following injunction: that no one shall become burdensome to his fellows by introducing strangers, or even near relatives, in order that the quiet or the rest may not be disturbed by these means, and so altercations and quarrels arise; but as they were admitted to the support of the College out of charity, so must they all live meekly in fellowship, without burdening each other, but sharing all things fairly.

Care and a diligent solicitude shall be taken that no persons be admitted but those who are of good conduct, chaste, peaceable, humble, indigent, of ability for study, and desirous of improvement. Among those, however, who are to be admitted and to receive this gratuitous support, those persons who are of my own kin are to be the chief and first.

If a Scholar shall be clearly found guilty, by competent witnesses or other conclusive evidence, before the Superior, assisted by six or seven of the Seniors of the House, of perjury, sacrilege, theft, or robbery, homicide, adultery, or other lapse of the flesh, or beating a Fellow, or the Warden, which is worse, the whole commission of such an offence, even for the first time, shall suffice to show him most worthy of expulsion from the House. But if a suspicion only of some grave crime shall arise against him, or if he shall commit any one of the lighter offences, as for instance, some trifling act of disobedience, he is to be reproved by an admonition, thrice repeated. But in case he slight this thrice repeated warning, he is to be expelled from the House without hope of restoration, and to remain for ever in that state of expulsion.

A Scrutiny shall be holden in the House by the Warden and the Seniors, and all the Scholars there present, three times in the year; a diligent enquiry is to be instituted into the life, conduct, morals, and progress in learning, of each and all; and what requires correction then is to be corrected, and excesses are to be visited with condign punishment.

The Warden, upon receiving notice from the Senior [or Vice-Warden] and the Scholars, is once a year, to convene, on a day certain, all the stewards and brethren of their manors and possessions, to some one of the manors or places: and then a diligent enquiry is to be instituted by the said Senior and Scholars into the life, conduct and morals of the Warden, stewards and brethren; and thereupon, after the accounts of each have been audited, delinquencies are to be severely punished, but all the persons who are found to have acted with prudence and fidelity are to be continued in their former administrations.

I also enjoin the Scholars above all things, in God's name and by their hopes of happiness both in this life and the next, that in all things, and above all things, they ever observe unity, and mutual charity, peace, concord and love.

THE MERTON SCHOLARS, 1285

Rad. de Oddyham, W. de Chelsam, Rob. de Riplingham, Peter de Insula, J. de Tinteshale, William de Bosco, Rob. de Albruwyc, William de Lee, J. de Wendover, R. de Cokeswell, Jo. de Mora, Rob. Scarle, Hen. de Fodrige, J. de Clive, R. de Haregrave, Galf. de Codington.

THE MERTON LIBRARY, 1300

❧ *The first twelve books in the college library were these, as recorded in the first catalogue:*

1. Half the psalter, glossed.
2. Job and the minor prophets, etc, glossed.
3. Canonical Epistles, *Song of Songs* and *Lamentations*, glossed.
4. Aristotle, *De Animalibus.*
5. Aquinas on second book of the Sentences.
6. Aquinas on third book of the Sentences.
7. Bonaventura on second book of the Sentences.
8. Bonaventura on third and fourth book of the Sentences.
9. Petrus Cormastor, *Historia Scholastica.*
10. Augustine on the *City of God.*
11. Gregory, Pastoralia.
12. Various works by Jerome and other fathers, with glosses on Aristotle's *Ethics.*

❧ *By 1360 the theological manuscripts alone were valued at about £132, estimated by F. M. Powicke* (The Medieval Books of Merton College, 1931) *to be about half the value of an ocean-going ship. The Merton Fellow's statutory annual allowance of 50s. could then have bought him, if he had spent it all on books, about a dozen cheap texts a year.*

ALFRED'S COLLEGE

❧ *Three Oxford colleges—Balliol, Merton, and University College—were founded in the thirteenth century. Of the three, University College consistently claimed seniority on the totally spurious grounds that it had really been founded centuries before by King Alfred [see page 4]. When in 1377 Edmund Francis and Ideonea, his wife, brought an action against its Master and Fellows for the possession of some property, they brazenly claimed the privileges of a royal foundation:*

To their most Excellent and most dread and most Sovereign Lord the King, and to his most Sage Council, Shew his poor orators, the Master and Scholars of his College, called Mickle University Hall in Oxenford, which College was first founded by your noble Progenitor, King Alfred (whom God assoil) for the maintenance of twenty-six Divines for ever.

That whereas one Edmond Francis, Citizen of London, hath in regard of his great Power, commenced a Suit in the King's Bench against some of the Tenants of the said Master and Scholars, for certain Lands and tenements with which the college was endowed, and from time to time endeavour to destroy and utterly disinherit your said College, of the rest of its Endowment . . . That it may please your most Sovereign and gracious Lord and King, since you are our true Founder and Advocate, to make the aforesaid parties appear before your their most Sage Council, to show in evidences upon the rights of the aforesaid matter, so that on account of the poverty of your said orators your said College be not disinherited, having regard, most gracious Lord, that the noble Saints, John of Beverley, Bede, and Richard Armacan, and many other famous Doctors and Clerks were formerly Scholars in your said College, and commenced Divines therein.

Though both John of Beverley and the Venerable Bede died a century before Alfred came to the throne, and though University College was actually founded four centuries later, the plea was accepted by the Crown: sustained by a number of forged charters, the Master and Scholars of University College stuck to their story until it was laughed out of serious discussion in the nineteenth century.

THE OXFORD MAN, *c.* 1400

A clerk ther was of Oxenford also,
That unto logyk hadde long ygo.
As lene was his hors as is a rake,
And he was nat right fat, I undertake,
But loked holwe, and therto soberly.
Ful thredbar was his overeste courtepy;
For he hadde geten him yet no benefice,
Ne was so worldly for to have office.
For hym was levere have at his beddes heed
Twenty bookes, clad in blak or reed,
Of Aristotle and his philosophie,
Than robes riche, or fithele, or gay sautrie.
But al be that he was a philosophre,
Yet hadde he but litel gold in cofre;
But al that he myghte of his freendes hente,
On bookes and on lernynge he it spente,
And bisily gan for the soules preye
Of hem that yaf hym wherwith to scoleye.
Of studie took he moost cure and moost heede.
Noght o word spak he moore than was neede,

And that was seyd in forme and reverence,
And short and quyk, and ful of hy sentence;
Sownynge in moral vertu was his speche,
And gladly wolde he lerne, and gladly teche.

Geoffrey Chaucer, *The Canterbury Tales, c.* 1390.

THE ASCENDANCY

Out of the tumults of the Middle Ages the University evolved an ascendancy over the city of Oxford, incorporated in a series of statutory privileges. Here are some items from a statement of them, probably codified in about 1520:

Item the townysmen (whan it shalbe necessarie to make inquisition for conservacion of the kyngs peace) muste appere before the chauncellar what hower so ever they be callid.

Item that no justice of the peace, the mayr, nor the bailiffs to medle in causis of the peace for transgression withyn the precincte of the universitie if a scolar be that one parte, but the chauncellar to have the heryng & determinacion thereof according to right.

Item if the officer of the universitie do first arest the brekars of the peace withyn the towne and suburbes of Oxforde, althoo nether parte be of the privilege yett ye correction and punyshment therof shall perteyne onlie to ye chauncellar.

Item the chauncellar in causis determinable before hym may punyshe obstinate persons and transgressors whether they be of thuniversite or of the towne, and that by incarceracion or banyshment from the universitie, the towne and suburbes.

Item the schirif and kepar of the castell are bownde to receive, kepe and delyver the chauncellars prisonars at his ordinacion and comaundment.

Item the chauncellar onlie hath custodie of the assisse of bredde, wyne and ale, correction and punyshment of the same, with fynys, amersments and other profects commyng thereof withyn the towne and suburbes of Oxford.

Item the chauncellar may compell both scolars and townysmen to pave the streets and kepe them cleyne, amovyng blocks, stonys, fyme, kyne, swyne, &c.

Item the chauncellar hath power to banyshe incontinent & viciose women, if they offende withyn the towne or the precincte of the same, so that such banysshid persons shall not dwell within x myles of the Universitie.

Item all privilegid persons are exemptid from all archbisshopps, legatts, bisshoppys and other ordinarye judgys for contracts enterid . . . for excessis, crymes, fawlts withyn the precincte of thuniversitie committed, and for all scolasticall acts, subjected to the jurisdiction of the chauncellar onlie.

Item the aldermen and LVIII of the towne to be sworne every yere in S. Mary chirch for conservacion of the privilegis of the Universitie:

The Othe

Ye shall swere that all liberties and fre customys which the chauncellar and scolars of this Universitie have by the chartres and frauntys of ower soveren lorde the kyng & his progenitours and all other customys which the said chauncellar & scolars have reasonablie usid, well and firmelie ye shall holde & feythfullie cause to be holden, saving your fidelitie gyven to our soveren lorde the kyng.

UNIVERSITY JUSTICE

The principal instrument of University dominance was the Chancellor's Court, whose powers were wide. As these translated extracts from its Latin proceedings show, it often dealt imaginatively with its cases:

1439: John Coventry agrees that if the Vicar of Wem shall not have paid his debt of 36s 8d before All Hallows day, he will take as a pledge for the same a certain piece of cloth dyed and embroidered with the history of King Robert.

1447: A farmer, indebted to the Warden of Merton College, undertakes to pay his baker's bill.

1448: Master Thomas Bysshopp, Principal of White Hall, having been falsely defamed . as being a Scotchman, by Master Thomas Elslake, produces witnesses to prove that he and his family are true Englishmen.

1450: John Martyn, a schoolmaster, being threatened with excommunication, formed a conspiracy with his scholars to the end that, when the sentence should be in reading, they should snatch the document from the hands of the priest, and drag him from the pulpit. He is therefore imprisoned, together with the other ringleaders.

1458: T. Bentlee, organ-player of All Souls, is imprisoned for adultery, repents with tears, and is released after three hours.

1461: One John Vincent, keeper of the inn *The Cardinal's Hatte*, asserts that a horse entrusted to him was stolen by two Welsh scholars, who rode off with it towards Wales. He is condemned to pay the price of the horse to its owner in three instalments.

1465: Richard Lancester, canon, and Simon Marshall, on the one side, and John Marshall, schoolmaster, and his wife on the other, shall not abuse or make faces at each other, shall forgive each other for all past offences, and at their joint charges provide an entertainment in St. Mary's College, the one party to provide a goose and measure of wine, the other bread and beer.

THE BORYS HEDE

A student of The Queen's College (founded 1340), encountering an inimical boar on Shotover Hill, was said to have fought it off by stuffing a volume of Aristotle down its snout, at the same time crying the triumphant words 'Graecum Est'. The story was remembered at the college every year at a feast, during which a boar's head was processed into the hall on a silver salver, and after two trumpet-calls the following carol was sung:

> Hey, hey, hey, hey, the borrys hede is army'd gaye;
> The borrys hede in hond I brynge,
> The borys hede ye furst mes.
>
> The borys hede, as I yow say,
> He takes his leyfe and gothe his way,
> Gone after ye xij tweyl ffyt day,
>   With hey.

OXFORD HUMOUR, 1420

One of the chief academic exercises of medieval Oxford was the disputation, a kind of stylized Latin debate by which the student demonstrated his fluency in argument and his command of language. Students had to perform in several public disputations before they qualified for their degrees, and by the end of the fourteenth century a pattern of performance had been developed. It was the duty of the Father of the Act, the Senior Proctor, to

*commend the candidates to the audience, and sometimes he did this comic-
ally, as in the case of Mr. Dobbys of Merton, a candidate in 1420:*

Mr. Dobbys's name denotes duplicity and fickleness, because firstly
D stands for Duplex; secondly, his name has two syllables; thirdly,
it has double B in the middle; and fourthly, *bis* at the very end. Mr.
Dobbys has a large head, a very low forehead, beetling eyebrows,
black staring eyes, a monstrous mouth, a large nose, a protruding
upper lip, and big ears; features which prove him undisciplined,
choleric, unsteady, impetuous, proud, feeble, fatuous, unvirtuous,
greedy, wicked, rough, quarrelsome, abusive, foolish and ignorant.
It is related of him that one night after a deep carouse, when on his
way from Carfax to Merton, he found it advisable to take his bearings.
Whipping out his astrolabe he observed the altitude of the stars, but,
on getting the view of the firmament through the sights, he fancied
that sky and stars were rushing down upon him. Stepping quickly
aside he quietly fell into a large pond. 'Ah, ah', says he, 'now I'm in
a nice soft bed I will rest in the Lord.' Recalled to his senses when the
cold struck through, he rose from the watery couch and proceeded to
his room where he retired to bed fully clothed. On the morrow, in
answer to kind inquiries, he denied all knowledge of the pond. Thus
were his feckless drunken ways amply proved.

🎗 *This is the first recorded example of the officially sanctioned humour later
to be expressed by Terrae-Filius, the licensed jester of the University. It
was translated by Strickland Gibson in* The Bodleian Quarterly Record,
1930.

THE WAR MEMORIAL, 1443

The Preface of Henry Chiechelee to the Statutes of his College,
namely, All Soulen College, Oxon

To all the faithful of Christ to whom the present letters shall come,
Henry by surname Chiechelee, by divine permission Archbishop of
Canterbury, Primate of all England, and Legate of the Apostolic See,
[wishes] health in Him, who is the true Health of all men. When we
weigh oftentimes, and dwell with the eyes of true compassion on the
state of the unarmed soldiery of the Chirch, which from lack of estate,
and the other miseries of this world that is day by day sinking to
worse, hath already in piteous sort decayed immeasurably; and when,
with no less touch of pity, We inly compassionate that general ailment
of the armed militia of the world, which hath been very much reduced

by the wars between the realms of England and France, We cannot choose but mourn, calling to mind the magnificence and honour, whereby both of the hosts aforesaid, in the struggle of a pious emulation to surpass each other, had rendered the kingdom of England, which was of yore, and in Our own times no less renowned, on just grounds formidable to its adversaries, and had proclaimed it illustrious and most glorious among foreign nations. We, therefore, pondering with tedious exercise of thought, how, after the short measure of our meanness, We may, even in the smallest degree, spiritually or temporally succour, by the medicine of some antidote, each soldiery aforesaid, albeit so grievously wounded, have at last made up Our mind to petition the Most Christian Prince Henry the Sixth, the illustrious King of England and France now being . . . that he himself would deign to found and establish for all future times, out of those goods of fortune which the overflowing bounty of the Creator hath most lavishly bestowed upon Us in this life, on a certain ground-plot of Our own, which We have lately purchased for this purpose within the town of the University of Oxford, one College of poor and indigent scholars, being Clerks, who are constantly bounded not so much to ply therein the various sciences and faculties, as with all devotion to pray for the souls of glorious memory of Henry the Fifth, lately King of England and France, his own illustrious progenitor, and the Lord Thomas, Duke of Clarence, and the other Lords and lieges of his realm of England, whom in his own, and in his said father's times, the havoc of that warfare so long prevailing between the said two realms hath drenched with the bowl of bitter death, and also for the souls of all the faithful departed; the same to be commonly called ALL SOULEN COLLEGE.

Archbishop Chichele was supposed to have dreamt that when the foundations of his new college were dug, there would be found 'a swapping mallard imprisoned in the sink or sewere, wele fattened and almost ybosten'. Sure enough the duck was found, and ever afterwards the Fellows of the college were to celebrate the fact in a Mallard Feast:

As touching the first institution of this Ceremony . . . I cannot give any account of it, but when they have a mind to keep it, the night is always within a night or two of All Souls. Then there are six Electors which nominate ye Lord of the Mallard, which Lord is to beare the expences of the Ceremony. When he is chosen, he appoints six officers, who march before him with white staves in their hands, and meddalls hanging upon their breasts tied with a large blew ribbond. Upon ye meddalls is cut on the one side the Lrd of the Mallard wth his officers, on the other ye mallard as he is carried upon a long Poll.

When y*e* Ld is seated in his chair with his officers of state (as above
sd) before him, they carry him thrice about the Quadrangle and sing
this song:

> Griffin Turkey Bustard Capon
> Let other hungry mortalls gape on
> And on their bones wth stomacks fall hard
> But let All souls men have the mallard
> Hough the bloud of King Edward, by ye bloud of King Edward
> It was a swapping swapping Mallard.
>
> Swapping he was from bill to eye
> Swapping he was from wing to thigh
> His swapping toole of generation
> Out swap'd all the winged Nation
> Ho the bloud &c.
>
> Then let us sing & dance a Galliard
> To the remembrance of the mallard
> And as the mallard does in Poole
> Let's dabble dive and duck in Bowle
> Ho the bloud &c.

When that is done they . . . go with 20 or 30 Torches (which are
allwayes carried before them) upon the Leads of ye Colledge where
they sing their song as before. This ended, they go into their common
rooms, where they make themselves merry with what wine every one
has mind to, there being at that time plenty of all sorts. . . . Then he
that bore the mallard chops of his head, dropping some of the bloud
into every tumbler, which being drunk off, every one disposeth of
himselfe as he thinks fit, it being generally day-brake.

Thomas Baskerville, c. 1675.

HUMPHREY THE BENEFACTOR

> At Oxenford thys lord his bookis fele
> Hath ev'ry clerk at work. They of hem gete
> Metaphysic; phisic these rather feele;
> They natural, moral they rather trete;
> Theologie here ye is with to mete;
> Him liketh loke in boke historical.
> In deskis xii hym selve as half a strete
> Hath boked their librair univ'al.

This fifteenth-century verse refers to the collection of books given to Oxford in the 1430s by Humphrey, Duke of Gloucester. By then exciting new ideas from the Continent—the New Learning of the Renaissance—were germinating in Oxford, and they were handsomely encouraged by the Duke's present. The University, decreeing that a new library should be named for him, sent a mellifluous report of the benefaction to Parliament in London:

To oure ryght worshypfull syres, the Speker, Knygthes and burges of this worshypfull parliment of oure sovereyn lorde Kyng.

Ryght worshypfull syres, grace, pece and prosperite be to yow, to Goddis worshyp and to the gode welth of the reme of England ever duryng. Worthy syres, for as moch that meny of yowr owne issu and also Kynnesmen hath be, beth now, and shall be in tyme comyng tenderly and bisely noryshed and avaunced with the rype frute of Konnyng in oure moder the Universite of Oxon; in to the glory and the worshyp of Godd in speciall, and to the mayntenaunce of Crysten fayth, causyng of wyse men in the reme, and to yow grete joy, confort and eternall mede, that causeth supporteth and forthefyth such studiers; Therfore we conceyvyth that your naturesses and benevolence shold enjoy with us of the fortheraunce of the sayde Universite; and because oure ryth speciall lorde and myghty prince, the Duke of Gloucestre, hathe late endoed and so magnified oure sayde Universite with a thousand pound worth and more of preciose bokes, to the lovyng of Godd, encrese of clergy and konnyng men; to the goode governaunce and prosperite of the Reme of Englond withoute end, before all other remes and contres of the world; Wherffore we beseche your sage discrecions to consider the gloriose yiftes of the graciose prince to oure sayde Universite, for the comyn profyte and worshyp of the reme, to thanke hym hertyly and also prey Godd to thanke hym in tyme commyng, wher goode dedys ben rewarded. And our lord Godd so inspire and governe yow to His plesaunce with helyth of soule and body. *Wryte at Oxon. . . .*

All the hole Universite of Oxon.

The Duke also agreed, it seems, to pay for the construction of a new building, above the Divinity Schools, to house the books he had given, and furthermore promised to give the University his own private library of Latin books. Unfortunately he died, suspiciously and in political eclipse, before his pledges could be fulfilled, and the University expended much ink, emotion, and flexible spelling in trying to get its rights:

To ye ryght hygh and myghty prince, oure most singuler good lord and protectour, ye Marqois of Suthfolk.

Right hyght and myghty prince, owr most singuler good lord and protectour, we recommend us unto yowr gud lordschip in ye most lowly wyse, thankyng yow . . . for ye gret favour and tendirness that ye have had at al tyms, and now in special, unto ye spede of owr maters; that in als muche as ye duk of Glowcestre, now late passid to God, a lytil be for his deth grantyd to owr Universite of Oxenford all his buks of study, also odyr boks longyng to scole mater, lyk as we be informyd be feyth and credible personys, and morover grantyd ye sam wt his onne mowghte, her in a tyme of a convocation in our semble howse, be fore diverse doctoes and maysters and other notable, many mo graduat men beyng in his presance, we myght rekever and have theys sayd boks; ye wyche was to us most special and singuler tresour; consayning ye gret wel disposid multitude of scolars and the gret penury of boks ye ben amang us. And we schalle pray God hertly for yowr gud & nobil estat: the wyche God longe màynteyne yn preperite and gud helth of soule and body. *Wrytn at Oxenford in owr semble howse ye viii day of Aprile* [1447]. Yowr poor oratrice Universite of Oxenford.

🦁 *But it did no good, and the Duke's splendid private library went to Cambridge.*

GRADUATION DAY

🦁 *In 1452 George Nevill of Balliol College, brother to the Earl of Warwick, became a Master of Arts. He celebrated by giving a banquet, spread over two days, for 900 guests, of which this is the menu:*

The first course: A Suttletee, the Borehead and the Bull. Brawne and Mustard. Frumenty and Venyson. Fesant in brase. Swan with Chawdne. Capon of Grece. Hernshew. Poplar. Custard Royall. Grant Flanpart deperted. Leshe damask. Frutor lumbert. A Suttletee.

The second course: Vian in brase. Crane in sawce. Yong Pocok. Cony. Pygeons. Byttor. Curlew. Carcell. Partrych. Venson baked. Fryed Meat in past. Lesh lumbert. A Frutor. A Sutteltee.

The third course: Gely Royal deperted. Hanch of Venson rosted. Wodecoke. Plover. Knottys. Styntis. Quayles. Larkys. Quynces baked. Viant in past. A Frutor. Lesh. A Sutteltee.

OXFORD PRINTING

This year [1464] that admirable invention of Printing came to Oxford to the end that learning might be encreased and encouragement given to Scholars to proceed in Letters. The manner how was thus. That art being invented, some say at Mentz, some at Harlem . . . Thomas Bouchier, Archbishop of Canterbury, moved King Hen. VI to use all possible means to get a printing mould to be brought into the Kingdom, and contributed 300 marks towards the fetching of a Workman from Harlem. . . . The King made up that sum a 1000 marks, and sent over to manage the business one Robert Turnour . . . who took to his assistance Will. Caxton, a Citizen and Trader in Holland. . . . These having received the money, went first to Amsterdam, then to Leyden, not daring to go to Harlem, that Town being jealous of their trade, they having imprisoned divers who came thither for the said purpose. . . .

At last with much ado they got off one of the under-workmen called Frederic Corsellis, who late one night stole from his fellows in disguise into a vessel prepared for that purpose, and so was conveyed safe to London: but being there some time, twas not thought fit to set him on work at that place but at Oxford, of which University the said Archbishop was sometime Chancellor. Corsellis therefore was carried with a guard to Oxford, which continually watched to hinder him from making his escape till he had made good his promise in teaching them to print.

Anthony Wood, *History and Antiquities of the University of Oxford*, 1674.

This story was pure invention, devised by Richard Atkyns of Balliol (he apparently wished to ingratiate himself with Charles II by pretending that printing was a royal prerogative). The first Oxford book seems in fact to have been printed by Theodoric Rood, who came freely to Oxford from Cologne and set up a press in the High Street. Here is the colophon of his first book:

EXPLICIT EXPOSICIO SANCTI JERONIMI IN SIMBOLO APOSTOLORUM AD PAPAM LAURENCIUM IMPRESSA OXONIE ET FINITA ANNO DOMINI, M.cccc.lxviij xvij die decembris.

This was misleading too, for Rufinus of Aquileia, not St. Jerome, was the true author of this Latin commentary on the Apostles' creed. Worse still, it is now thought that Herr Rood contrived to make a misprint in the date: the first Oxford book was printed not in 1468, which would make it the first book printed in England, but in 1478.

THE KING, THE UNIVERSITY, AND THE BISHOP OF BATH

The religious orthodoxy of Oxford was often suspect, especially after the teachings of John Wyclif, Master of Balliol in the 1360s, gave rise to the subversive Lollard movement. In the following letter from Henry VII to the University, written in 1487, we may sense a continuing doubt in the royal mind about Oxford's loyalty to Crown and Church:

Trusty and welbelovyde, we grete you wele. And where as we late ffore certen grete and urgent caases touchynge owre personne and the quietenesse off thys owre realm, sent fore the bysshoppe of bathe, resaunt wtyn owre Universite there, to come unto us; he, obstynatly leyng aparte hys naturall deuty and observaunce, refused to obey owre commaundement yn that party: wherewyth we nethere were nere yett be content or pleased. And for as much as we be enformyd that he there contynieth, usyng certan practyses prohybyte by the lawes off holy church, and other damnabyll conjurecies and conspiraes, as it ys probabylly shewyd, as wele agenst us as to ye subversione of the universall wole and tranquillyte off thys owre realme; and, as we be assured, he makyth not hys abode there as a scolare and student . . . but, to theffect he myght be taken and reputyd as a student there, schamfully lurkyth and hydythe hymmselff . . . we send for him ayene atte thys tyme by oure trusty and wolbeloved chaplayn, master Edward Wylloughby; and desire and pray you that ye wyll se oure sayd chapelayne may have the conveyance off hym to us, withowte any lett or interruptione. . . . Where for we wol and straytly commande you, that ye suffyr oure servant, brynger hereoff, to take the sayde Bisshop and to bring hym unto us, wt owte any resistence lett or contradiction or eny personne or personnes of our sayde Universitie; assuring yow that, yff ye of obstynacye refuse to obeye thys our commaundment, we shall not only sende thyder suche power as oure entent in thys partie shalbe undowtly executed and fulfylled, but also provide for the punnishment off your disobeissaunce in suche sharpe wyse as shalbe to the ferfull example of them so presumyng or attemptyng heraftyr.

ERASMUS IN OXFORD

The Renaissance burst energetically in Oxford, and under the influence of progressive scholars like William Grocin, Thomas Linacre and John Colet, the forms of medieval scholarship began to wither. The great Dutch humanist Desiderius Erasmus came to Oxford in 1499, living for three months at St. Mary's College, and was delighted with what he found (he writes his letters in Latin):

He goes to a college feast: Nothing was wanting. A choice time, a choice place, no arrangements neglected. The good cheer would have satisfied Epicurus; the table-talk would have pleased Pythagoras. The guests might have peopled an Academy, and not merely made up a dinner party.

He replies to a complaint that he has given no lectures in Oxford: In our day Theology, which ought to be at the head of all literature, is mainly studied by persons who from their dullness or lack of sense are scarcely fit for any literature at all. This I say, not of learned honest professors of Theology, to whom I look up with the greatest respect, but of that sordid and superstitious crowd of divines, who think nothing of any learning but their own . . . You exhort me, or rather you urge me with reproaches, to endeavour to kindle the studies of this University—chilled, as you write, during these winter months—by commenting on the ancient Moses or the eloquent Isaiah . . . But I . . . do not think that I possess the strength of mind to sustain the jealousy of so many men, who would be eager to maintain their own ground.

Immediately after leaving Oxford, he sums up his impressions: How do you like our England, you will say? Believe me, when I assure you that I have never liked anything so much before. I find the climate both pleasant and wholesome; and I have met with so much kindness, and so much learning, not hacknied and trivial, but deep, accurate, ancient, Latin and Greek, that but for the curiosity of seeing it, I do not much care for Italy. . . .

🦥 *In later years Erasmus returned to England and settled in Cambridge, but he always preferred Oxford all the same.*

THE RENAISSANCE COLLEGE

🦥 *By the first years of the sixteenth century the humanist studies of the New Learning were strongly in the ascendant. When Richard Fox, the blind bishop of Winchester, thought of founding a new monastic establishment in the city, he was discouraged (in Latin) by his friend Hugh Oldham, Bishop of Exeter:*

What, my Lord, shall we build houses and provide livelihoods for a company of bussing monks whose end and fall we ourselves may live to see? No, no, it is more meet that we should have care to provide for the increase of learning, and for such as by their learning shall do good in the Church and the Commonwealth.

🐝 *So Fox instead founded Corpus Christi, the first true Renaissance college in Oxford, with Greek and Latin lecturers on the establishment, and an enlightened emphasis on liberal studies. Fox's Statutes for the college, drawn up in Latin in 1517, defined its ideals in a variety of quaint metaphor and idiom. They begin with the fancy that the best way to heaven is by a ladder, of which one side is called virtue, the other knowledge:*

We, therefore, Richard Fox, by Divine Providence Bishop of Winchester, being both desirous ourselves of ascending by this ladder to heaven and of entering therein, and being anxious to aid and assist others in a similar ascent and entrance, have founded, reared, and constructed, in the University of Oxford, out of the means which God of his bounty hath bestowed on us, a certain bee garden, which we have named the College of Corpus Christi, wherein scholars, like ingenious bees, are by day and night to make wax to the honour of God, and honey, dropping sweetness, to the profit of themselves and of all Christians. We appoint and decree by these presents, that in this bee garden there shall dwell for ever a President, to hold authority over the rest, twenty Scholars, or Fellows, the same number of Disciples, three Lecturers to be therein employed, each in his office and order; and, moreover, six Ministers of the Chapel, of whom two must be Priests, two not Priests, but Clerks and Acolytes, or at the least initiated by the primary tonsure, and the two remaining Choristers.

🐝 *The domestic affairs of the college are minutely regulated, down to the time-table for the washerwomen—'to come on Monday or Tuesday when dinner is over, to the College gate'—and stern rules of behaviour are decreed:*

From the Temple pass we to the Hall, for we would sow honey-bearing flowers there also, that on them the studious and ingenious bees may diet both in body and in soul. Wherefore we enact, that every day at dinner-time, the President and all the Fellows, Scholars, Students and Ministers of the Chapel then present in the Hall, shall listen to a portion of the Bible to be read before them, . . . All the aforesaid persons, as they hear in silence, shall earnestly and reverently listen and attend to the reading, and not engage in any talk, story-telling, din, laughter, disturbance, noise, or other enormities, while the reading goes on, but ruminate on the words in reading, as their ghostly food. . . .

Abuse, detraction, strife, buffoonery, prating, and the other vices of the tongue, seldom wait on an empty stomach, but often on one that is bloated and gorged; wherefore, in order that we may make stand against the causes of such misconduct, we command and enact, that in our College every day, after dinner and supper . . . every one of the Seniors, of whatsoever degree or estate they may be, shall forthwith, without any interval, betake themselves to their studies, or elsewhere, and not allow the other younger sort to loiter there any longer, except when either debates of the house, or other arduous business regarding the College, is required to be immediately taken in hand . . . or when, for the sake of reverence towards God, his glorious Mother, or some other saint, a fire is made there for the comfort of all the indwellers; for then the Fellows and Scholars of our College may be allowed to stay in the Hall for the purpose of temperate recreation, by means of songs and other reputable sports, as becomes Clerks, subsequently to the above meals and potations; and to compare one with another, read, and recount poems, histories, and the wonders of the world, and other things of the same kind.

As to the intellectual curriculum of the college, it is defined horticulturally:

The bees make not honey of all flowers without choice, but from those of all the sweetest and best scents and savours, which are tasted and distinguishable in the honey itself; hence the kinds of honey in different regions are various according to the diversity of the flowers, and neither Britain, Attica, nor Hybla can produce honey, so long as the honey-bearing flowers are far away. We therefore are resolved to constitute within our bee-garden for ever, three right skilful herbalists, therein to plant and sow stocks, herbs and flowers of the choicest, as well for fruit as thrift, that ingenious bees swarming hitherward from the whole gymnasium of Oxford, may thereout suck and cull matter convertible not so much into food for themselves, as to the behoof, grace, and honour of the whole English name, and to the praise of God, the best and greatest of beings. Of the above three, one is to be the sower and planter of the Latin tongue, and to be called the Reader or Professor of the Arts of humanity; who is manfully to root out barbarity from our garden, and cast it forth, should it at any time germinate therein. . . . The second herbalist of our apiary is to be, and to be called, the Reader of the Grecists and of the Greek Language; whom we have placed in our bee-garden expressly, because the Holy Canons have established and commanded, most suitably for good letters, and Christian literature especially, that such an one should never be wanting in this University of Oxford, in like manner as in some few other most famous places of learning. . . . Lastly, a third gardener, whom it behoves the other gardeners to obey, wait on, and serve, shall be called and be the Reader in sacred Divinity, a study which we have ever holden of such importance, as to have constructed this our apiary for its sake, either wholly, or most chiefly; and we pray, and in virtue of our authority command all the bees to strive and endeavour, with all zeal and earnestness, to engage in it. . . .

🐝 *Bishop Fox concludes with the warning that members of the college who break the rules must be 'punished, harrassed, and most bitterly afflicted with the penalties of the Statutes, without pardon':*

But whosoever shall keep them without offence and steadfastly, and, so far as he can, in their integrity, and shall procure their observance, shall dwell and be fed in safety, immunity, peace and honour in our Hive for a season; and shall, after no long delay, having obtained his discharge, arrive at, and take upon himself that most resplendent illumination for which he has been so industriously storing wax; so that at the last he shall clearly descry the most precious Body of Christ, incomparably sweeter than all honey, to which we have dedicated our Hive, and to enjoy that sight, with the highest bliss, for evermore. .

🐝 *Erasmus, the Dutch scholar, was much impressed by the conception of Corpus Christi, and especially by Fox's plan for a library of books in Latin, Greek, and English:*

Not a few places have been made famous by some popular monument. Rhodes is celebrated for its Colossus; Caria for the tomb of Mausolus; Memphis for its pyramids; Cnidus for its sculptured Venus; Thebes for its magic Mammon. My mind foretells that in the future this college, like some holy temple dedicated to good learning, will be accounted among the chief glories of Britain in all countries of the world, and that the spectacle of that tri-lingual library will in the future draw more persons to Oxford than Rome drew to herself of old, though her many marvels made her a place of pilgrimage.

From a letter in Latin, 1519.

A VALUE JUDGEMENT

A Mayster of Arte
Is not worthe a Farte

From *The Jestes of Scoggin*, attr.
to Andrew Boorde, d. 1549.

DIRTY CUPS AND MIDNIGHT COOKING

🦌 *Magdalen College, founded in 1458, was inspected by a commissary from its Visitor, the Bishop of Winchester, in 1507. It was found to be in a terrible state, its functions in disorder and its Fellows spitefully at odds. Among the complaints made to the inquiry were these:*

Stokes was unchaste with the wife of a tailor.

Stokysley baptized a cat and practised witchcraft.

Gregory climbed the great gate by the tower, and brought a Stranger into College.

Kendàll wears a gown not sewn together in front.

Pots and cups are very seldom washed, but are kept in such a dirty state that one sometimes shudders to drink out of them.

Gunne has had cooked eggs at the Taberd in the middle of the night.

Kyftyll played cards with the butler at Christmas time for money.

Smyth keeps a ferret in College, Lenard a sparrow-hawk, Parkyns a weasel, while Morcott, Heycock and Smyth stole and killed a calf in the garden of one master Court.

From the Episcopal Register, Winchester, 1507.

🦌 *All the errant Fellows were repentant, and having shamelessly told tales about each other, drank a loving-cup together and were forgiven.*

THE CARDINAL'S COLLEGE

🦌 *By 1525 there were twelve Oxford colleges. Thomas Wolsey, Cardinal of York and Lord Chancellor to Henry VIII, planned to surpass them all with the thirteenth, which he financed partly by suppressing various religious houses, and made room for by demolishing most of a monastery and all of a church:*

Hee stiled [it] by the name and title of *Thomas Wolsey Cardinall of Yorke his Colledge in the Universitie of Oxford*. This hee builded to the praise, glory and honor of the holie and undevided Trinitie, the most holy Virgin Mary, the blessed Virgin St Friswide, and All Saints. . . .

This Colledge the Cardinall ordeined to be a perpetuall Nurcery of Learning of the faculties of Divinitie, the Cannon and Civill Law, of Humanity alsoe and Phisick, and for a perpetuall observance of God's worshipp in and of the number of one Deane and 60 Cannons Seculars, more or lesse, augmenting and diminishing, according to the abilities and exigences of this Foundation. . . . Of sett purpose, hee made certaine pauses and delaies, that hee might make choise of the sharpest and quickest witts, among whom William Tindall, that translated the Bible into the English Tongue, was one, and Taverner, the worthy Musitian, was the Organist. . . . All the time of his life, hee kept, perpetually in himselfe, power to manage the lands of his Colledge. . . .

Leonard Hutten, *Antiquities of Oxford, c.* 1600.

🦋 *The foundation provided for the following members:*

1 Dean
60 Canons primi ordinis
40 Canons secundi ordinis
13 Chaplains
12 Lay Clerks
16 Choristers
1 Teacher of Music
6 Professors
4 Legal Officers
23 Servants

🦋 *The first parts of the building to be completed were the kitchen and dining-hall, among the most magnificent in Europe, giving rise to the quip:*

Cardinalis iste instituit collegium, et absolvit popinam. (*Your cardinal began a college, and produced an eating-house.*)

🦋 *Though Wolsey was dedicated to the New Learning, and saw his college as a truly Renaissance foundation, he was vehemently opposed to the Protestant ideas then arriving in England from Europe. Unfortunately some of the original scholars of his college proved to be infected with Lutheran heresies, and the Cardinal ordered the arrest of the chief activist, Thomas Garrett. In February 1528 Garrett escaped from the custody of Dr. Cottisford, Rector of Lincoln College, and here one of his associates, Anthony Dalaber, describes how the disquieting news reached the Dean and Chapter of Cardinal College during a service in the still unfinished college chapel:*

Evensong was begun; the Dean and the canons were there in their grey amices; they were almost at 'Magnificat' before I came thither. I stood in the choir door, and heard Master Taverner play, and others of the chapel there sing, with and among whom I myself was wont to sing also; but now my singing and music were turned into sighing and musing. As I there stood, in cometh Dr. Cottisford, the Commissary, as fast as ever he could go, bare-headed, as pale as ashes (I know his grief well enough); and to the Dean he goeth into the choir, where he was sitting in his stall, and talked with him very sorrowfully; what, I know not; but whereof, I might and did truly guess. I went aside from the choir door to see and hear more. The Commissary and Dean came out of the choir, wonderfully troubled as it seemed. About the middle of the church met them Dr. London, puffing, blustering, and blowing like a hungry and greedy lion seeking his prey. They talked together awhile; but the Commissary was much blamed by them, insomuch as he wept for sorrow. The doctors departed, and sent abroad their servants and their spies everywhere.

After consulting an astrologer they found and imprisoned six of the heretics. Taverner was excused on the grounds that he was 'but an Organist': three others, imprisoned and fed for five months on nothing but salt fish (according to John Foxe's Actes and Monuments, *1563) died of their treatment before they could be brought to the stake. But in the following year Wolsey fell from power anyway. He appealed to the Crown to spare his college, complaining of*

. . . indyssposycion of body and mynde by the reason of suche gret hevynes as I am yn, being put from my slep, and mete for such advertysments as I have had from yow of the dyssolucion of my College.

Rumour said that the King would demolish at least part of the unfinished building, the French Ambassador reported to his King in November 1529,

. . . were it for no other purpose than that of removing the Cardinal's escutcheon, which will be no easy work as there is hardly a stone from the top of the building to the very foundations where his blazoned armorial is not sculptured.

In the event, though Henry suppressed Cardinal College absolutely, he preserved its buildings for his own purposes, intending presently to found a college of his own:

. . . Not so great or of such magnificence as my Lord Cardinal intended to have, for it is not thought meet for the common weal of our realm. Yet we will have a College honourably to maintain the service of God and literature.

The requiem of the Cardinal's college was written by Shakespeare, in Henry VIII *(Act IV, Scene 2):*

> though unfinished, yet so famous,
> So excellent in art, and still so rising,
> That Christendom shall ever speak his virtue. . . .

SEEKING A DIVORCE

And in caas ye do not upryghtly accordinge to divine lernynge handle your selfe herin, ye may be assurede, that we, not without great cause, shall soo gwykely and sharpely loke to your unnaturall misdemeanure therin, that it shall not bee to your gwietnesse and ease hereafter.

🦋 *With these threatening words in 1530 Henry VIII invited the opinion of the University of Oxford upon his proposed divorce from Queen Catherine, the cause of his break with Rome and so of the English Reformation. Cambridge and several European universities had already tactfully pronounced in his favour, but Oxford was less amenable. The Masters of Arts refused to allow the Faculty of Theology to deal with the matter, and demanded to be represented themselves. A commission of persuasion was sent to Oxford, advised by a learned Italian friar, Nicholas de Burgo, but got a rough welcome:*

> Women (that season) in Oxforde were busye
> Their harts were goode, it appeared no lesse:
> as Fryer Nicholas chanced to come by
> Halas (saide some) that we might this knave dresse
> for his unthankefull daylye busyness
> againste our Queen, good [Catherine] deare,
> he should evyl to cheeave, hee sholde not sure mysse.
>
> Withe that a woman (I saw it trulye)
> a lumpe of Osmundys let harde at him flying;
> which myste of his noddle the more pytie,
> and on his Fryers heelys it came trycelynge,
> who sodanly, as he it perceavynge
> made his complaynte upon the women
> that thirtye the morowe, were in Buckerdo.

Will Forrest, The Second Gresyld, 1558.

['*Buckerdo*'—Bocardo—*was the town gaol.*]

🦋 *The King thereupon wrote to the University again, still more ominously:*

Trustie and welbeloved we grete you wele, and of late bee enformed to our no litill merveille discontentacion, that a great parte of the youth of that Universitie with contemptuous faction and maner, dayli contynynge togeders neither regardynge their dutie to us their soveraigne Lorde, nor yet conformynge them to the opinions and ordres of the vertuous, wise, sadde and profound lerned men of that Unyversitie, wilfullie do styke upon the opinions to have a great

nombre of the Regents and non Regents, to be associate unto the
Doctours, Proctors and Bachelors of Divinitie, for the determination
of our question, whiche we beleve hath not often been seen, that such
a greate nombre of ryght smalle lernynge in regarde to the other,
should be wyned with soo famous a soorte, or in a maner staye ther
seniours in so wayghtie a cause whiche as we thinke shulde be no
smalle dishonnour to ower Universitie there, but most speciallye to
you the seniours and rewlers of the same: assurynge you that this
their unnaturall and unkynd demeanoure is not only ryght muche to
oure displeasure, but also muche to bee merveyled of, upon what
growndes and occasions, they beinge our mere subgietts sholde show
them self more unkind and wylfull in this matier, then all other
Universities bothe in this and all other Regions doo. Finally we
trustinge in the dexteritie and wisdome of you and other the sadde,
discrete and substanciall lernede men of that Universitie bee in perfite
hope, that ye will conduce and frame the saide yonge personnes unto
good ordre and conformitie, as it becomythe you to doo: whereof we
bee desyrous to here with conveniente diligence. And doubte yee not
that we shall regarde the demeanure of every oon of that Universitie
according to their merits and demerits. And iff the youth of that
Universitie will play maistres as they begynne to doo, we doubte not,
but that they shall wele perceyve that *non est bonum irritare crabrones*
[it is not good to stir up hornets].

Geven undre our Signet at our Castell of Windesore.

🐝 *This time, helped along by bribes and further threats, the University
succumbed: the divorce was approved, the King's marriage was dissolved,
and the authority of the Pope was abolished in England.*

SCOURING THE UNIVERSITY

🐝 *The Reformation scoured Oxford. The monastic houses of Osney, Rewley,
and Godstow were all abolished, and their assets appropriated, but the
King, for all his difficulties with Oxford about his divorce, had no wish to
harm the University. When it was suggested to him that its colleges might
profitably be dissolved too, this is how he replied:*

Ah, sirha, I percieve the abbeie lands have fleshed you and set your
teeth on edge to aske also those colleges. And whereas we had a
regard onelie to pull downe sinne by defacing the monasteries, you
have a desire also to overthrow all goodnesse by subversion of
colleges. I tell you, sirs, that I judge no land in England better

bestowed than that which is given to our universities, for by their maintenance our realme shall be well governed when we be dead and rotten. As you love your welfares therffore, follow no more this veine, but content your selves with that you have alreadie, or else seeke honest meanes whereby to increase your livelods, for I love not learning so ill, that I will impaire the revenues of anie one house by a penie, whereby it may be upholden.

Raphael Holinshed, Chronicles, 1577.

🦁 *On the other hand the learning of the University was forcibly modernized. This is William Tyndale, translator of the Bible, looking back on the scholasticism which had so long dominated Oxford teaching:*

Remember ye not how within this thirty years *and far* less, and yet dureth to this day, the old barking curs, Duns' disciples and like draff called Scotists, the children of darkness, raged in every pulpit against Greek, Latin and Hebrew; and what sorrow the schoolmasters, that taught the true Latin tongue, had with them; some breaking the pulpit with their fists for madness, and roaring out with open and foaming mouth, that if there were but one Terence or Virgil in the world, and that same in their sleeves, and a fire before them, they would burn them therein, though it should cost them their lives; affirming that all good learning decayed and was utterly lost, since men gave them unto the Latin tongue. . . . They have ordained that no man shall look at the scripture until he be noselled in heathen learning eight or nine years, and armed with *false principles*, with which he is clean shut out of the understanding of the scripture. And at his first coming unto University he is sworn that he shall not defame the University, whatsoever he seeth. And when he taketh first degree he is sworn that he shall hold none opinions condemned by the church; but what such opinions be, that he shall not know.

🦁 *And this is Richard Layton, sent to Oxford in 1535 as a Visitor from the King, on the symbolic end of the Duns Scotus tradition of learning:*

Wee have set Dunce in Bocardo, (a prison so called) and have utterly banished him Oxford for ever, with all his blynd glosses, and is now made a common servant to every man, fast nayled up upon posts in all common howses of easement, *id quod oculis meis vidi*. And the second time wee came to New College, after wee had declared your injunctions wee fownd all the great Quadrant Court full of the Leaves of Dunce the wind blowing them into every corner; and there wee fownd one Master Greenefeld a Gentleman of Buckinghamshire gathering up part of the said book leaves (as he said) therewith to

make him Scwells or Blaunchers to keepe the Deere within the wood,
and thereby to have the better crye with his hounds.

Richard Layton, in a letter to Thomas Cromwell, 1535.

THE KING'S COLLEGE

*As part of his rejuvenation of Oxford Henry VIII re-founded, in 1532,
the abolished Cardinal College, combining it with the episcopal see of
Oxford, making its chapel a cathedral, and calling it Christ Church. It
was naturally a stronghold of the new Protestant orthodoxy. Its first
Dean, Richard Cox, appointed the Italian Protestant Peter Martyr to
be a Canon of Christ Church, and the two men scandalized Catholic thinking by marrying (their wives were the first married women ever to live in
an Oxford college). It was a divine of different views, William Forrest,
who anathematized Cox in the following furious but incoherent stanzas:*

> Abhorringe his Order of sacrede preeistehod
> a whoare he tooke hym, wife cowlde he take none,
> for contrarye vowe, he made unto God,
> when, of his mynysters, hee tooke to bee one:
> but, for he wolde not to the Dyvyl alone
> he wrought (by all meanys) others to entrappe
> with hym (for eaver) to cursse their mishappe.

> He wrought by his holy stynkeinge Martyr
> Peter, that Paule his breath coulde not abyde,
> (for that, like Sathans true knyght of the Gartyr
> his holy doctryne, hee heere falcyfide)
> that whoe (of Preeists) in maryage was not tyde,
> he was afflicted, tormoyled and toste
> to loss of lyvynge or some other coste.

> This was a worthye famous Doctor,
> this was a man worthie of preamynence
> this was a Christian true Professor,
> this was a man of right intelligence:
> the Dyvyl hee was; I say my conscience
> he was (I saye) an erraunt cursed Theeif
> his actys declare, yee neade no ferdre preeif.

*When the Catholic Queen Mary came to the throne in 1553 Cox lost his
Deanery and was imprisoned, but the accession of the Protestant Elizabeth
five years later saw him Bishop of Ely, and it was to him, when he*

declined to co-operate in a building scheme, that the Queen is supposed to have sent her famous letter:

PROUD PRELATE

You know what you were before I made you; if you do not immediately comply with my request, by G-d I will unfrock you.

ELIZABETH

Peter Martyr fled to the Continent when Mary came to the throne. His wife had died in 1552, and her remains were exhumed from the Cathedral burial ground and thrown on a dungheap in the Deanery yard: fortunately the bones of St. Frideswide, the patron saint of the Cathedral, were about that time rediscovered in a remote corner of the church, so when the Protestants again came to power the relics of the two ladies were buried in one grave, 'permixta et confusa'. What became of the 'whoare' Mrs. Cox, nobody knows.

AN OXFORD INVENTION, *c.* 1540

Alexander Nowell of Brasenose College is said to have invented bottled ale:

Without any offence it may be remembered, the leaving of a bottle of ale, when fishing, in the grass, he found it some days after, no bottle but a gun, such the sound at the opening thereof and this is believed (casualty is mother of more inventions than industry) the original of bottled ale in England.

Thomas Fuller, *Worthies of England*, 1662.

TWO CHRIST CHURCH ORDERS, *c.* 1550

Concerning Dogges:

No student, scholar, chaplain nor servant or any belonging to the House shall lodge any dogg except the porter to dryve oute cattell and hogges out of the House.

What Everie Scholler Ought to Have:

1. Imprimis a Tutor, one of the Divines or of the Philosophers.
2. Honeste apparell and cumblye for a scholler.
3. Psalterium of Leo Juda translation.
4. His catechisme sett forthe in the Kyng's booke by harte.
5. Grace accustomed to be said at meales by harte.
6. Theie must also tayke an othe to the kyng.

OXFORD MARTYRS

The accession to the throne of the Catholic Queen Mary plunged Oxford once again into religious uncertainty, and it was in Oxford that Thomas Cranmer, Archbishop of Canterbury, Nicholas Ridley, Bishop of London, and Hugh Latimer, Bishop of Worcester, were examined for their Protestant heresies by a commission of Oxford and Cambridge divines. After lengthy disputations in the Divinity Schools they were condemned to death by burning. Lord Williams of Thame, the Queen's Commissioner, and the Vice-Chancellor of the University presided over the burnings of Latimer and Ridley in Broad Street on 16 October 1555, when one Bishop is supposed to have addressed the other in the famous words:

Be of good comfort, Master Ridley, and play the man. We shall this day light such a candle, by God's grace, in England, as I trust shall never be put out.

Archbishop Cranmer, whose execution had to be sanctioned by the Pope, had longer to wait, and in the interval he recanted his alleged heresies, recognizing the authority of the Pope in England, and the truth of Roman doctrines. He was condemned to be burnt nevertheless, and John Foxe, in his Book of Martyrs, *tells the story of his public humiliation at the hands of Henry Cole, Warden of New College, in St. Mary's Church, and his subsequent fate:*

In this so great frequency and expectation [on 25 March 1556] Cranmer at the length cometh from Bocardo prison into St. Mary's Church (the chief church in the university) because it was a foul and rainy day, in this order: the mayor went before; next him the aldermen in their place and degree; after them was Cranmer brought between two Friers which, mumbling to-and-fro certain Psalms in the streets, answered one another untill they came to the Church door and there they began the Song of Simeon, Nunc dimittis, and entring into the Church, the Psalm-saying Friers brought him to his landing and there left him. There was a stage set over against the Pulpit, of a mean height from the ground, where Cranmer had his standing, waiting until Cole made him ready to his Sermon.

The lamentable case and sight of that man gave a sorrowful spectacle to all Christian eyes that beheld him. He that late was Archbishop, Metropolitan, and Primate of England and the King's Privy Councellor, being now in a bare and ragged gown and ill favouredly cloathed, with an old square Cap, exposed to the contempt of all men, did admonish men not only of his own calamity but also of their state and fortune . . . In this habit, when he had stood a good space upon the stage, he lifted up his hands to Heaven and prayed unto God once or twice, till at length Dr. Cole coming into the Pulpit began his Sermon.

Cole's sermon, quoting the examples of St. Laurence on the grill, St. Andrew on the cross, and the three children to whom God made the flame to seem like a pleasant dew, ended with the promise that immediately after Cranmer's death, 'Dirges, Masses, and Funerals would be executed for him in all the Churches of Oxford'.

Cranmer in all this mean time, with what great grief of mind he stood hearing this Sermon, the outward shews of his Body and countenance did better express than any man can declare, one while lifting up his hands and eyes unto Heaven and then again for shame letting them down to the Earth . . . More than twenty several times the tears gushed out abundantly, dropping down marvellously from his Fatherly Face. They which were present do testify that they never saw in any child more tears.

Cole then demanded that Cranmer repeat the terms of his recantation, 'that all men may understand that you are a Catholick indeed', but Cranmer, after a few innocuous pietisms, stunned the congregation by recanting his recantation.

And now I come to the great thing which so much troubleth my conscience more than any thing that ever I did or said in my whole life, and that is the setting abroad of a writing contrary to the truth; which now I renounce and refuse, as things written with my hand, contrary to the truth which I thought in my heart, and written for fear of death and to save my life if it might be, and that is, all such bills and papers which I have written or signed with my hand since my degradation; wherein I have written many things untrue. And forasmuch as my hand offended, writing contrary to my heart, my hand shall first be punished thereofe: for may I come to the fire it shall be first burned. And as for the Pope, I refuse him as Christ's enemy and Antichrist. . . .

Here the standers by were all astonied, marvelled, were amazed, did look one upon another, whose expectations he had so notably deceived. Some began to admonish him of his recantation and to accuse him of falsehood. Briefly, it was a World to see the Doctors beguiled of so great an hope.

Cranmer was led away to the waiting stake in Broad Street, harried along the way by the two Spanish friars, 'vexing, troubling, and threatening him most cruelly'. There, barefoot, bareheaded, fastened with an iron chain, in a long shirt down to his feet, he was committed to the flames.

And when the wood was kindled and the fire began to burn near him, stretching out his Arm, he put his right hand into the flame, which he held so stedfast and immovable (saving that once with the same hand he wiped his face) that all men might see his hand burned before his Body was touched. His eyes were lifted up to heaven, and oftentimes he repeated 'this unworthy right hand', as long as his voice would suffer him; and using often the words of St. Stephen, 'Lord Jesus receive my spirit', in the greatness of the flame he gave up the ghost.

One of the Spanish friars, unaccustomed it seemed to such fortitude in death, ran to Lord Williams of Thame crying that 'the Archbishop was vexed in mind and died in great desperation'. But Lord Williams, Foxe tells us, merely smiled, 'and (as it were) by silence rebuked the Friers folly'. Three centuries later the Oxford Protestant Martyrs were to be commemorated in the pinnacle of Gilbert Scott's Martyrs' Memorial in St. Giles, popularly alleged by generations of undergraduates to be the spire of an otherwise submerged church; but a more haunting monument to the occasion, to the age, and perhaps to the first tumultuous years of Oxford University history, is to be found in the account-book of the City Bailiffs for 21 March 1556:

	s	d
For an 100 of wood-fagots	6	0
For an 100 and half of furs-fagots	3	4
For the carriage of them	0	8
For two labourers	1	4

Peace and War
1558 – 1700

Granted a breathing-space under the reign of Elizabeth I, and settled at last into some sort of legal cohesion, the University of Oxford was plunged into confusion again by the English Civil War and its aftermath, emerging in a state of demoralized exhaustion from the various contradictory challenges of the seventeenth century.

RESTORING ORDER

Queen Mary's Catholic interregnum had tragically disturbed the University of Oxford. As the antiquarian Anthony Wood wrote in retrospect, 'two religions being now as it were on foot, divers of the chiefest of the University retired and absented themselves till they saw how affairs would proceed'. Queen Elizabeth's accession restored stability. On the one hand, a Royal Commission clamped down on the last remnants of Roman Catholicism at Oxford:

Whereas by credible report we are inform'd that as yet there are remaining in your College divers Monuments of superstition undefac'd: These be by virtue of the Queen's Majesty's Commission to us directed to wylle and commande you forthwith upon the syght herof utterlye to deface, or cause to be defac'd, so that they may not hereafter serve to any superstitious purpose, all Copes, Vestments, Albes, Missals, Books, Crosses and such other idolatrous and superstitious Monuments whatsoever, and within eight days after the receipt herof to bringe true certificate of their whole doinge herin to us or our Colleagues, whereof fayle you not as you will answere to the contrarye at your perill.

On the other hand the University was disciplined by a forcible new Chancellor, the Earl of Leicester, who took office in 1564 and soon let his views be known to the authorities at Oxford:

The disorders, not muttered of, not secretly informed here and there in corners, but openly cried out uppon continually and almost in every place, are such, as touch no less then your religion, your lives and conversation, and the whole estate of your Universitye, Professions and Learning. . . . The chefest points are the want of instructing your youth in the Principles of Religion, the little care that Tutors have that waye, and most especially the suffering of secret and lurking Papists amongst you, which seduce your youth and carry them over by flockes to the Seminaries beyond Seas. This is so evident, that it cannot be denied, so heynous both in the publique Estate and yours, both to GOD, Church, Prince and Cuntrye, that it cannot be excused. . . .

In your Conversation and Life are these things noted. Excesse in apparell, as silke and velvet, and cutt dubbletts, hose, deepe ruffs and such like, like unto, or rather exceeding, both Inns of Courte men and Courtiers.

The Haunting of the Towne, that the streets are every daye and all day longe more full of Schollers then Townsmen.

That Ordinary Tables and Ale-houses, growen to great number, are not yet so many as they may be full fraight all daye and much of the night, with Schollers tipling, dicing, carding, and I will not say worse occupied. . . .

Is this the antient discipline of that Universitye for Schollers that are sent thether to be brought up in all modestye to go thus disguised, that are alowed frugally by their freinds and founders to be thus wastfull in apparell and expenses, that are sent thether as it were to a Mart of good Learning and good Education? to learne indeed nothing ells but to jelt in the stretes and to tipple in Tavernes, returning to their freinds (as I heare many of them of good sort complaine) lesse learned then when they came thether, and worse mannered than if they had been so long conversant amongst the worst sort of people?

Noe this is not the old Universitye order. . . .

On the whole the prescription worked: it took time for the University to settle down, but from the days of Elizabeth I Oxford never seriously wavered again in its loyalty to the established church and crown of England.

THE QUEEN IN OXFORD

Queen Elizabeth I made two State visits to Oxford. In 1566 she came attended by Robert Cecil, her Lord Treasurer, the Spanish Ambassador, and a train of nobles and maids of honour:

The one and thirtieth of August the queenes majestie in hir progress came to the universitie of Oxford and was of all the students which had looked for hir comming thither two yeares, so honorablie and joifully received as either their loialness towards the queenes majestie or the expectation of their friends did require.

Raphael Holinshed, *Chronicle*, 1577.

Oxford was hung all about with honorific verses in Latin, Greek, and Hebrew, and the Queen, met by a mounted posse of academics, was greeted by the assembled student body:

One Robert Deale of New Coll. spake before her at the North Gate, called Bocardo, an Oration in the name of all the Scholars, that stood one by one on each side of the street from that place to Quatervois: which being finished, she went forward, the Scholars all kneeling and unanimously crying, 'Vivat Regina'; which the Queen taking very kindly, answered oftentimes with a joyful countenance, 'Gratias ago, gratias ago'.

At her coming to Quatervois (commonly called Carfax) an Oration was made in the Greek Tongue by Mr. Lawrence the King's Professor of that language in the University; which being finisht, she seemed to be so well pleased with it, that she gave him thanks in the Greek Tongue, adding, that it was the best Oration that ever she heard in Greek, and that 'we would answer you presently, but with this great company we are somewhat abashed: we will talk more with you in our Chamber'.

Anthony Wood, *History and Antiquities of the University of Oxford*, 1674.

The Queen attended disputations in philosophy, civil law, divinity and physick, and in St. Mary's Church she herself made a much admired Latin oration. A play, Palaemon and Arcyte, *was performed for her at Christ Church (always a great place for plays). During the first part the stage fell down, killing three people and injuring five more:*

Which disaster coming to the Queen's knowledge [*says Wood*] she sent forthwith the Vicechancellor and her Chirurgeons to help them, and to have a care that they wanted nothing for their recovery. Afterwards the Actors performed their parts so well, that the Queen laughed heartily thereat, and gave the Author of the Play [Richard Edwards] great thanks for his pains.

During the second part of the performance, staged the following night, a fox hunt was enacted in the quadrangle, so convincing, Wood says, that some of the students thought it to be real:

They cried out 'Now now—there there—he's caught, he's caught.' All which the Queen merrily beholding said, 'O excellent! those boys in very troth are ready to leap out of the windows to follow the hounds.'

The play was so brilliant that its author, Mr. Edwards, was advised to write no more, in case he went mad from sheer genius: but it was his last production anyway, for he died within a few months.

At the end of her visit the Queen was escorted out of Oxford by 'certeine doctors of the universitie riding before in their scarlet gowns and hoods'. Wood continues:

When she came to the forest of Shotover, about two miles from Oxford, the Earl of Leycester, Chancellor of the University, told her

that the University liberties reached no further that way; whereupon
Mr Roger Marbeck spake an eloquent oration to her, containing
many things relating to learning and the encouragement thereof by
her; of its late eclipse, and of the great probability of its being now
revived under the government of so learned a Princess etc., which
being done she gave him her hand to kiss with many thanks to the
whole University.

🐉 *Queen Elizabeth's second visit, in 1592, seems to have been a more
emotional occasion (she was now fifty-eight). Once again, Wood reports,
she attended the statutory disputations:*

One of the questions was,

   'Whether it be lawful to dissemble in cause of Religion?'

Which being looked upon as a nice question caused much attention
from the courtly Auditory. One argument more witty than solid,
that was urged by one of the opponents, was this—'It is lawful to
dispute of Religion, therefore 'tis lawful to dissemble': and so going
on, said, 'I myself now do that which is lawful; but I do now dissemble: ergo, it is lawful to dissemble.' At which her Majesty and
all the auditory were very merry. The Bishop [of Hereford] in his
Oration concerning the said question, allowed a secresy, but without
a dissimulation; a policy, but not without piety, lest men taking too
much of the serpent, have too little of the dove. All that then was
disliked in him, was the tediousness in his concluding Oration; for
the Queen, being something weary of it, sent twice to him to cut it
short, because herself intended to make a publick speech that evening;
but he would not, or as some told her, could not put himself out of a
set methodical speech for fear he should have marred all, or else
confounded his memory. . . .

   Next morning . . . she proceeded to her Oration; and when she
was in the midst thereof, she cast her eye aside, and saw the old
Lord Treasurer Burleigh (Cecil) standing on his lame feet for want
of a stool; whereupon she called in all haste for a stool for him; nor
would she proceed in her speech till she saw him provided of one.
Then fell she to it again, as if there had been no interruption. Upon
which one that knew he might be bold with her, told her after she
had concluded, that she did it of purpose to shew, that she could
interrupt her speech, and not be put out, although the Bishop durst
not adventure to do a less matter the day before.

🐉 *Her departure this time (she never came to Oxford again) was to give rise
to one of the most frequently quoted of all Oxford quotations:*

In the afternoon she left Oxford, and going through Fishstreet to Quatervois, and thence to the East Gate, received the hearty wishes (mixt with tears) of the people; and casting her eyes on the walls of St. Mary's Church, All Souls, University and Magdalen Colleges, which were mostly hung with Verses and emblematical expressions of Poetry, was often seen to give gracious nods to the Scholars. When she came to Shotover Hill (the utmost confines of the University) accompanied with those Doctors and Masters that brought her in, she graciously received a farewell Oration from one of them, in the name of the whole University. Which being done, she gave them many thanks, and her hand to kiss; and then looking wistfully towards Oxford, said to this effect in the Latin Tongue: 'Farewell, farewell, dear Oxford, God bless thee, and increase thy sons in number, holiness and virtue, &c.' And so went towards Ricote.

OXFORD BELLS

Mr. Wirdescue told me that the use of ringing uppon the Coronation day was never used here in Englande before the time of Queene Elizabeth, in whose fortenth yeare of her raigne or thereabouts, it began first at Oxford, thus. St. Hugh's day beinge a gaudy day in Lincolne College, the masters and the other company after their gaudies and feastinge went to ringe at Allhallowes, for exercise sake. Mr. Waite beinge then mayor of Oxford and dwellinge thereabouts, being much displeased with their ringinge (for he was a great precisian) came to the Church to knowe the cause of the ringinge. And at length beinge let in by the ringers, who had sut the doores privately to themselves, he demanded of them the cause of their ringinge, charginge them with popery, that they rang a *dirige* for Queen Mary, etc., because she died upon that day. The most part answered that they did it for exercise; but one, seeinge his fellowes pressed by the mayor so neere, answered that they runge not for Queen Marie's *dirige* but for joy of Queen Elizabeth's coronation and that that was the cause of the ringinge. Whereuppon the mayor goinge away, in spite of that answer, caused Karfox bells to be runge, and the rest as many as he could command, and so the custom grewe.

Brian Twyne (1579?–1644), manuscript in the Bodleian Library, 1610.

TROUBLES AT CORPUS CHRISTI

🐝 *One of the more unsettled colleges in Queen Elizabeth's time was Corpus Christi, Richard Fox's 'hive of blissful bees' (pages 36–38), which had a strong Roman Catholic interest. In 1566 a disgruntled Fellow, Hieronymus Reynolds, brought a series of charges against the college President, Thomas Greenway, which were heard in the college chapel before the Bishop of Winchester's Chancellor, representing the college Visitor. They included the following:*

He spoyleth the Colledge wodes, as the common report is, and maketh in every sale a part of mony unto himself. He is noted of many men to have had connexion with viii Infamous women, ii at Heyford (of which parish he was Rector), whereof one he brought from Warminster, another from London, one at Exeter, one in St. Allbones an olde acquaintance of his when he dwelte there, and fower at London, as Barbara his Ostes at the Cock, Margaret Burton, Johane Townsende and Alice of the Cock. . . . He is accompted a Whoremonger, a common drunkard, a mutable papist and an unpreching prelate and one of an Italian faith. He resorteth to bull-beytinge and bearebeyting in London. In Christmas last past he, comming drunke from the Towne, sat in the Hall amonge the Schollers until i of the clock, totering with his legge, tipling with his mouth, and hering bawdy songes with his eares as, My Lady hath a prety thinge, and such like. He kepeth vi horse continually in the stable, whereas the Colledge nedeth and alloweth but five. He ys a faithfull frende to all the papistes and a mortall enemy to all the protestants in this house, and therefore ys reported to study Jacke Maicher, a wicked boke written in the italian tongue. He calleth prestes sonnes prestes Brattes. He admitted Mr Belly without an oathe. He hath lefte in our fine Box but iis vid, in which, at his cumming, he found cccccli.

🐝 *The President denied all charges (not altogether convincingly, especially about the Infamous women) and was apparently confirmed in his office. Two years later he retired anyway, withdrawing to the varied pleasures of his parish, and the college was temporarily without a President:*

In the vacancy the Queen Elizabeth I commended to the choice of the Society one William Cole, sometime Fellow of that College, afterwards an exile in Queen Mary's Reign, suffering then very great hardships at Zurich. But, when the prefixed time came, the Fellows, who were most inclined to the R. Catholic persuasion, made choice of one Rob. Harrison, Master of Arts, not long since removed from the

College by the Visitor for his (as 'twas pretended) Religion, not at all taking notice of the said Cole, being very unwilling to have him, his wife, and children, and his Zurichian Discipline introduced among them.

🔖 *The Queen overruled them: Cole was forcibly installed, and three Roman Catholic Fellows were expelled. Two remained quietly in Oxford. The third, George Napier, went abroad:*

[After] spending some time in one of the English Colleges, that was about these times erected, [he] came again into England and lived as a seminary Priest among his relations, sometimes in Holywell near Oxford, and sometimes in the country near adjoining, among those of his profession. At length, being taken at Kertlington, and examined by one Chamberlaine, Esq., a Justice of the Peace, was sent Prisoner to the Castle of Oxford, and, the next Sessions after, being convicted of Treason, was on the 9 Nov. 1610 hanged, drawn, and quartered in the Castle yard. The next day his head and quarters were set upon the 4 Gates of the City, and upon that great one belonging to Ch.Ch. next to St. Aldate's Church, to the great terror of the Catholics that were then in and near Oxford. He was much pitied for that his grey hairs should come to such an end, and lamented by many that such rigour should be shewn on an innocent and harmless person. . . . Some, if not all, of his quarters were afterwards conveyed away by stealth, and buried at Sandford near Oxford, in the old Chapel there.

🔖 *As for Cole, he proved such an unpopular President that he was taken to task by the College Visitor, the Bishop of Winchester, himself a former exile in Zurich, where the two men had lived together in great poverty:*

After long discourses on both sides, the Bishop plainly told him— 'Well well, Mr. President, seeing it is so, you and the College must part without any more ado, and therefore see that you provide for yourself.' Mr. Cole therefore, not being able to say any more, fetcht a deep sigh and said—'What, my good Lord, must I then eat mice at Zurich again?' meaning that must he endure the same misery again that he did at Zurich, when he was an exile in Queen Mary's. . . . At which words the Bishop being much terrified, for they worked with him more than all his former oratory had done, said no more, but bid him be at rest and deal honestly with the college.

Anthony Wood, *History and Antiquities of the University of Oxford*, 1674.

🦎 *The Bishop of London, informed by letter of these various proceedings,
expressed his approval forcibly:*

My Lords, I like this letter very well, and think, as the writer, if by
some extraordinary ready means that house or school be not purged,
those godly foundations shall be but a nursery of adder's brood, to
poison the Church of Christ.

Edm. London.

STAYING TOO LONG

From our entrance into the universitie unto the last degree received
is commonlie eighteene, or peradventure twentie yeeres, in which
time, if a student hath not obteined sufficient learning, thereby to
serve his own turne and benefit his common wealth, let him not looke
by tarieng longer to come by anie more. For after this time and
40 yeeres of age, the most part of students doo commonlie give over
their woonted diligence and live like drone bees on the fat of colleges,
withholding better wits from the possession of their places and yet
dooing little good in their own vocation and calling. . . . Long continuance in those places is either a signe of lacke of friends or of
learning or of good and upright life, as bishop Fox sometimes noted,
who thought it sacrilege for a man to tarrie anie longer at Oxford
than he had a desire to profit.

William Harrison, Description of England, 1577.

PLAGUE STRIKES THE UNIVERSITY

🦎 *There are repeated records of pestilence in Oxford. Here Anthony Wood*
(History and Antiquities of the University of Oxford, *1674*) *describes
the plague of 1577:*

The 15, 16 and 17 day of July [1577] sickened . . . above 300 persons,
and within 12 days space died an hundred Scholars, besides many
Citizens. . . . The time without doubt was very calamitous and full
of sorrow; some leaving their beds, occasioned by the rage of their
disease and pain, would beat their keepers and nurses, and drive
them from their presence. Others like mad men would run about the
streets, markets, lanes, and other places. Some again would leap
headlong into deep waters. . . . The Physicians fled, not to avoid
trouble, which more and more came upon them, but to save themselves
and theirs. The Doctors and Heads of Houses all almost to one fled,

and not any College or Hall there was, but had some taken away by this infection, either in their respective Houses, or else in the country where they depended on safety.

Some thought that this Oxford mortality . . . was devised by the Rom. Catholics, who used the Art Magick in the design. . . .

🦶 *Whether it was or not, several Oxford colleges acquired country houses to which their members could flee in times of epidemic. Wood himself attributed the infection to the stink of the prisoners to which Oxford had been exposed at a recent Assizes, but he was no admirer of the Oxford climate, either:*

Colds without coffing or running at the nose [*he wrote in his diary one day*], onlie a languidness and faintness. Certainly Oxford is no good aire.

A PRODIGAL VISITOR

🦶 *A Polish prince, Albertus de Alasco, visited Oxford in 1583 with recommendations from Queen Elizabeth herself. The University did him proud, welcoming him with banquets, fireworks, plays, disputations, and Latin orations, but according to Anthony Wood* (History and Antiquities of the University of Oxford, 1674) *the poor prince over-reached himself in response:*

Such an entertainment it was, that the like before or since was never made for one of his Degree, costing the University with the Colleges (who contributed towards the entertainment) about 350*l*. And indeed considering the worthiness of the person for whom it was chiefly made, could not be less. He was one 'tam Marti quam Mercurio': a very good Soldier and a very good Scholar, an admirable Linguist, Philosopher and Mathematician. His deportment was very winning and plausible, his personage proper, utterance sweet, nature facile and wit excellent. But that which was in him most observable, was his prodigality, for so far did he exceed his abilities, that being not able to keep within bounds (notwithstanding he had 50 Castles of great value with a wife) was forced at length to quit England (after he had tarried there 4 Months) to prevent the coming on of Creditors, and retiring to his own Country, was afterwards seen at Crakow by an English Gentleman very poor and bare.

COLLEGE PROPERTY

A note of the white game belonginge unto the Collegge of Christ Church in the Universitie of Oxford, which have been found this yeare 1594 at the Swan uppinge time, And what broodes there are of the same and where they doe lye:

17 old swans: 4 at the high Bridge at Stanton, one each at Stanton Moor, Bynssy Church, Broadford, Sandford, Abingdon, Sutton, Sparsy Bridge, Yefley Mill, Ottmoor, Ruely Lock, Wytam Steene, Hincksy Mill, and another at Abingdon. Also 9 cygnetts.

Christ Church archives, 1594.

THE BIRTH OF THE BODLEIAN

By the end of the sixteenth century squalor had fallen upon the splendid library founded by Duke Humphrey of Gloucester in the previous century: its books were dispersed, its bookshelves had been sold, its buildings stood empty and forlorn.

This was the State of Things when Sir Thomas Bodley Knt. considered the Damage which Learning had sustained, and the great Use that a publick Library would be to the Students. . . . Sir *Thomas* had all the qualities of a *Mecenias*, he was an excellent Scholar himself, a Lover of Learning in others, and the Proprietor of a very plentiful Estate. After a mature Deliberation, he desir'd Leave of the University to furnish Duke *Humphrey's* Library once more, with Desks, Seats, and Books, at his own Costs and Charges; which being gain'd, he acquitted himself beyond all Expectation.

John Ayliffe, *The Antient and Present State of the University of Oxford*, 1714.

AN OFFER

To the Vice-Chancellor, the Universitie of Oxford.

Sir,

Although you know me not, as I suppose, yet for the farthering of an offer of evident utilitie to your whole Universitie, I will not be too scrupulous in craving your assistance. I have been alwaies of a mind, that yf God of his goodness should make me able to doe any thing for the benefit of posteritie, I would shew some token of affection that I have evermore boarne to the Studies of good Learning. I know my portion is too slender to perform for the present any answerable act to my willing disposition, but yet to notifie some part of my desire in that behalf, I have resolved thus to deale. Where there hath bin hertofore a publicke Library in Oxford, which you know is apparent by the rome itself remayning and by your Statute Records, I will take the charge and cost upon me to reduce it againe to his former use, and to make it fit and handsome with seates and shelves and deskes and all that may be needfull, to stirre up other mens benevolence to help to furnish it with bookes. And this I purpose to begin, assoone as timber can be gotten, to the intent that you may reape some speedie profitt of my project. And where before as I conceave, it was to be reputed, but a store of bookes of diverse Benefactors, because it never had any lasting alowance for augmentation of the number or supplie of bookes decaied, wherby it came to passe, than when those that were in being were either wasted or embeziled, the whole foundation came to ruin: To meet with that inconvenience, I will so provide hereafter (if GOD do not hinder my present designe) as you shall be still assured of a standing annual rent to be disbursed every yere in buying of bookes, in Officers' Stipends, and other pertinent occasions; with which provision, and some order for the preservation of the place and of the furniture of it from accustomed abuses, it may perhaps, in tyme to come, prove a notable Treasure for the multitude of volumes an excellent benefit for the use and ease of Students, and a singular ornament in the University. . . .

From London Feb. 23, 1597

Your affectionate frend
THO: BODLEY

A COMMITMENT

December 12, 1610, at Stationers Hall, in Ave Mary Lane in London: Out of their zeale to the advancement of learnings, and at the request of the right worshipfull Sir Thomas Bodley, Knight . . . [this Companye] did by their Indenture . . . for them and their successors, graunte and confirme unto the Chancellor, Maisters and Schollers of the Universitie of Oxford, and to their successors for ever, That of all bookes after that from time to time to be printed in the said Company of Stationers, beinge new books and coppies never printed before, or thoughe formerly printed yet newly augmented or enlarged, there should be freelie given one perfect Booke of every such booke (in quyers) of the first ympression thereof, towardes the furnishinge and increase of the said Library.

Ever afterwards, first under this agreement, later as a Copyright Library, the Bodleian was to have a right to a copy of every book printed in Britain.

[63]

AN OATH

You promise, and solemnly engage before God, Best and Greatest, that whenever you shall enter the public library of the University, you will frame your mind to study in modesty and silence, and will use the books and other furniture in such manner that they may last as long as possible. Also that you will neither in your own person steal, change, make erasures, deform, tear, cut, write notes in, interline, wilfully spoil, obliterate, defile, or in any other way retrench, ill-use, wear away, or deteriorate any book or books, nor authorise any other person to commit the like; but, so far as in you lies, will stop any other delinquent or delinquents, and will make known their ill-conduct to the Vice-Chancellor or his deputy within three days after you are made aware of it yourself: so help you God, as you touch the Holy Gospels of Christ.

From the Latin of the Bodleian Statutes, 1610.

A MEMORIAL

Sir Thomas Bodley died Jan. 28. 1612 after he had made fit Statutes for the Government of the Place, and they had been confirm'd in *Convocation*, and he declared by the University to be the Founder of the Library: But with him the Genius of the Place did not seem to fall, since there are now more than double or treble the Number of Books in it than were there at the Time of his Death. . . . Upon the whole, this Library is much larger than that of any University in *Europe*; nay, it exceeds those of all the Sovereigns in *Europe*, except the Emperor's and the *French* Kings. . . .

John Ayliffe, *The Antient and Present State of the University of Oxford*, 1714.

TWO REFUSALS

1645: Bodley's Librarian received, via the Vice-Chancellor, an instruction from Charles I: *Deliver unto the bearer hereof for the present use of his Maiesty, a Book Intituled* Histoire Universelle du Sieur d'Aubigné: *and this shall be your warrant.* Since the Founder's Statutes forbade the sending of books out of the building, the order was refused, and King Charles cancelled it.

1654: My Lord Protector Oliver Cromwell sent his letter to Mr. Vice-Chancellor to borrow a M.S. (Joh. de Muris) for the Portugal Ambassador. A copy of the Statute was sent (but not the book), which when his Highness had read, he was so satisf'd, and commended the prudence of the Founder, who had made the place so sacred.

From the Annals of the Bodleian Library.

THE BIRTH OF THE BODLEIAN

Authors seek ye? 'Ready before your eyes!'
Each classic author in his bookcase cries.
Of this great work, scarce paralleled on earth,
Seek ye the Founder? Bodley gave it birth.

> John Owen, *Epigrams*, 1612 (from the Latin).

What ever happy Booke is chained here,
No other place or people needs to feare:—
His Chaine's a Passport to goe everywhere!

> Abraham Cowley, written in the copy of
> his Poems which he presented to the
> Bodleian, 1656.

Most noble Bodley! we are bound to thee
For no small part of our eternity.

Thou has made us all thine heirs; whatever we
Hereafter write, 'tis thy posterity.

This is thy monument! here thou shalt stand
Till the times fail in their last grain of sand.

Thou canst not die! Here you are more than safe,
Where every book is thy large epitaph.

> Henry Vaughan (1622–95), 'The Silurist'.

A GRIEVANCE

To the Worshipful Mr. Vice-Chancellor and to all heads and governors of Colleges and Halls within the famous University of Oxon.
The humble petition of William Snoshill of East Lockinge in the county of Berks, labourer, and of Jane the wife of Thomas Hatton of Childrey in the county aforesaid, labourer, sister of the said William Snoshill,

Humbly sheweth,

That your Petitioners being the grand-children of the sister of Sir Thomas Bodley, the munificent founder of the Bodleian Library in your University, being now reduc'd to a poor and low estate, do with all humility make bold to represent their distrest condition to your consideration, hoping that out of your tender pity and commiseration, and that regard you have for the pious memory of so great a benefactor to your University, to whom your poor Petitioners are so nearly allied, you will be pleas'd to consider them as real objects of your charity and compassion, and thereby you will lay an eternal obligation on them of praying for your present and future happiness.

William Snoshill
Jane Hatton.

🙕 *The Curators of the Library allowed Mr. Snoshill and Mrs. Hatton £4 out of the library funds, while Dr. Altham, Professor of Hebrew, and Dr. Hudson, librarian, gave each of them ten shillings more.*

HARMONY

🙕 *The last years of Queen Elizabeth's reign were relatively tranquil ones for Oxford, and visitors were struck by the orderly calm of the University:*

[Oxford] students lead a life almost monastic; for as the monks had nothing in the world to do, but when they had said their prayers at stated hours, to employ themselves in instructive studies, no more have these. They are divided into three Tables: the first is called the Fellows Table, to which are admitted Earls, Barons, Gentlemen, Doctors and Masters of Arts, but very few of the latter; this is more plentifully and expensively served than the others; the second is for Masters of Arts, Bachelors, some Gentlemen, and eminent Citizens: the third is for people of lower condition. While the rest are at dinner or supper in a great Hall, one of the Students reads aloud the Bible, which is placed on a desk in the middle of the Hall . . . as soon as Grace is said after each meal, every one is at liberty, either to retire to his own chambers, or to walk in the College garden, there being none that has not a delightful one. Their habit is almost the same as that of the Jesuits, their gowns reaching down to the ancles, sometimes lined with furr. . . .

> Paul Hentzner, *A Journey into England in 1598*, translated by Horace Walpole, 1757.

Oxford separates its scholars into three ranks or orders, Masters, Bachelors, Disciples. That these ranks should be confused one with another is regarded in the light of a crime. Thus it comes to pass that Masters do not rashly mingle with Bachelors, nor Bachelors lower themselves to the level of the disciples; much less do the disciples, the lowest order of all, dare to place themselves on terms of equality with their superiors. To these latter it is not even permitted to go outside their college unless accompanied by a comrade. And so a pleasing harmony is engendered between these orders, one with another, and also a great reverence on the part of all towards the heads of colleges. . . . The colleges of Oxford approach more nearly to well-ordered cloisters of religious and monks than to an assembly of young men and youths congregated in their respective halls.

Nicholas Fitzherbert, *Oxoniensis in Anglia Academia Descriptio*, 1602 (from the Latin).

ON THE RIVER, SEVENTEENTH CENTURY

In summer time to Medley,
   My love and I would go;
The boatmen there stood ready,
   My love and I to row.
For cream there would we call,
   For cakes and for prunes too;
But now; alas! she's left me,
   Falero, lero, loo.

As we walked home together
   At midnight through the Town,
To keep away the weather
   O'er her I'd cast my Gown:
No cold my Love should feel,
   Whate'er the heavens could do:—
But now alas! she's left me,
   Falero, lero, loo.

George Wither, 1604.

JACOBITE OCCASIONS

King James I, himself a bookish man, thought highly of Oxford University, and gave it the Parliamentary representation which it was to enjoy for nearly 350 years. He figures often in its annals. In 1605, for instance, he attended disputations in Physick and Philosophy during a four-day visit to the city, and according to the antiquarian Anthony Wood, enjoyed himself:

While the aforesaid Exercises were performing, the King shewed himself to be of an admirable wit and judgement, sufficiently applauded by the Scholars by clapping their hands and humming: which though strange to him at first hearing, yet when he understood, upon enquiry, what that noise meant (which they told him signified applause) was very well contented.

🐝 *On the other hand when in 1617 a University play was performed for him, it was not altogether a success:*

The Comedy of *Barten Holyday*, student of Christ Church, called the Marriage of Arts, was acted publickly in Christ Church Hall with no great applause, and the wits now of the University being minded to show themselves before the King, were resolved to act the said Comedy at Woodstock; wherefore the Author making some foolish alterations in it, was accordingly performed on a Sunday night, 26 Aug. But it being too grave for the King and too Scholar-like for the Auditory (or as some say that the Actors had taken too much wine before), His Majesty after two Acts offered several times to withdraw; but being perswaded by some of those that were near him, to have patience till it was ended, lest the young men should be discouraged, adventured it, though much against his will.

Anthony Wood, *The History and Antiquities of the University of Oxford*, 1674.

🐝 *In 1603 King James exposed an Oxford charlatan:*

Richard Haydock, of New College, practiced Physick in the day and preached in the night in bed. His practice came by his profession and his preaching (as he pretended) by Revelation; for he would take a text in his sleep and deliver a good sermon upon it; and tho' his auditory were willing to silence him, by pulling, hauling, and pinching, yet would he pertinaceously persist to the end, and sleep still. The fame of this Sleeping Preacher flies abroad with a light wing; which coming to the King's knowledge he commanded him to the Court, where he sate up one night to hear him. And when the time came that the Preacher thought it was fit for him to be asleep, he began with a prayer, then took a text of Scripture, which he significantly enough insisted on a while; but after made an excursion against the Pope, the Cross in Baptism, and the last Canons of the Church of England, and so concluded sleeping. The King would not trouble him that night, letting him rest after his labours; but sent for him the next morning and in private handled him so like a cunning Chirurgeon, that he found out the sore; making him confess not only his sin and error in his act, but the cause that urged him to it, which was, that he apprehended himself as a buried man in the University, being of a low condition, and if something eminent and remarkable did not spring from him, to give life to his reputation, he should never appear anybody, which made him attempt this novelty. The King, finding him ingenious in his confession, pardoned him.

Arthur Wilson, Life and Reign of James I, 1653.

Haydock left Oxford after this unsettling experience, but lived happily ever afterwards as a respected physician in Salisbury.

In 1620 the King presented to the University a new edition of his own works:

The King's Booke was receaved with a great deale of solemnitie, and in a solemne procession was carried from St. Marie's (where the Convocation was) by the Vice-Chancellor, accompanied with some 24 Doctors in scarlett, and the rest of the bodie of the Universitie, unto the Publick Librarie where the Keeper, one Mr Rows, made a verie prettie Speech, and placed it *in archivis, intuentibus nobis et reliquis academicis*, with a great deale of respect.

Patrick Young, in a letter, 1620.

The 'prettie Speech' said that the only thing that had been wanting to complete Sir Thomas Bodley's happiness during his lifetime was a gift from the King himself of his 'miraculously perfect' works, but that doubtless in the other world Sir Thomas was sharing in the glory of that day. In the following year the King, during a visit to the Bodleian, repaid the fancy with an often-quoted wish:

Were I not a king, I would be a university man; and if it were that I must be a prisoner, if I might have my wish, I would have no other prison than this library, and be chained together with these good authors.

PRIDEAUX OF EXETER

John Prideaux (1578–1650), Rector of Exeter College, began life in humble circumstances in Devon.

A good gentlewoman of the Parish took some compassion on him— and kept him sometime at school until he had gotten some smattering in the Latin Tongue and School learning. Thus meanly furnished, his Genius strongly inclined him to go to Oxford, and accordingly he did so, in very poor habit and sordid (no better than leather Breeches) to seek his Fortune. Being thus come out of the West, a tedious Journey on Foot, to this noblest Seat of the Muses, whither should he first apply himself for succour but to that Society therein where most of his Countrymen resided? I mean Exeter College. Here he is said at the beginning to have lived in very mean Condition and to have gotten his Livelyhood by doing servile offices in the kitchen: yet all this while he minded his Book, and what leisure he could obtain from the Business of the Scullery, he would improve it all in study. . . . Fair Blossoms of Learning promising a future good Encrease, appearing upon this young Man, the College began to take notice of him, and at length admitted him a Member of their House. . . . In the Rectorship of his College, he carried himself so winning and pleasing by his gentle Government and Fatherly Instruction, that it flourished in his time more than any House in the University, with many Scholars, as well of great as mean Birth; Yea, many Foreigners of illustrious Families led by the fame of his Learning and Wisdom, as he had been another Solomon, came over purposely to sit at his Feet, and to gain Instruction.

John Prince, *Worthies of Devon*, 1701.

A LOSS OF DIGNITY

In 1634 the foundation stone was laid of an extension to Bodley's library:

The Vicechancellor, Doctors, Heads of Houses and Proctors, met at St. Mary's Church about 8 of the clock in the morning; from thence each having his respective formalities on, came to this place, and took their seats that were then erected on the brim of the foundation. Over against them was built a scaffold where the two Proctors with divers Masters stood. After they were all settled, the University Musicians who stood upon the leads at the west end of the Library sounded a lesson on their wind music. Which being done the singing men of Christ Church, with others, sang a lesson, after which the Senior Proctor Mr. Herbert Pelham of Magdalen College made an eloquent Oration; that being ended also the music sounded again, and continued playing till the Vicechancellor went to the bottom of the foundation to lay the first stone in one of the south angles. But no sooner he had deposited a piece of gold on the said stone, according to the usual manner in such ceremonies, but the earth fell in from one side of the foundation, and the scaffold that was thereon broke and fell with it, so that all those that were thereon to the number of an hundred at least, namely the Proctors, Principals of Halls, Masters, and some Bachelaurs fell down all together one upon another into the foundation, among whom the under Butler of Exeter College had his shoulder broken or put out of joint, and a Scholar's arm bruised, as I have been informed.

The solemnity being thus concluded, with such a sad catastrophe, the breach was soon after made up, and the work going chearfully forward, was in four years space finished.

Anthony Wood, *History and Antiquities of the University of Oxford*, 1674.

JEALOUSIES

Though by the sixteenth century riots between town and gown were rare and generally mild events, old jealousies still alienated the communities, the University scorning the city's pretensions, the townspeople resenting the academics' legal privileges—sometimes, as the first of these two reports seems to suggest, with reason:

Tuesday, Aug. 26, 1634, an assises held at Oxford for the tryall of John Dunne, M.A., of Christchurch, for supposinge to have killed a little boy called Humfrey Dunt, a basket maker's son of Grampoole; and of John Goffe, M.A., and fellow of Magdalen College, for being supposed to have killed one Joseph Boys, a chaplaine of that house. Mr. Unton Croke, under stewarde of the Universitie, did sett judge by commission. There were two several commissions for those two tryalls; which being read in the Lower Gild Hall where the assises were kept, the Vice-Chancellor and the doctors being there present in their robes and some masters in their white minied hooddes, first the said steward made a short speech unto the bench in Latin by way of congratulation that the Universitie liberties were so well preserved. Then afterwards he gave the charge to the jury; and then he proceeded to the arraignment of the prisoners at the barre; and then they broke up and went all to dinner to the signe of the Starre. After dinner they came to the halle againe in their robes as before, about two o'clock, and proceeded further to the tryall first of John Dunne, who was acquitted; then of John Goffe, who was likewise acquitted. . . .

When Dr. Kinge, Dean of Christ Church, was Vice-Chancellor, there happened to be a quarter sessions about Christmas, 1609, whereat Sir David Williams, who was one of the circulating judges for Oxfordshire, was present in the upper Gildhall; unto which also came the mayor of the towne, one Alderman Harris, whom the judge placed at his right hand; and the Vice-Chancellor, Dr. Kinge, cominge in, a while after, did offer allso to set at the judge's right hand, and would have displaced the mayor; which the judge would not suffer, allowinge of the Mayor's placings, and that it was due unto him and not to the Vice-Chancellor. Whereupon the Vice-Chancellor made no stirre about it then, but sate there all the while below the mayor. And when they rose from the bench and were come down into the street, goeinge up towarde Carfax with a purpose to dine all together at the Starre, the judge did again cause the mayor to take the hand of the Vice-Chancellor; where uppon, about Alderman Wright, his house, beinge a corner house at Carfoxe, the Vice-Chancellor would goe no further, but called back the Bedell and turned downeward to Christ Church. Whereupon the judge asked him, sayinge, 'What! Mr. Vice-Chancellor, will you not dine with us?' unto which the Vice-Chancellor replied that he could have a dinner at home at Christ Church, and takinge no other leave, departed and went home. . . .

Brian Twyne (1579?–1644), Manuscript in the University Archives.

SANDERSON OF LINCOLN

Isaak Walton affectionately portrays the logician Robert Sanderson, Fellow of Lincoln College (1587–1662):

Though he was blest with a clearer judgement than other men; yet he was so distrustful of it, that he did usually over-consider of consequences, and would so delay and reconsider what to determine, that though none ever determin'd better, yet, when the bell toll'd for him to appear and read his Divinity Lectures in Oxford, and all the Scholars attended to hear him, he had not then, or not till then, resolv'd and writ what he meant to determine; so that that appear'd to be a truth, which his old dear friend Dr. Sheldon would often say of him, namely, 'That his judgement was so much superiour to his fancy, that whatever this suggested, that dislik'd and controul'd; still considering and reconsidering, till his time was so wasted, that he was forced to write, not (probably) what was best, but what he thought last.'

In 1616 Sanderson was elected Senior Proctor for the year, but he used his authority gently:

If in his night-walk he met with irregular Scholars absent from their Colleges at University hours, or disordered by drink, or in scandalous company, he did not use his power of punishing to an extremity; but did usually take their names, and a promise to appear before him unsent for next morning: and when they did, convinced them, with such obligingness, and reason added to it, that they parted from him with such resolutions, as the man after God's own heart was possessed with, when he said, 'There is a mercy with thee, and therefore thou shalt be feared': Psal. cxxx. 4. And by this and a like behaviour to all men, he was so happy as to lay down this dangerous employment, as but very few, if any, have done, even without an enemy.

❧ *His learning was his joy:*

His memory was so matchless and firm, as 'twas only overcome by his bashfulness; for he alone, or to a friend, could repeat all the Odes of Horace, all Tully's Offices, and much of Juvenal and Persius, without book: and would say 'the repetition of one of the Odes of Horace to himself, was to him such music, as a lesson on the viol was to others, when they played it to themselves or friends'.

Life of Robert Sanderson, 1678.

CORBET OF CHRIST CHURCH

Richard Corbet [Dean of Christ Church from 1620 to 1629] was very facetious, and a good fellowe. One time he and some of his acquaintance being merry at Fryar Bacon's study (where was good liquor sold), they were drinking on the leads of the house, and one of the schilars was asleepe, and had a pair of good silke stockings on. Dr. Corbet (then M.A., if not B.D.) gott a paire of cizers and cutt them full of little holes, but when the other awaked, and percieved how and by whom he was abused, he did chastise him, and made him pay for them. After he was D. of Divinity, he sang ballads at the Crosse at Abingdon on a market-day. He and some of his camerades were at the taverne by the crosse. . . . The ballad singer complaynd, he had no custome, he could not putt-off his ballades. The jolly Doctor putts-off his gowne, and putts-on the ballad singer's leathern jacket, and being a handsome man, and had a rare full voice, he presently vended a great many, and had a great audience.

John Aubrey, *Brief Lives*, 1669–96.

THE BOBARTS AND THE PHYSICK GARDEN

❧ *The Oxford Physick Garden, later called the Botanic Garden, was founded in 1621 on the site of the former Jewish cemetery. Its first gardener was a German, Jacob Bobart (1599–1679), whose son Jacob*

Bobart the Younger (1641–1719) later took charge of the garden too as Professor of Botany. Both were celebrated Oxford figures:

Amongst ye severall famous structures & curiosities wherewith ye flourishing University of Oxford is enriched, that of ye Publick Physick Garden deserves not ye last place, being a matter of great use & ornament, prouving serviceable not only to all Physitians, Apothecaryes, and those who are more imediately concerned in the practise of Physick, but to persons of all qualities serving to help ye diseased and for ye delight & pleasure of those of perfect health, containing therein 3000 severall sorts of plants for ye honor of our nation and Universitie & service of ye Comonwealth. . . . Old Jacob Bobart father to this present Jacob may be said to be ye man yt first gave life & beauty to this famous place, who by his care & industry replenish'd the walls, wth all manner of good fruits our clime would ripen, & bedeck the earth wth great variety of trees plants & exotic flowers, dayly augmented by the Botanists, who bring them hither from ye remote Quarters of ye world. . . .

Here I may take leave to speake a word or two of old Jacob who is now fled from his Earthly Paradise. As to Country he was by birth a German born in Brunswick that great Rum-Brewhouse of Europe: In his younger dayes as I remember I have heard him say he was sometime a Soldier by which Imploy and travail he had opportunitie of Augmenting his knowledge, for to his native Dutch he added the English Language, and he did understand Latine pretty well. As to fabrick of body he was by nature very well built, (his son in respect of him but a shrimp) tall straite and strong with square shoulders and a head well set upon them. In his latter days he delighted to weare a long Beard and once against Whitsontide had a fancy to tagg it with silver, which drew much Company in the Phisick-Garden.

Thomas Baskerville, c. 1675.

AN ACCUSATION

Mark Coleman, a melancholy distracted man, sometime a singing-man of Ch. Ch, walking in the Garden caught fast hold of J. Bobart senr's long beard, crying, 'Help! Help!' Upon which people coming in and enquiring of the outcry, Coleman made reply that Bobart had eaten his horse and his tayle hung out of his mouth.

Anthony Wood, The History and Antiquities of the University of Oxford, 1674

Poor Coleman is described by Thomas Baskerville, about the same time, as 'ye distracted man, one well skill'd in Musick, when hee was in his right minde. . . .'

THE BOBARTS AND THE PHYSICK GARDEN

EXPERTISE

In [New] Colledge, the house of office or Bog-house is a famous pile of building, the dung of it computed by old Jacob Bobart to be worth a great deal of money, who said this Compost when rotton was an excellent soil to fill deep holes to plant young vines.

Thomas Baskerville, c. 1675.

R.I.P.

February 4 [1679] Jacob Bobart died: servant to the University: an understanding man: the best gardener in England: hath a book extant.

Anthony Wood, Life and Times, 1632–95.

BOBART'S DRAGON

Mr. Jacob Bobart [the Younger] did about forty years ago find a dead rat in the Physic Garden, which he made to resemble the common picture of dragons, by altering its head and tail, and thrusting in taper sharp sticks, which distended the skin on each side till it mimicked wings. He let it dry as hard as possible. The learned immediately pronounced it a dragon, and one of them sent an accurate description of it to Dr. Magliabechi, Librarian of the Grand Duke of Tuscany. Several fine copies of verses were wrote upon so rare a subject, but at last Mr. Bobart owned the cheat. However it was looked upon as a masterpiece of art; and as such, deposited in the Museum, or Anatomy School, at Oxford.

Zachary Grey, notes upon Hudibras, 1744.

THE PROFESSOR

I was greatly shocked by the hideous features and generally villainous appearance of this good and honest man [Bobart the Younger]. His wife, a filthy old hag, was with him, and although she may be the ugliest of her sex he is certainly the more repulsive of the two. An unusually pointed and very long nose, little eyes set deep in the head, a twisted mouth almost without upper lips, a great deep scar in one cheek and the whole face and hands as black and coarse as those of the poorest gardener or farm-labourer. His clothing and especially his hat were also very bad. Such is the aspect of the Professor, who would most naturally be taken for the gardener.

Z. C. von Üffenbach, Oxford in 1710, ed. W. H. and W. J. C. Quarrell, 1928.

EPITAPH TO THE FATHER

To the Pious Memory of Jacob Bobart, a native German. A man of great integrity, chosen by the founder to be keeper of the Physic Garden. He dyed Feb. 4, 1679, in the 81st year of his age.

On his memorial tablet at St. Peter's-in-the-East, Oxford.

EPITAPH TO THE SON

Here lies Jacob Bobart
Nailed up in a cupboard.

In *Terrae-Filius*, 1721.

EXPELLED—WITH EXPENSES

MEMORANDUM, That after many yeares patience in which we
had laboured and expected the reformation of Matthias Watson (a
Fellow of our Societie) from a notorious lewd and deboscht course of
lyfe, after soe long patience and many yearely and almost dayly
admonitions, with all tenderness & compassion given him, both in
publicke & in private, by ech of the societie aparte, and by all of us in
generall at our severall chapter dayes, at which publicke meetings
he had for diverse yeares many & sundry peremptory warninges either
to reforme himself from such scandalous deboshtness or els to provide
himself elsewhere; and having now usd all other possible means to
reclaime him, by injoyneinge of him to absent himself from the
Colledge to see if by a more private lyfe he might recover himself,
since he hath of a longe time both in the Colledge and elsewhere soe
misbehaved himself that he hathe long been a discredite and shamefull
burden to the Colledge as also a notorious and insufferable scandall
in the office of his ministrie & callinge . . . we were inforced this 8 of
August in the yeare 1625 to proceed to his finall expulsion, when
upon our meetinge in this Chappell being found incorrigible by a
major Parte according to Statute, his place was pronounced voyd and
three dayes libertie only given him to provide for his remoovall to
his freinds; yet soe charitable & compassionate did we desire to
approove ourselves in this forced acte of expulsion that by the consent
of the major parte we agreed to buy him a new sute of apparell and
to hire a messenger and horse to carrie him downe to his freinds and
to supply him with sufficient money for his expences by the waye;
which money layd forth for him should be taken up out of the allowance of a fellowship untill the end of the quarter before Christmass and
the divident which should have been due to him at Ester followinge.

Minutes of a college meeting, Lincoln College, 1625.

AMBITION

From Banbury, desirous to add knowledge
To zeal, and to be taught in Magdalen College,
The River Cherwell doth to Isis runne
And bears her company to Abington.

John Taylor, *Thames and Isis*, 1632.

LAUD'S COLLEGE

*Archbishop Laud, Chancellor of the University and reformer of its
Statutes, was educated at St. John's College, and financed the erection of
some splendid new buildings there. King Charles I and Queen Henrietta
Maria attended the opening ceremony in 1636, and Laud, who arrived
himself in Oxford in a coach and six, attended by fifty mounted re-
tainers, described the event in his diary:*

Dinner being ready, they passed from the old into the new library,
built by myself, where the King and Queen and the Prince Elector
dined at one table, which stood cross at the upper end. And prince
Rupert with all the lords and ladies present, which were very many,
dined at a long table in the same room. All the several tables, to the
number of 13 besides these 2, were disposed in several chambers of
the college, and had several men appointed to attend them; and I
thank God I had that happiness, that all things were in very good
order and that no man went out at the Gates, courtier or other, but
content, which was a happiness quite beyond expectation.

*The puddings were especially magnificent, being modelled to represent
a full Convocation of the University in session—'masters were set in paste,
scholars in jellies': though other proud sons of St. John's contributed to
the feast 7 stags, 63 bucks and does, 5 oxen, 74 wethers, 2 lambs and a calf,
nevertheless the festivities cost the Archbishop £2,666. 1s. 7d., or rather
more than half the cost of the new buildings themselves.*

AN OXFORD INVENTION, 1636

That night, after the King [Charles I], Queen and two Princes had
supped, they saw a Comedy acted in Christ Church Hall. . . . Therein
was the perfect resemblance of the billows of the Sea rolling, and an

artificial island, with Churches and Houses waving up and down and floating, as also rocks, trees and hills. Many other fine pieces of work and Landscapes did also appear at sundry openings thereof, and a Chair also seen to come gliding on the Stage without any visible help. All these representations, being the first (as I have been informed) that were used on the English stage . . . originally due to the invention of Oxford Scholars.

Anthony Wood, *The History and Antiquities of the University of Oxford*, 1674.

LAUD'S CODE

In 1636 William Laud, Archbishop of Canterbury and Chancellor of the University, codified the jumbled mass of rules and statutes by which Oxford confusedly governed itself. 'This work', said he, 'I hope God will soe blesse as that it may much improve the honour and good government of that place, a thing very necessary in this life both for Church and Commonwealth.' The Laudian Code, which was written in Latin, and was to survive until 1864, concerned itself with every detail of University life:

PERSONAL APPEARANCES:

It is enacted that all the heads, fellows and scholars of colleges, as well as all persons in holy orders, shall dress as becomes clerks. Also that all others (except the sons of barons having the right of voting in the Upper House of Parliament, and also of barons of the Scotch and Irish peerages) shall wear dresses of a black or dark colour, and shall not imitate anything betokening pride or luxury, but hold themselves aloof from them. Moreover they shall be obliged to abstain from that absurd and assuming practice of walking publicly in boots. There must be, also, a mean observed in the dressing of the hair; and they are not to encourage the growth of curls, or immoderately long hair.

DIVINE TRUTH:

It is enacted, that the lecturers in philosophy shall, as often as they happen to treat of questions regarding God, the eternity of the world, the immortality of the soul, and others of the same kind, always follow the opinion of those persons who, on such points, dissent the least from Christian truth. But if the opinions of the philosophers are in any other respects altogether contrary to godliness, the lecturers shall earnestly remind their scholars or hearers of the feebleness of human sense to comprehend those things, the truth of which we know for certain by divine revelation.

DRINK AND TOBACCO:

It is enacted, that scholars of all conditions shall keep away from inns, eating-houses, wine-shops, and all houses whatever within the city, or precinct of the University, wherein wine or any other drink, or the Nicotian herb, or tobacco, is commonly sold; also that if any person does otherwise, and is not eighteen years old, and not a graduate, he shall be flogged in public.

SHOWMEN:

Neither rope-dancers nor players (who go on the stage for gain's sake), nor sword-matches, or sword players are to be permitted within the University of Oxford. All stage-players, rope-dancers and fencers transgressing are to be incarcerated.

SPORT:

It is enacted that scholars of all conditions shall abstain from every kind of game in which there is a money stake, as for instance, the games of dibs dice and cards, and also ball-play in the private yards and greens of the townsmen. Also, they must refrain from every kind of sport or exercise, whence danger, wrong or inconvenience may arise to others, from hunting wild animals with hounds of any kind, ferrets, nets or toils; and also from all parade and display of guns and cross-bows, and, again, from the use of hawks for fowling. In like manner, no scholars of any condition (and least of all graduates) are to play foot-ball within the University or its precinct.

IDLING ABOUT:

It is enacted, that scholars (particularly the younger sort, and undergraduates) shall not idle and wander about the City, or its suburbs, nor in the streets, or public market, or Carfax (at Penniless Bench as they commonly call it), nor be seen standing or loitering about the townsmen or workmens' shops; a description of offenders this which the old statutes of our University denominated scouts and truants.

FREQUENTING HARLOTS:

It is enacted, that scholars and graduates of all conditions are to keep away during the day, and especially at night, from the shops and houses of the townsmen; but particularly from houses where women of ill or suspected fame or harlots are kept or harboured, whose company is peremptorily forbidden to all scholars whatever, either in their private rooms or in the citizens' houses.

UNUSUAL FASHIONS:

It is enacted, that if any persons shall introduce new and unusual fashions in dress, the Vice-Chancellor and heads of colleges and halls shall, after debate among themselves, publish their opinions on the subject. Then the Vice-Chancellor is to inhibit the cutters-out or tailors of clothes from the power of making up such dresses.

CARRYING ARMS:

It is enacted that no student or other person shall by day or night carry either offensive or defensive arms, such as swords, poignards, daggers (commonly called stilettos), dirks, bows and arrows, guns, or warlike weapons or implements, within the verge of the University, unless when he happens to make a journey to parts remote, or to return therefrom, excepting parties who carry bows and arrows for fair amusement's sake.

TWO UNIVERSITY FUNCTIONARIES:

The duty of the Clerk of the University is to call the members together by the ringing of the usual bell, and to see that the places, schools churches, houses, chairs, and cushions are clean, and to garnish them with their ornaments. To look after the University clock; but if he is slovenly in his attention, or else on purpose retards its going, and then makes it too fast, the Vice-Chancellor is to set a fine of ten shillings upon him. At the command of the Vice-Chancellor or Proctors, to give the boys a public flogging, if any there be who deserve blows.

The duty of the Tintinnabulary is, at the death of doctors, masters, scholars, and other privileged persons, to put on the clothes of the deceased and give notice of their burial by ringing the bell which he carries in his hand.

EPITAPH FOR AN ORGANIST

Here lies one blown out of breath,
Who lived a merry life and dy'd a merry death.

Written for William Meredith, d. 1637,
organist of New College.

I kept both servants and horses at Oxford, and was allowed what expense or recreation I desired, which liberty I never much abused; but it gave me the opportunity of obliging by entertainments the better sort and supporting divers of the activest of the lower rank with giving them leave to eat when in distress upon my expense, it being no small honour amongst those sort of men, that my name in the buttery book willingly owned twice the expense of any in the University. This expense, my quality, proficiency in learning, and natural affability easily not only obtained the good-will of the wiser and older sort, but made me the leader even of all the rough young men of that college, famous for the courage and strength of tall, raw-boned Cornish and Devonshire gentlemen, which in great numbers yearly came to that college. . . .

It [was] a foolish custom of great antiquity that one of the seniors in the evening called the freshmen (which are such as came since that time twelvemonth) to the fire and made them hold out their chin, and they with the nail of their right thumb, left long for that purpose, grate off all the skin from the lip to the chin, and then cause them to drink a beer glass of water and salt. The time approaching when I should be thus used, I considered that it had happened in that year more and lustier young gentlemen had come to the college than had done in several years before, so that the freshmen were a very strong body. Upon this I consulted my two cousin-germans, the Tookers, my aunt's sone, both freshmen, both stout and very strong, and several others, and at last the whole party were cheerfully engaged to stand stoutly to defence of their chins. We all appeared at the fires in the hall, and my Lord of Pembroke's son calling me first, as we knew by custom it would begin with me, I according to agreement gave the signal, striking him a box on the ear, and immediately the freshmen fell on, and we easily cleared the buttery and the hall, but bachelors and young masters coming in to assist the seniors, we were compelled to retreat to a ground chamber in the quadrangle. They pressing at the door, some of the stoutest and strongest of our freshmen, giant-like boys, opened the doors, let in as many as they pleased, and shut the door by main strength against the rest; those let in they fell upon and had beaten very severely, but that my authority with them stopped them, some of them being considerable enough to make terms for us, which they did, for Dr. Prideaux being called out to suppress the mutiny, the old Doctor, always favourable to youth offending out of courage, wishing with the fears of those we had within, gave us articles of pardon for what had passed, and an utter abolition in that college of that foolish custom.

Anthony Ashley Cooper, 1st Lord Shaftesbury, 1621–83.

KETTELL OF TRINITY

He was a very tall well growne man. His gowne and surplice and hood being on, he had a terrible gigantique aspect, with his sharp grey eies.

He had a very venerable presence, and was an excellent governour. One of his maximes of governing was to keepe-downe the *juvenilis impetus*.

Mr. —. one of the fellowes, was wont to say, that Dr. Kettel's braine was like a hasty-pudding, where there was memorie, judgement, and phancy all stirred together. He had all these faculties in great measure, but they were all just so jumbled together.

A neighbour of mine told me he heard him preach once in St. Marie's Church. I know not whether this was the only time or no that he used this following way of conclusion:—'But now I see it is time for me to shutt up my booke, for I see the doctors' men come-in wiping of their beardes from the ale-house'—(He could from the pulpit plainly see them, and 'twas their custome in sermon to go there, and about the end of sermon to returne to wayte on their masters).

The Doctor's fashion was to goe up and down the college, and peepe in at the key-holes to see whether the boyes did follow their books or no.

He observed that the howses that had the smallest beer had most drunkards, for it forced them to goe into the town to comfort their stomachs; wherfore Dr Kettle alwayes had in his College excellent beer, not better to be had in Oxon.

He was irreconcileable to long haire; called them hairy scalpes, and as for periwigges (which were then very rarely worne) he beleeved them to be the scalpes of men cutt off after they were hanged, and so tanned and dressed for use. When he observed the scolars' haire longer than ordinary (especially if they were scholars of the howse), he would bring a pair of cizers in his muffe (which he commonly wore), and woe be to them that sate on the outside of the table. I remember he cutt Mr. Radford's haire with the knife that chipps the bread on the buttery-hatch.

He dragg'd with one (i.e. right) foot a little, by which he gave warning (like the rattlesnake) of his comeing.

Upon Trinity Sunday he would commonly preach at the Colledge, whither a number of the scholars of other howses would come, to laugh at him.

He was a person of great charity. In his college, where he observed diligent boyes that he ghessed had but a slender exhibition from their friends, he would many times putt money in at their windowes; that his right hand did not know what his left did.

Dr. Kettle was wont to say that 'Seneca writes, as a boare does pisse', scilicet, by jirkes.

He sang a shrill high treble; but there was one (J. Hoskyns) who had a higher, and would play the wag with the Dr. to make him straine his voice up to his.

When he scolded at the idle young boies of his colledge, he used these names, viz. *Turds*, *Tarrarags* (these were the worst sort, rude raskells), *Rascal-Jacks*, *Blindcinques*, *Scobberlotchers* (these did no hurt, were sober, but went idleing about the grove with their hands in their pocketts, and telling the number of the trees there, or so).

Tis probable this venerable Dr. might have lived some yeares longer, and finisht his century, had not those civill warres come on: which much grieved him, that was wont to be absolute in the colledge, to be affronted and disrespected by rude soldiers. The dissolutenesse of the times, as I have sayd, grieving the good old Doctor, his dayes were shortned, and dyed (July) anno Domini 1643.

John Aubrey, *Brief Lives*, 1669–96.

PURITANS

In 1640 the magnificent Chancellor of the University, the High Church Archbishop Laud, was arrested by order of Parliament and impeached for treason. He resigned the Chancellorship, feeling that he could be 'no farther usefull'. Until then the Puritans who had made their appearance at Oxford had been hardly more than figures of fun:

Those people called Puritans . . . being now numerous and observing their private meetings in Oxford, were not wanting certain Scholars that made it their recreation to scoff at, and jeere, them. These last were a company of boone Fellows, stiled themselves 'the College or Society of Wormes', and appointed Readers from among them that should lecture it at their merry meetings against the Puritans. They imitated them in their whining Tones, with the lifting up of eyes, in their antick actions, and left nothing undone, whereby they might make them ridiculous.

Anthony Wood, *The History and Antiquities of the University of Oxford*, 1674.

🦋 *Very soon the Puritan movement was to challenge the very style and purpose of Oxford: but by then the 'boone Fellows' had fought and lost a war, and Archbishop Laud had been executed.*

WAR

🦋 *When in 1642 civil war broke out between Charles I and Parliament, there was no doubt which side Oxford would support. Though many of the townspeople supported Parliament, and though there were many Puritans within the University, Oxford remained overwhelmingly royalist from start to finish. Before the fighting began the King wrote to the Vice-Chancellor, from York, asking for money:*

July 7, 1642:

Charles R

Reverend father in GOD, right trusty and well beloved we greet you well. Whereas upon a false and scandalous pretence, and which we have sufficiently made appeare to be such by our actions and declarations, and by the declaration of our Lords and Councellours here present with us, that we intended to make Warre upon our Parliament; Horse is still levied, and Plate and Money is still brought in against us, notwithstanding our Declarations and Proclamations to the contrary. . . . And whereas our University of Oxford is not only involved in the consequences of such dangerous and illegall proceedings equally with the rest of our Subjects, but by our perpetuall care and protection of such nurseries of Learning, we have especiall reason to expect the particular care of us, and their extraordinary assistance to our defence and preservation: These are therefore to will and require you to signifie to that our University in such manner as shall appear to you best for our service, that any sums of money that either any of our Colleges, out of their Treasuries, or any person thereof out of their particular fortunes, shall pay to this bearer Dr. Richard Chaworth, and receive his receipt for the same, shall be

Corpus Christi College, 1726: in the left-hand quadrangle, the chained fox which commemorated the name of the college founder, Bishop Fox

The Physic Gardens, 1766

received by us as a very acceptable service to us, and repaid by us with interest of 8*li* per centum, justly and speedily as it shall please GOD to settle the distractions of this poore Kingdome, of which our conscience beares us witness that we are not the cause.

🐾 *Convocation agreed to lend whatever money the University had, and Dr. Chaworth went away with £860. In the following week Parliament, hearing of this 'wicked purpose and intention', warned the colleges not to follow suit:*

Wee do hereby declare that the said Colleges are not bound by any such order of Convocation, being in itself unlawfull and injurious to the foundations of the Colleges, and in regard of the end to which it is designed treacherous to GOD and the Commonwealth, and that all the parties, actours and contrivers therof, are thereby liable to severe punishment, and shall for the same be questioned according to Law. Wherefore for the preservation of the Kingdome and preventing of the ruine and destruction of that famous Universitie, we do order and command, that the Heads and Fellowes of the said Colleges respectively, do surcease and forbeare that wicked and unlawfull course, and do forthwith put their plate and money into some safe place under good security, that it be not employed against the Parliament, certifying us in whose custody it doth remaine.

🐾 *The colleges disregarded the warning, and presently handed over to the King most of their plate, which was used to make coinage in a mint set up at Oxford. These were their respective contributions:*

	lb	oz	d
The Cathedral Church of Christ	172	3	14
Jesus Coll	86	11	5
Oriel Coll	82	0	19
Queen Coll	193	3	1
Lincoln Coll	47	2	5
University Coll	61	6	5
Brazen Nose Coll	121	2	15
St. Mary Magdalen Coll	296	6	15
All Souls Coll	253	1	19
Balliol Coll	41	4	0
Merton Coll	79	11	10
Trinity Coll	174	7	10
	1610	1	18

John Gutch, *Collectanea Curiosa*, 1781.

🐾 *The city was first occupied by a Parliamentary force, who removed many*

*pictures and 'Papist ornaments' from the churches, and mutilated the
statue of the Virgin Mary over the porch of St. Mary's, but at the end of
1642 the King set up his headquarters in Oxford, and the court remained
there for four years. The King lived at Christ Church, Queen Henrietta
Maria at Merton, and the courtiers and their ladies were scattered
through the colleges:*

Our grove [at Trinity College] was the Daphne for the ladies and
their gallants to walke in, and many times my lady Isabella Thynne
would make her entry with a theorbo or lute played before her. I
have heard her play on it in the grove myself, which she did rarely;
for which Mr. Edmund Waller hath in his Poems for ever made her
famous. . . . She was most beautifull, most humble, charitable, etc.
but she could not subdue one thing. I remember one time this lady
and fine Mris. Fenshawe (her great and intimate friend, who lay at
our college—she was wont, and my lady Thynne, to come to our
Chapell, mornings, halfe dressd, like angells), would have a frolick
to make a visitt to the President [Dr. Ralph Kettell]. The old Dr.
quickly perceived that they came to abuse him; he addresses his discourse to Mris. Fenshawe, saying, 'Madam, your husband and father
I bred up here, and I knew your grandfather; I know you to be a
gentlewoman, I will not say you are a whore; but gett you gonne for
a very woman.'

John Aubrey, Brief Lives, 1669–96.

*Though there was fighting in the outskirts of the city, Oxford was almost
unscathed, and when the war was lost surrendered itself quietly to the
Parliamentarians. Many Oxford royalists were heartbroken. This is an
epitaph in the church at North Hinksey, in the western purlieus of the city:*

Reader, look to thy feet. There lies Wm. Fynmore . . . who in the
year of our Ld. 1646, when loyalty and the Church fainted, lay down
and died.

*And this is a stanza from a poem addressed to Oxford in 1650 by Thomas
Vaughan of Jesus College:*

> Give my soul leave to studie a degree,
> Of sorrow, that may fit my fate and thee,
> And till my eyes can weep what can I think,
> Spare my fond teares, and here accept my ink.

*When the Parliamentarians took over the city, Anthony Wood tells us,
the University was 'exhausted of its treasure and deprived of its number of
Sons':*

Lectures and Exercises for the most part ceased, the Schools being employed as Granaries for the Garrison, which was some reason why so many Scholars were superannuated at the Surrender. . . . Those few also that were remaining, were for the most part, especially such that were young, much debauched, and become idle by their bearing Arms and keeping company with rude Soldiers. Much of their precious time was lost by being upon the guard night after night, and by doing those duties which appertained to them as bearers of Arms, and so consequently had opportunities, as Lay-Soldiers had, of gaming, drinking, swearing, &c. as notoriously appeared to the Visitors that were sent by the Parliament to reform the University. . . . I have had the opportunity (I cannot say happiness) to peruse several songs, ballads, and such like frivolous stuff, that they were made by some of the ingenious sort of them, while they kept guard at the Holybush and Angel, near Rewley, in the West Suburbs; which, though their humour and chiefest of their actions are in them described, yet I shall pass them by, as very unworthy to be here, or any part, mentioned.

The Colleges were much out of repair by the negligence of Soldiers, Courtiers and others that lay in them, a few Chambers that were the meanest (in some Colleges none at all) being reserved for Scholars use. Their treasure and plate was all gone. . . . the books of some Libraries imbeziled, and the number of Scholars few, and mostly indigent. . . . The Halls (wherein as in some Colleges, ale and beer were sold by the penny in their respective Butteries) were very ruinous, occasioned through the same ways as the Colleges were, and so they remained, except Magdalen Hall and New Inn, (which were upon the Surrender replenished with the Presbyterian faction) for several years after. . . . In a word, there was scarce the face of an University left, all things being out of order and disturbed.

PEACE

The Puritans who took control of Oxford after the civil war were not, at least to loyal Oxford eyes, very attractive:

As to manner; factious, saucy and some impudent and conceited, morose (incident to most that are sedentary and studious), false, factious in college, and delighting in petty plots. . . . Scorning at anything that seemed formall; laughing at a man in a cassock or canonicall coat or long cloak to the heels, at those praying with hats before their eyes when they come into the church or kneeling down against a pillar or form. . . . Never stiled any church by the name of 'St.'

as 'St. Marie's' 'S. Peter's' etc; but 'he preached at Marie's, Peter's', etc.

As to discipline; By constant preaching and praying they worked very much upon the affections of people, and some in so great manner that they proved no better than crazed people, or such that are dreamers of dreams, that pretend to revelations, to be instructed by visions. . . . Disputing constantly, and many good disputants then bred up, especially in philosophy. . . . Fighting in the streets (to the great scandall of the gowne), frequent. The sale of books very much, practicall divinity and quaint discourses, and money plenty. . . .

They used to love and encourage instrumental musick; but did not care for vocall, because that was used in church by the prelaticall partie. They would not goe to ale-houses or taverns, but send for their liquors to their respective chambers and tiple it there. Some would go in publick; but then, if overtaken, they were so cunning as to dissemble it in their way home by a lame leg or that some suddaine paine there had taken them. . . .

Many also of them that were the sons of upstart gentlemen, such that had got the good places into their hands belonging to the law-courts and had bought the lands of the clergy and gentry, were generally very proud, saucy, impudent, and seldom gave respect to any but the leading person. As for any of the old stock, they laughed and flouted at them, scarse gave them the wall, much less the common civility of a hat: and so it was that the antient gentry of the nation were dispised.

Anthony Wood, Life and Times, 1632–95.

🙚 *A corps of Puritan preachers took over the pulpit of St. Mary's Church, and were said to have given the Oxford scholars some wry amusement by describing their predecessors as 'Dumb Dogs, idle Drones, blind Seers, &c'. Wood lists a number of reasons why they were not taken very seriously:*

1. Their Prayers and Sermons were very tedious.
2. They made wry mouths, squint eyes, and scru'd faces, quite altering them from what GOD and Nature had made them.
3. They had antick behaviours, squeaking voices and puling tones, fit rather for Stage players and Country beggars to use, than such that were to speak the Oracles of God.
4. The truth is, they and the generality of their profession did so frame their countenances at the entrance into the pulpit, as also their pronunciation both in their Prayers and Sermons, and used the Scripture phrase, (whether understood by the people or not) as that 'no Tragedian in the world could have acted the part of a right godly man better than they did.'

❧ *In 1647 Parliament ordered a Visitation to 'reform and regulate' the University. It began badly. The University authorities were ordered to appear before its members between nine and eleven in the morning of 4 June, but the Visitors went first to hear a Visitation Sermon in St. Mary's Church, and this proved so long that by the time they moved on to the Schools eleven o'clock had come and gone, to the delight of the Vice-Chancellor, Doctors, and Proctors waiting there:*

This fair advantage being unexpectedly offered them (for which they had none to thank but the Preacher, one of the Visitors) they thought they had no reason but to embrace it. And therefore upon the striking of the clock, they . . . entred immediately into the Convocation House and there declared by the mouth of one of the Proctors (whereunto they had the attestation of a Public Notary) that 'whereas by virtue of an Ordinance of the two Houses, they had been cited to appear there that day, viz. 4 June between the hours of 9 and 11 in the forenoon, they had obeyed, and because the time limited was then expired, they held themselves not obliged to further attendance'. Whereupon the Vice-chancellor [Samuel Fell of Christ Church] gave command that they should every man forthwith repair home to their several Colleges.

In their return (the Vicechancellor and Doctors marching in a full body with the Bedells before them) they met the Visitors just in the Proscholium by the Divinity School door, where the passage being somewhat narrow, one of the Bedells (W. Ball a bold fellow) called to them 'Room for Mr Vicechancellor'—whereupon they were pleased to deny self and gave the way. The Vicechancellor very civilly moved his cap to them, saying 'Good morrow Gentlemen, 'tis past eleven of the clock'—and so passed on without taking further notice of them. Upon this there followed a great Humme from the Scholars, and so they parted, they holding on to their place of Visitation (which was the Convocation House or Apoditerium) and the Scholars to their dinner.

Anthony Wood, *The History and Antiquities of the University of Oxford*, 1674.

❧ *The Visitors, momentarily baffled, presently responded fiercely, removing Fell from the Vice-Chancellorship and from his Deanery at Christ Church (though his wife refused to budge). In April 1648 the newly installed Chancellor of the University, Lord Pembroke (described in a royalist pamphlet as 'a long-legged peece of impertinency') arrived to supervise the proceedings, and as Anthony Wood's strongly prejudiced records show, vigorously set about bringing the colleges to heel:*

In the morn. about 9 of the clock, the Chancellor, Visitors, and a strong guard of Musqueteers went to Magdalen College, and entring

into the Common Hall in expectation of finding the Members according to summons, not one but Mr. John Dale junior was found there.
No sooner they were settled, but up comes Mr. Thom. Smyth . . .
(one of the Members of the said House) and very boldly asked the
Chancellor 'By what Authority he sat there?' The Chancellor upon
this seemed to be so much troubled that he could not attend the business in hand; at length Mr. Cheynell perceiving it, told him, 'My
Lord, be not troubled, for that man' (pointing to Smyth) 'is mad'.
But Mr. Smyth overhearing said, 'Sir, I would have you to know
that I am not so mad as you &c.' which being heard by the freshmen
and rabble caused great laughter among them, because they well
knew that Cheynell was an hot-headed crazed person.

In the afternoon they go to All Souls College, and finding none of
the Fellows in the Hall there, were much troubled. At length they
sent for Dr. Sheldon the Warden (then walking in his garden) who
appearing before them, did with great moderation of mind ask them
'by what authority they summoned him?' Upon which the Authority
was shewn and read. Dr. Sheldon told them that it concerned not
him at all, for it was dated March the 8, and gave the Chancellor
and Visitors power to give possession to those which were then
voted into the places of such that had been removed by them. Also
that he was not so much as then questioned, nor voted out of his
place till March 30 &c. This puzzled the Chancellor and Visitors
very much, nor was there any answer for the present given.

At 8 of the clock in the Morn. [the] Visitors went to Christchurch with
an intent to enter the Dean's Lodgings to receive the Members of
that House according to order. But they finding them shut and none
within who would open them, sent for Andrew Burrough Provost
Marshal of the Garrison of Oxford and a Guard of Musqueteers and
others, who, being come with hammers and sledges, break open the
said doors, wherein finding Mrs. Fell and her Children, said 'that
they came in a fair way to her and desired her to quit her house'—
But she refusing, they set a guard of Soldiers in the Rooms into which
they had entred, wherein remaining for some time, endeavoured
(as 'tis said) to weary her out with noise, rudeness, smell of Tobacco,
&c. But being not dismayed with those matters continued as a Cap-
tive in her own house till she was carried forth.

🦋 *She and the other women in the house were carried out in chairs, while
the children got planks. The Visitation proceeded, empowered now to sum-
mon and suspend any uncooperative scholars, and the Fellows, students, and
servants of the colleges were separately asked whether or not they would*

submit to the authority of Parliament. Some of the answers they got at Magdalen College, taken from the Register of the Visitors, may stand for the rest:

Lodovicus Mason: I am not of the understandinge (my yeares beinge so tender) to hold your Thesis which you propose, either affirmative or negative.

Jo. Drake: To this Question whether I will submitt to the authoritie of Parliament in this Visitation, I Answere: that if the word Submitt signifie that the 2 Houses of Parliament without and against his most excellent Majestie, have a lawfull power to visite this Universitie, either by themselves or others: That then I cannot in conscience and in regard of my Oathes made to my Soveraigne and Leige Lord the Kinge, and of the Oathes made to this Universitie, without perjury submitt and acknowledge such a power.

William Sydenshaw: The Question beinge soe sublime, it passeth my weake apprehension to give any positive Answere to it.

Hugh Holden: The matter required of mee concerninge my livelyhood, I doe desire tyme to consider with myselfe, and to give in a full Answere in a matter soe much concerninge mee.

Daniell Jones, Chiefe Buttler: I must and doe submitt to the authoritie of Parliament in this Visitation.

William Hearne, Second Cooke: My conscience will not allow mee to conforme myselfe to this Visitation.

Jo: Tutchin, Junior Butler: I Jo: Tutchin to this Question am not able to resolve.

Hugh Phillips, Chorister and but a Schooleboy 14 yeare old: I confesse that I am not scholler sufficient to give an Answere to this Question propounded.

Thomas Ellis of Jesus College responded to the Visitors in the high Welsh way:

After a seriouse and diligent consultation had with my owne Conscience, I have at length pitched upon this resolution: That I cannot submit to this your Visitation, without the hazard of shipwrackinge of my soule: how pretiouse a thinge that is to every man, I neede not insist to tell you: I beseech God the Father of Mercies to strengthen mee with his grace for the mentayninge of good conscience while I am

THO. ELLIS

L. Smith of All Souls replied in the English kind:

I ever thought the high Court of Parliament the supreame Power of England, and shall alwayes submitt to that Power and authoritie soe farre as lawfully I may.

🐾 *And Lyonell Pine of Wadham relied upon opacity:*

I hope noe man, since hee cannot finde in my life past whence to censure me, greedy to find faults that hee will rake my own brest to confesse that which noe man accuse me of, neither doe I myself yet know, viz: what I possibly shall doe hereafter this when I shall be commaunded that which I yet never heard of.

🐾 *Most of the colleges were given new heads; most of the professors were replaced; Oliver Cromwell himself became Chancellor of the University; piety was rampant.*

You shall not want my prayers [*wrote Cromwell to the University*] that that seed and stock of Piety and Learning so marvellously springing up amongst you may be useful to that great and glorious Kingdom of our Lord Jesus Christ.

🐾 *As George Trosse, a puritanical student at Pembroke College declared in retrospect,*

I thank my GOD from the Bottom of my Heart, that I went to Oxford where there were so many *Sermons* preach'd, and so many *Excellent, Orthodox,* and *practical Divines,* to preach them.

🐾 *Even so, the Visitors' efforts at moral reform were not uniformly successful, and finding that students were still failing to attend sermons and lectures, in 1657 they resorted to drastic means:*

An Order For Catechizing

For the better instruction of youth in the principles of true religion, and saving knowledge of Jesus Christ: It is agreed and Ordered by the Visitors of this University: That there be catechising weekly in every Colledge and Hall in this University, upon Saturday in the afternoone, betweene the houres of five and six: to be performed by the Head of the House, or upon his necessary absence (or some other just cause of hindrance) by some other meete person therunto appoynted by him: All undergraduates are hereby enjoyned to attend in the place appoynted to be instructed: And if the said season appoynted above shall appeare inconvenient to any particuler Colledge, the Governor or Governors of the said Colledge are hereby desired to signify to the Visitors within the space of 14 daies what other day and houre they have fixed on.

🐾 *Three years later Charles II was restored to the throne, and Puritan Oxford vanished for ever.*

SCENES OF UNIVERSITY LIFE

Despite the rumpus of the civil war, University life somehow continued, as these extracts demonstrate:

AN INITIATION AT MERTON, 1647

December.—At that time Christmas appearing, there were fires of charcole made in the common hall. . . . At all these fires every night, which began to be made a little after five of the clock, the senior under-graduats would bring into the hall the juniors or freshmen between that time and six of the clock, and there make them sit downe on a forme in the middle of the hall, joyning to the declaiming desk: which done, every one in order was to speake some pretty apothegme, or make a jest or bull, or speake some eloquent nonsense, to make the company laugh. But if any of the freshmen came off dull, or not cleverly, some of the forward or pragmatical seniors would '*tuck*' them, that is, set the nail of their thumb to their chin, just under the lower lopp, and by the help of their other fingers under the chin, they would give him a mark, which sometimes would produce blood.

Now for a diversion and to make you laugh at the folly and simplicity of those times, I shall entertaine you with part of a speech which A. Wood spoke while he stood on the forme placed on the table, with his gowne and band off and uncovered.

Most reverend Seniors,

May it please your Gravities to admit into your presence a kitten of the Muses, and a meer frog of Helicon to croak the cataracts of his plumbeous cerebrosity before your sagacious ingenuities. Perhaps you may expect that I should thunder out demicannon words, and level my sulphurious throat against my fellowes of the Tyrocinian crew: but this being the universal judgement of the wee fresh water Academians, behold, as so many stygian furies or ghosts risen out of their winding sheets, wee present ourselves before your tribunal, and therefore I will not fulminate nor tonitruate words nor swell into gigantick streins: such towring ebullitions do not exuberate in my Aganippe, being at the lowest ebb . . .
&c.

Thus he went forward with smart reflections on the rest of the freshmen and some of the servants, which might have been here set downe, had not the speech been borrowed of him by several of the seniors who imbezel'd it.

Anthony Wood, *Life and Times, 1632–95.*

A GUIDE TO OXFORD CHARACTERS, 1650

A Pretender to Learning is a kind of Scholler-Mountebank. You find him in his Slippers, and a Pen in his eare, in which formality he was asleep. His Table is spred wide with some Classick Folio, which is as constant to it as the carpet, and hath lain open in the same Page this halfe yeere. He walks much alone in the posture of Meditation, and has a Book still before his face in the fields. He is a great Nomenclator of Authors, which hee has read in generall in the Catalogue, and in particular in the Title, and goes seldome so farre as the Dedication. Hee never talkes of any thing but learning, and learnes all from talking.

A Young Gentleman of the University is one that comes there to weare a gown, and to say hereafter, he has beene at the University. His Father sent him thither, because he heard there were the best Fencing and Dancing schooles. His main loytering is at the Library, where he studies Armes and Bookes of Honour, and turnes a Gentleman-Critick in Pedigrees. Of all things hee endures not to bee mistaken for a Scholler, and hates a black suit though it bee of Satin.

A Plodding Student is a kind of Alchymist or Persecuter of Nature, that would change the dull lead of his brain into finer mettle. His Study is not great but continuall, and consists much in the sitting up till after midnight in a run gowne and a Nightcap, to the vanquishing perhaps of some sixe lines. His Invention is no more than the finding out of his papers, and his few gleamings there, and his disposition of them is just as the book-binder's, a setting or glewing of them together. Hee may with much industry make a breach into Logicke, and arive at some ability in an Argument: but for politer Studies, hee dare not skirmish with them, and for poetry, accounts it impregnable.

A Downe-right Scholler is one that has much learning in the Ore, unwrought and untryde, which time and experience fashions and refines. The time has got a veine of making him ridiculous, and men laugh at him by tradition; but his fault is onely this, that his mind is somewhat too much taken up with his minde, and his thoughts not loaden with any carriage besides. The hermitage of his study makes him somewhat uncouth in the world, and men make him worse by staring on him. But practise him a little in men, and brush him over with good company, and he shall outbalance those glisterers as much as a solid substance does a feather, or gold gold lace.

From John Earle (d. 1665), *Microcosmography*.

A STUDENT'S EXPENSES, 1655

Charges upon my sonne William his going to Oxford July 1655. For a College pott, 10*li*. 2s. 9d. Deposited with his Bursar, 10*li*.

Deposited with Mr Gale his tutor, 10*li*. For lether chaires and cyrtanes, 2*li*. 11s. For a pewter cesterne, 10s. For a bason, chamberpot, snuffers and candlesticks, tinderbox, 12s. For andirons, fireshovell, tongs, bellowes, &c., 13s. 6d. For a table, lock and key, &c., 18s. 6d. For diet and horsemeat at Wickham, 1*li*. 13s. 8d. For diet and horsemeat at Oxford, 6*li*. 8s. 2d. For a tawny gowne, 6*li*. To the officers of Magdalen Colledge, 1*li*. 5s. Spent in the journey, in lace, knifes, seing the Colledges and Library, &c., 6*li*. Total 56*li*. 14s. 7d.

From the papers of Sir William Roberts in the Bodleian Library.

William jr., a gentleman-commoner of Magdalen, left Oxford without taking a degree, but became a Baronet and Member of Parliament nevertheless.

A PURITAN PRESIDENT

Thomas Goodwin was installed as President of Magdalen College by the Parliamentary Visitation that followed the Civil War. He was very Puritan, and was known as Nine-caps, supposedly because 'having a cold head he was forced to wear so many'. An essay in The Spectator, *1712, describes an encounter with this eccentric fundamentalist:*

A Gentleman, who was lately a great Ornament to the Learned World, has diverted me more than once with an Account of the Reception which he met with from a very famous Independent Minister, who was Head of a College in those times. This Gentleman was then a young Adventurer in the Republic of Letters, and just fitted out for the University with a good Cargo of *Latin* and *Greek*. His Friends were resolved that he should try his Fortune at an Election, which was drawing near in the College, of which the Independent Minister whom I have before mentioned, was Governor. The Youth, according to Custom, waited on him in order to be examined. He was received at the Door by a Servant, who was one of that gloomy Generation that were in Fashion. He conducted him, with great Silence and Seriousness, to a long Gallery which was darkened at Noonsday and had only a single Candle burning in it. After a short Stay in this melancholly apartment, he was led into a Chamber hung with Black, where he entertained himself for some time by the glimmering of a Taper, till at length the Head of the Colledge came out to him, from an inner Room, with half-a-Dozen Night-Caps upon his head, and a religious Horror in his Countenance. The young man trembled; but his Fears encreased, when, instead of being asked what Progress he had made in Learning, he was examined how he abounded in Grace. His *Latin* and *Greek* stood him in little stead; he was to give an account only of the State of his Soul, whether he was of the Number of the Elect; what was the Occasion of his Conversion; upon what Day of the Month, and Hour of the Day it happened; how it was carried on, and when compleated. The whole Examination was summed up with one short question, namely, *Whether he was prepared for Death*? The Boy, who had been bred up by honest Parents, was frighted out of his Wits at the Solemnity of the Proceeding, and by the last dreadful Interrogatory; so that upon making his escape out of this House of Mourning he could never be brought a Second time to the Examination, as not being able to go through the Terrors of it.

🦁 *A different impression is given of Goodwin by his contemporary Thomas Woodcock:*

He was somewhat whimsycall, in a frolic pist once in old Mr. Lothian's pocket (this I suppose was before his trouble of conscience and conversion made him serious). . . . He prayed with his hatt on and sitting.

CIVILITIES OF OXFORD

🦁 *Many remarkable men flourished in the apparently unpromising climate of Cromwellian Oxford, notably the group of savants led by John Wilkins at Wadham College—himself appointed to his Wardenship by the Parliamentary Visitors:*

We all din'd at that most obliging and universally-curious Dr. Wilkins's, at Wadham College. He was the first who shew'd me the transparent apiaries, which he had built like castles and palaces, and so order'd them one upon another as to take the hony without destroying the bees. These were adorn'd with a variety of dials, little statues, vanes, &c. and he was so abundantly civil, as finding me pleas'd with them, to present me with one of the hives which he had empty. . . . He had also contriv'd an hollow statue which gave a voice and utter'd words, by a long conceal'd pipe that went to its mouth, whilst one speaks through it at a good distance. He had above in his lodgings and gallery variety of shadows, dyals, perspectives, and many other artificial, mathematical, and magical curiosities, a way-wiser, a thermometer, a monstrous magnet, conic and other sections, a ballance on a demi-circle, most of them of his owne and that prodigious young scholar Mr. Chr. Wren, who presented me with a piece of white marble, which he had stain'd with a lively red very deepe, as beautiful as if it had been natural.

Thus satisfied with the civilities of Oxford, we left it. . . .

John Evelyn, *Diary*, 1654.

A PRODIGY

The 'prodigious' Christopher Wren's first Oxford building was the Sheldonian Theatre, begun in 1664, the University's ceremonial place of assembly, and originally its printing works too. Its ceiling was one of the architectural wonders of the day, prompting this stanza by 'C.S.':

> Near Earth's deep centre the Foundation lies;
> While the Roof bids Good Morning to the Skies,
> Whose unsupported Arch floats in the air
> As if no Buildings, but a Bird hung there.

As for the ceiling decoration, an ample allegory of Truth and wholesome Sciences by the otherwise mostly forgotten Robert Streater, the contemporary poet Robert Whitehall wrote of it that

> Future ages must confess they owe,
> To Streater more than Michelangelo.

CASUALTIES OF WAR

Among the casualties of the civil war, it seems, were a hive of bees which had for more than a century obligingly lived at Corpus Christi College— the 'holy bee-hive' founded by Richard Fox in 1517 (pages 36–38). These hadappeared there first to welcome the eminent rhetorician Ludovicus Vives, appointed in 1520 to be Professor of Rhetoric, but according to Robert Plot, in his Natural History of Oxfordshire (1677), they were disturbed by the times:

In the year 1630, the Leads over Vives his Study being pluckt up, their Stall was taken, and with it an incredible Mass of Honey: but the Bees, as presaging their intended and imminent Destruction (whereas they were never known to have swarmed before) did that Spring (to preserve their famous kind) send down a fair Swarm into the President's Garden, which in the Year 1633 yielded two Swarms; one whereof pitched in the Garden for the President; the other they sent up as a new Colony to preserve the Memory of this Mellifluous Doctor, as the University stiled him in a Letter to the Cardinal.

And there continued, as I am informed by several ancient Members of that Society that knew them, till by the Parliament Visitation, in Anno 1648, for their Loyalty to the King, they were all, but two, turned out of their Places, at what time, with the rest of the Inhabitants of the College, they removed themselves, but no further than the East End of the same Cloister, where as if the Feminine sympathized with the Masculine Monarchy, they Instantly declined, and came shortly to nothing. After the Expiration of which ancient Race, there came 'tis true, another Colony to the East Corner of the Cloyster, where they continued till after the Return of his most Sacred Majesty Charles II: but it not being certain that they were any of the Remains of the ancient Stock (though 'tis said they removed thence to the first place) nor any of them continuing long there, I have chose rather to fix their Period in the Year 1648, than to give too much Credit to Uncertainties.

And thus unhappily, after above Six-Score Years Continuance, ended the famous stock of Vives his Bees, where 'tis pity they had not remained, as Virgil calls them, an Immortale Genus.

🐝 *Corpus Christi was not, however, deprived of all its livestock. In David Loggan's print of the college, drawn in 1675, a fox is to be seen chained in a courtyard, preserving the memory of the founder's name, and in the same year Thomas Baskerville reported three other familiars:*

Mr. Taylor yet resident [in Corpus Christi] as I remember told me that one Ouldame was a good Benefactor to this Colledge, who has in his Escuchion Three Owls, and they have a Yew tree by ye Colledge seldome without 3 Owles. The people it seemes out of respect to their Benefactor taking care to preserve and Cherish these Birds.

THE SCHOLAR-GYPSY: AN OXFORD LEGEND

There was very lately a Lad in the *University of Oxford*, who being of very pregnant and ready parts, and yet wanting the encouragement of preferment; was by his poverty forc'd to leave his studies there; and to cast himself upon the wide world for a livelyhood. Now, his necessities growing dayly on him, and wanting the help of friends to relieve him, he was at last forc'd to joyn himself to a company of *Vagabond Gypsies*, whom occasionally he met with, and to follow their Trade for a maintenance. Among these extravagent people, by the insinuating subtilty of his carriage, he quickly got so much of their love and esteem; as that they discovered to him their *Mystery*: in the practice of which, by the pregnancy of his wit and parts he soon grew so good and proficient, as to be able to out-do his Instructors. After he had been a pretty while well exercis'd in the Trade; there chanc'd to ride by a couple of *Scholars* who had formerly bin of his acquaintance. The Scholars had quickly spyed out their old friend, among the *Gypsies*; and their amazement to see him among such society, had wellnigh discover'd him: but by a sign he prevented their owning him before that Crew; and taking one of them aside privately, desired him with a friend to go to an Inn, not far distant thence, promising there to come to them. They accordingly went thither, and he follows: after their first salutations, his friends enquire how he came to lead so odd a life as that was, and to joyn himself with such a *cheating beggerly* company. The *Scholar-Gypsy* having given them an account of the necessity, which drove him to that kind of life; told them, that the people he went with were not such *Impostours* as they were taken for, but that they had a *traditional* kind of learning among them, and could do wonders by the power of *Imagination*, and that he himself had learnt much of their Art, and improved it further than themselves could. And to evince the truth of what he told them, he said he'd remove into another room, leaving them to discourse together; and upon his return tell them the sum of what they had talked of: which accordingly, he perform'd, giving them a full account of what had pass'd between them in his absence. The *Scholars* being amaz'd at so unexpected a discovery, earnestly desir'd him to unriddle the *mystery*. In which he gave them satisfaction, by telling them, that what he did was by the power of *Imagination*, his Phancy binding theirs; and that himself had dictated to them the discourse, they held together, while he was from them: That there were warrantable wayes of heightening the *Imagination* to that pitch, as to bind another's; and that when he had compass'd the whole *secret*, some parts of which he said he was yet ignorant of, he intended to leave their company, and give the world an account of what he had learned.

Joseph Glanvill, *The Vanity of Dogmatizing*, 1661.

OXFORD RESTORED

Oxford greeted the restoration of the monarchy exuberantly, and when Charles II came to the city in 1681 it was like old times for the delighted royalists:

All the way the king passed were such shoutings, acclamations, and ringing of bells, made by loyall hearts and smart lads of the layetie of Oxon, that the aire was so much peirced that the clouds seemed to divide. The generall cry was 'Long live King Charles', and many drawing up to the very coach window cryed 'Let the king live, and the

devill hang up all roundheads': at which his majestie smiled and
seemed well pleased. . . . The youths were all on fire, and when love
and joy are mixed, cannot but follow rudeness and boysterousness.
Their hats did continually fly, and seriouslie had you been there, you
would have thought that they would have thrown away their verie
heads and leggs. . . . ('Twas observed by some of our *curiosi* that as
the king passed westward up the High Street, the small raine that
then fell which was driven by the west wind, was returned back all
the way in that street at least a man's length by the verie strength of
voices and hummings. . . .)

Anthony Wood, Life and Times, 1632-95.

🐾 *But the University over-reacted to the relaxation of Parliament's sober
controls:*

Of all places the University being fast to the monarchy, suffering most
and being most weary of the usurpation . . . when the King came in,
nay, when the King was but *voted* in, they were not onely like them
that dream, but like them who are out of their wits, mad, stark,
staring mad. To study was *fanaticism*, to be moderate was downright
rebellion, and thus it continued for a twelvemonth; and thus it would
have continued till this time, if it had not pleased God to raise up some
Vice-Chancellours who stemmed the torrent which carried so much
filth with it, and in defiance of the loyal zeal of the learned, the drunken
zeal of dunces, and the great amazement of young gentlemen who
really know not what they would have, but yet made the greatest
noise, reduced the University to that temperament that a man might
study and not be thought a dullard, might be sober and yet a con-
formist, a scholar and yet a Church of England man.

Stephen Penton, The Guardian's Instruction, 1688.

Divers things are desired by most sober men to be reformed that
now, anno 1666, and divers years before (especially from 1660)
are crept in among us.—Baudy houses and light huswifes giving
divers yong men the pox soe that desease is very common among
them and some obscure pocky doctors obtaine a living by it. And
whereas it was notorious formerly to those that had it, it is now soe
common (especially in Exeter College [and] Xt. Ch . . .) that they
glory of it.—Corrupters of youth, such that live obscurely and lurke
in the towne taking all advantages to make pray of scollers . . .—
Multitudes of ailhouses . . . keeping dice, cards, sketells, shuffle-
boords, billiard tables.—Extravigancie in apparell, having their suits
and hats dect with colored ribbons, and long haire periwiggs: and
extravigancie in their gownes.—Lying and swearing much used.—

Atheisme.—Disrespecte to seniors, sawciness; occasioned by Masters their accompaniing and courting undergraduates.

Why doth solid and serious learning decline, and few or none follow it now in the university? Answer: because of coffee-houses, where they spend all their time; and in entertainments at their chambers, where their studies are become places for victualers, also great drinking at taverns and ale-houses (Dr. Lampshire told me there were 370 in Oxford), spending their time in common chambers whole afternoons, and thence to the coffee-house.

Apr. 26 [1666], the new proctors took their places, but such rudeness performed by the undergraduates that never before was heard. They houted and hum'd all the way from the Scooles to Xt. Ch.; went even with the Proctors and Masters, staring them in the faces, then through all the quadrangles of Xt. Ch., nay, up into the hall, in so much that proctor Hodges was faigne to goe up and turne them out. They not contented with that; but they come againe, staring upon the Masters while they scambles, and laughing and hooting at them, and the like. After that houted and hum'd downe Mr. Bayley to Magd. (i.e. Magd. Coll.) crying 'Hum Bury but Hum as and Thom as'. This is for want of strict governors and orders to be made.

Anthony Wood, Life and Times, 1632–95.

🐾 *Nor were the Oxford Puritans altogether subdued by the turn of events:*

The first matter . . . that the restored persons looked after was to put themselves in the most prelaticall garbe that could be, and the rather, that they might encourage others, especially those of the intervall, to doe the like; to restore all signes of monarchy in the Universitie, the Common prayer, surplice and certaine customes.

But the Presbyterians, whos number was considerable, seeing their disciples daily fall off, endeavoured to make these matters ridiculous either in their common discourses, libells, or some idle pamphlets that they caused to be dispersed. They compared the organ to the whining of pigs; their singing, to that of a joviall crew in a blind ale-house. . . . They brought it to pass also to make them ridiculous in severall speeches spoken in the Act next year by such that had been initiated in their discipline, and to make the auditory believe that the devill used to walk in a surplice severall nights in Magdalen College cloyster.

Nay, some varlets of Christ Church were so impudent . . . to goe on the 21 January this year about 11 or 12 of the clock at night to a chamber under the common hall (where the choiresters learne their grammar) and then to take away all such surplices that they could find: and being so done, to throw them in a common privy house belonging to Peckwater Quadrangle, and there with long sticks to thrust them downe into the excrements. The next day being discovered, they were taken up and washed; but so enraged were the deane and canons, that they publickly protested, if they knew the person or persons that had committed that act, they should not onlie lose their places and be expelled the Universitie but also have their eares cut off in the market place. The Presbyterians were wonderfully pleased at this action, laughed hartily among themselves, and . . . soone after came out a ballad or lampoone . . . the beginning of which was this:—

> Have pity on us all, good Lairds,
> For surely wee are all uncleane;
> Our surplices are daub'd with tirds,
> And eke we have a shitten Deane.

🦍 *Even the staunchly royalist Wood, when the court returned to London after a prolonged stay in Oxford, viewed its departure with relief:*

The greater sort of the courtiers were high, proud, insolent, and looked upon scolars noe more than pedants, or pedagogicall persons: the lower sort also made no more of them then the greater, not suffering them to see the king or queen at dinner or supper or scarce at cards or at masse, never regarding that they had parted with their chambers and conveniences.

The townsmen, who were gainers by the court, grew rich and proud, and cared not for scolars; but when the court was gone they sneaked to them againe.

To give a further character of the court, they, though they were neat and gay in their apparell, yet they were very nasty and beastly, leaving at their departure their excrement in every corner, in chimneys, studies, cole-houses, cellars. Rude, rough, whoremongers; vaine, empty, carelesse.

DONS, 1666

20 Oct., S., Dr. John Wall, D.D. and canon of Xt Ch., departed this life and was buried . . . on the north side of the choire. He left 2000 and 20 li. to the city of Oxon for the maintenance of 30 poore widows, each to have 4 li. per annum a peice. His money left to veterans. . . .

He left the college nothing, though they and the University stood much in need of it at this time towards their buildings. But the reason was evident, because that Jasper Mayne, Dr. Richard Gardiner, and the deane also would be alwaies gibing him at meales when they meet togeather, especially after he had left the first 1000 li. to the towne, 1664; for which also the students jeered him, and some of the officers of the Act (of whom one or two were of Xt Ch.). and this without controule from the deane or others. None of the college followed him to church; only the choire going before him. . . .

'Tis said that Dr. Richard Gardiner came to Dr. Wall's lodgings the day he died and offered his service to pray with him, but Dr. Wall refused it. Upon which old Gardiner being inraged said that he was 'a mudde wall, a tottered wall, a toren wall, nay! a towne wall'; and broke his windows with his staff.

Anthony Wood, Life and Times, 1632–95.

MIGHTY TOME

O! The bonny Christ Church Bells
One, two, three, four, five, six,
That ring so mighty sweet,
So wonderous great,
And trowle so merrily, merrily.

O the first and second Bell,
Which every day at four and ten
Cry come, come, come, come, come,
Come to prayers,
And ye verger troops before ye Deane.

Tingle, tingle, tingle,
Says the little bell att 9
To call the beerers home;
But the devill a man
Will leave his Can
Till hee hears the mighty Tome.

Henry Aldrich, c. 1670

'Mighty Tome' was Great Tom, which had once hung in Osney Abbey, and was now installed in Tom Tower at Christ Church, designed for it by Christopher Wren.

THE VICE-CHANCELLOR AND THE FLYING COACH

🦌 *The return of the monarchy strengthened once again the academic grip on Oxford. In 1670 the University licensed its official carrier to run a flying coach service to London. Not to be outdone, the City authorized Messrs. Dye and Fossett, Carriers, to run a rival service. Instantly came the academic riposte:*

February 13th, 1671.

These are to give notice that whereas Thomas Dye and John Fossett have, without licence from me, and in contempt of the Chancellor, Masters and Scholars of the University (to whom the ordering and giving of all carriers of what kind soever, trading to or with the University and City of Oxford doth of right belong) presumed to set up a flying-coach to travail from hence to London. These are to require all scholars, privileged persons and members of the University not to travail in the said coach set up by Thomas Dye and John Fossett, nor to send letters of any goods whatsoever by the aforesaid flying-coach.

P. MEWS, Vice-Chancellor.

TWO DRINKING STORIES

There is over against Baliol College, a dingy, horrid, scandalous alehouse, fit for none but draymen and tinkers, and such as by going there have made themselves equally scandalous. Here the Baliol men continually ly, and by perpetuall bubbeing add art to their natural stupidity to make themselfes perfect sots. The head, being informed of this, called them togeather, and in a grave speech informed them of the mischiefs of that hellish liquor cold ale, that it destroyed both body and soul, and advised them by noe means to have anything more to do with it; but one of them, not willing soe tamely to be preached out of his beloved liquor, made reply that the Vice-Chancelour's men dranke ale at the Split Crow, and why should not they too? The old man, being nonplusd with this reply, immediately packeth away to the Vice-Chancelour, and informd him of the ill example his fellows gave the rest of the town by drinking ale, and desired him to prohibit them for the future; but Bathurst the Vice-Chancelour, not likeing his pro-posall, being formerly an old lover of ale himselfe, answered him roughly, that there was noe hurt in ale, and that as long as his fellows did noe worse he would not disturb them, and soe turnd the old man goeing; who returneing to his colledge, calld his fellows again and told them he had been with the Vice-Chauncelour, and that he told him there was noe hurt in ale; truely he thought there was, but now, being informed of the contrary, since the Vice-Chauncelour gave his men leave to drink ale, he would give them leave to; soe that now they may be sots by authority.

Van Tromp [the great Dutch admiral] came hither on Tuesday night and immediately waited on our Dean, by whom he was treated at dinner the next day; he desired he might have salt meat, he never useing to eat any other, which put Mr Dean much to it to find that which would please his pallet. He had much respects shown him here, and the University presented him with a Drs degree, but the seaman thinkeing that title out of his element would have nothing to doe with it. He was much gazed at by the boys, who perchance wondred to find him, whom they had found so famous in Gazets, to be at last but a drunkeing greazy Dutchman. . . . To drinke . . . is the only thing he is good for; and for fear he should loose soe commendable a quality he dayly exerciseth it, for wont of better company, with Price our butler and Rawlins the plumber, with whom he spendeth al the time he is here either in the brandy shop or tavern. . . .

We got a greater victory over [him] here then all your sea captaines in London, he confesseing that he was more drunke here then anywhere else since he came into England, which I thinke very little to the honour of our University. Dr. Speed was the chiefe man that encountred him, who mustering about five or six more as able men as himselfe at wine and brandy got the Dutchman to the Crown Tavern, and there soe plyed him with both that at 12 at night they were fain to carry him to his lodgeings.

Humphrey Prideaux, Letters to Dr. Ellis, 1675.

PORNOGRAPHY AT THE PRESS

Pornography was very nearly printed on the University's press, then housed in the Sheldonian Theatre, in 1675, when there was a plot to produce a surreptitious edition of 'Aretino's postures', Giulio Romano's celebrated engravings to illustrate Pietro Aretino's sonnets. Humphrey Prideaux of Christ Church reports in a letter to his friend John Ellis:

We were like to have had an edition of them from thence were it not that last night the whole worke was mard. The gentlemen of All

Souls had got them engraved, and had imployed our presse to print them of. The time that was chosen for the worke was the evening after 4, Mr Dean [the printer] after that time never useing to come to the theator, but last night, beeing imployed the other part of the day, he went not thither till the work was begun. How he tooke to find his presse workeing at such an imployment I leave you to immagine.

✤ *The project was abandoned.*

SECOND BEST

BENEATH THIS MARBLE IS BURIED
THO. WELSTED
WHO WAS STRUCK DOWN BY THE THROWING OF A STONE
HE WAS FIRST IN THIS SCHOOL
AND WE HOPE IS NOT LAST IN HEAVEN
WHITHER HE WENT
INSTEAD OF TO OXFORD
JANUARY 13, 1676
AGED 18

epitaph, Winchester College (from the Latin).

THIS NICER PIT

London likes grossly; but this nicer pit
Examines, fathoms, all the depths of wit;
The ready finger lays on every blot;
Knows what should justly please, and what should not.

But by the sacred genius of this place,
By every Muse, by each domestic grace,
Be kind to wit, which but endeavours well,
And, where you judge, presumes not to excel.

If his ambition may those hopes pursue,
Who with religion loves your arts and you,
Oxford to him a dearer name shall be
Than his own mother-university.
Thebes did his green unknowing youth engage;
He chooses Athens in his riper age.

John Dryden, *Prologue to the University of Oxford*, 1681

LOCKE OF CHRIST CHURCH

🕮 *John Locke the revolutionary philosopher (spelt in his own time 'Lock')
combined politics with the intellectual life and became a personal adviser
to the statesman Lord Shaftesbury (page 81). When Shaftesbury fell from
power in 1681, and was obliged to flee to Holland, Locke too, then a
Student (Fellow) of Christ Church, was suspected of treasonable activity,
and closely watched. His colleagues were mystified by his comings and
goings, as Humphrey Prideaux, also of Christ Church, reported to a
friend:*

Where J. L. goes I cannot by any means learn, all his voyages being
so cunningly contrived; sometimes he will goe to some acquaintances
of his near ye town, and then he will let anybody know where he is;
but at other times when I am assured he goes elsewhere, noe one
knows where he goes, and therefore the other is made use of only for
a blind. He hath in his last sally been absent at least 10 days, where I
cannot learn. Last night he returned; and sometimes he himselfe goes
out and leaves his man behind, who shall then to be often seen in ye
quadrangle to make people believe his master is at home, for he will
let noe one come to his chamber, and therefore it is not certain when he
is there or when he is absent. I fancy there are projects afoot.

🕮 *So unsafe did Locke feel that he too fled to Holland, and was named as a
traitor wanted by the British Government. In 1684 Lord Sunderland,
Secretary of State to Charles II, suggested to John Fell, Dean of Christ
Church, that Locke should be expelled from the college, but Fell replied that
there were no reasonable grounds for doing so—as Prideaux had observed:*

[He] lives very quietly with us, and not a word ever drops from his
mouth that discovers any thing of his heart within. . . . He seems to be
a man of very good converse, and that we have of him with content;
as for what else he is he keeps it to himselfe, and therefore troubles
not us with it nor we him.

🕮 *Fell himself said there had never been 'such a master of taciturny', but
Sunderland soon came back with a more peremptory letter:*

CHARLES R.

Right Reverent Father in God, and trusty and well-beloved, we greet
you well. Whereas we have received information of the factious and
disloyall behaviour of — Lock, one of the Students of that our
Colledge, we have thought it fit hereby to signify our will and pleasure

to you, that you forthwith remove him from his said Student's place and deprive him of all the rights and advantages thereunto belonging; for which this shall be your warrant. And so we bid you heartily farewell. Given at our Court at Whitehall the 11th day of November 1684, in the six and thirtieth yeare of our reigne.

By his Majestie's command
SUNDERLAND

This time the Christ Church authorities, seldom sympathetic anyway to visionary free-thinkers, complied at once, thus ending Locke's Oxford career and depriving themselves of their most eminent contemporary:

By the Dean and Chapter of the Cathedral Church of Christ in Oxon, 15 November 1684. The day and year above written his Majesty's mandate for the removal of Mr. Lock from his Student's place and deprivation of him from all rights and advantages thereto belonging was read in Chapter and ordered to be put in execution, there being present Jo: Oxon, Dean, Dr. Ed: Pocock, Dr. Henry Smyth, Dr. Jo: Hammond, Dr. Henry Aldrich.

After the Glorious Revolution of 1688 Locke returned in honour to England, but understandably declined to ask for his place at Christ Church again. It was not until 1754 that the Dean and Chapter made amends by placing a full-length statue of him, by Roubillac, on the staircase of their new college library, and by referring to him in a Chapter minute as 'John Lock, Esq., formerly Student of Christ Church and an ornament of this society'.

MR. WILDING'S EXPENDITURES

James Wilding took his degree at Merton College in 1687. Here are some of his expenditures:

For Paper, Inkhorn, and Lead pen	£ 00	01	10
For Candlesticke & Lanterne	00	01	11
For declaiming in ye Schools	00	04	00
For a purge	00	01	00
For being let blood	00	01	00
A dyet drink	00	02	06
For my letters of Ordination	00	01	00
To ye Bishop's Men	00	00	06

For carrying my surplice	00	01	06
To ye poor	00	00	02
For Venus Treacle	00	00	01
For Lobsters, 2	00	00	04
For mending my shoose	00	00	02
To Mary	00	00	06

published in *Collectanea*, Oxford Historical Society, 1890.

'Venus Treacle' I surmise to have been the panacea called Teriaca which originated in Venice.

THE TUTOR'S ADVICE

Stephen Penton, Principal of St. Edmund Hall from 1675 to 1683, published anonymously in 1688 The Guardian's Instructor, or the Gentleman's Romance, *written for the diversion and service of the gentry. Here is some of the advice it contains, offered by an imaginary Oxford tutor to the father of a prospective undergraduate:*

– That he observe the duties of the House for prayers, exercise, etc. as if he were the son of a *beggar*; for when a young boy is plumed up with a *new suit*, he is apt to fancie himself a fine thing.

– That he writes no letter to come home for the first *whole year*. It is a common and a very great inconvenience, that soon after a young gentleman is settled, and but *beginning to begin* to study, we have a tedious ill spell'd letter from a dear sister, who languishes and longs to see him as much almost as she doth for a husband. And this, together with rising to prayers at six a clock in the morning, softens the lazy youth into a fond desire of seeing them too. . . . The next news of him is at home; within a day or two he is invited to a *hunting match*, and the sickly youth, who was scarce able to rise to prayers, can now rise at four of the clock to a fox-chase; then must he be treated at an ale-house with a rump of beef seven miles from home, hear an uncle, cousin or neighbour rant and swear; and after such a sort of *education* for six or eight weeks, full of tears and melancholy, the sad soul returns to *Oxford*; his brains have been so *shogged*, he cannot think in a fortnight.

– That he frequent not *publick places*, such as are bowling green, racket court, etc. for beside the danger of firing his bloud by a *fever*, heightning passion into *cursing* and *swearing*, he must unavoidably grow acquainted with *promiscuous* company, whether they are or are not *vertuous*.

– Whatever letters of *complaints* he writes home I desire you to send me a copy; for ill-natured, untoward boys, when they find discipline sit hard upon them, they then will learn to lie, complain, and rail against the university, the college, and the tutour, and with a *whining* letter, make the mother, *make the father*, believe all that he can invent, when all this while his main design is to leave the university and go home again to spanning farthings.

– I understand by one of your daughters that you have brought him up a *fine padd* to keep here for his health's sake; now I will tell you the use of an horse in *Oxford*, and then do as you think fit. The horse must be kept at an *ale-house* or an *inn*, and he must have leave to go once *every day* to see him eat oats . . . and it will not be *genteel* to often to an house and spend nothing; and then there may be some danger of the horse growing *resty* if he be not used often, so that you must give him leave to go to *Abingdon* once every week to look out of the tavern window, and see the maids sell turnips; and in one month or two come home with a surfeit of poisoned wine, and save any *farther charges* by dying, and then you will be troubled to send for your horse again.

THE TRADESCANTS' ARK

In 1683 the Ashmolean Museum was opened, named after its patron the antiquary Elias Ashmole, and housed in a brand new building in Broad Street. Its nucleus was the collection of rarities made during the previous century by the two John Tradescants, father and son, which had been shipped to Oxford by barge, and included the following curiosities:

A Babylonian Vest.
Diverse sorts of Egges from Turkie; one given for a Dragons egge.
Easter Egges of the Patriarchs of *Jerusalem*.
Two feathers of the Phoenix tayle.
The claw of the bird Rock: who, as Authors report, is able to trusse an Elephant.
Dodar, from the Island *Mauritius*; it is not able to flie being so big.
Hares head, with rough horns three inches long.
Toad fish, and one with prickles.
Divers things cut on Plum-stones.
A Cherry-stone, upon one side *S. Geo*: and the Dragon, perfectly cut: and on the other side 88 Emperours faces.
The figure of a Man singing, and a Woman playing on the Lute, in 4° paper; The shadow of the worke being *David's* Psalms in Dutch.
A Brazen-balle to warme the Nunnes hands.

Blood that rained in the *Isle of Wight*, attested by *Sir Jo: Oglander*.
A bracelet made of thighes of Indian flies.
Turkish toothbrush.

FARCE AT TRINITY

One bright moonshine night two young gentlemen [of Trinity College], now both living, were drest up like Roman Statues, & by means of a ladder, placed in the two niches of the new Quadrangle, which were left vacant for the statues of some eminent benefactors. As soon as these had got their cue, H. got into the Fellows' Room, & with a great deal of counterfeited surprise and horrour tells them that he feared the College was haunted; for that there were human figures standing in the niches of the New Buildings. Two of the Fellows went out with him; saw them standing there to their very great surprise; but they had not courage enough to go near them without more company. But while they were gone back to fetch this out of their common room, a young fellow-commoner happened to go thro the Court with a Violin in his hand, on which he had been playing in the Grove; he had been that evening drinking in company, and had got himself pot valiant. As he pass'd by, one of the *Statues* hemm'd to him, and by that means discover'd himself to this Gentleman, who, after having stared at him some time, flies to a heap of rubbish that lay hard by, & snatching up some pieces of bricks, soon pelted one of the statues down . . . the statue had no remedy but to leap down upon him, which it did, breaking his fiddle, & demolishing him very much, & withal bestowing many kicks and cuffs upon him.

By this time the then Dean & some more of the Fellows had muster'd up courage enough to make a visit in consort to these statues, but when they come into that Quadrangle, they found the niches empty & no statues there. The rest ridiculed the Dean, who said he had actually seen the figures himself; & in the truth of that he was supported by Harris . . . while the dispute held, the groans of *Croom* affected them, for that was the name of the Hero that had stoned the statue till he had fetched him down upon his head: they approached him, took him up, & examin'd him nicely, but he finding the violin broke fell into such passion that they cou'd get nothing out of him but curses against the Devil who, he said, had *leap't upon his head*, beat him, and broke his violin.

The next morning *Croom* was Examin'd again before the President and Fellows, as also was Harris. . . . The Presdt. (D. Bathurst) saw into the whole affair: & telling the Dean and Fellows that they were either Drunk or Mad, invited Harris to dine with him, & made him tell the whole contrivance; & he express'd himself to be as sorry as the contrivers were that Croom's drunken valour had disappointed them of the pleasure of frightening the Fellows (if they shou'd have ventured to exorcise the statues) by dreadfull voices and speeches, which were ready prepared for that purpose.

John Harris, 1684.

JAMES II AND THE FELLOWS OF MAGDALEN

Oxford, Sept. 4th 1687

Gentlemen

The King Commands me to acquaint you, that He would have you attend Him in ye Deans lodgings in Xt Ch Colledge at three of ye Clock this afternoon.

I am Gentlemen Yr Most Humble Servant
Sunderland P.

To ye fellows of Magd. Colledge.

For all its loyalty to the Stuarts, Oxford overwhelmingly opposed the attempts of James II to Catholicize the University again, and the above letter to the Fellows of Magdalen College was an opening shot in the most famous of all clashes between Universitas Oxoniensis *and the English Crown. The King wished to place a Catholic in the Presidency of the college: this being against the college statutes, the Fellows defied him, electing instead their own candidate, John Hough, and in the autumn of 1687 James arrived in person to force upon them, as his second choice, the compliant Bishop of Oxford, Samuel Parker. As Dr. Aldworth, vice-president of the college, recalled it in his* Impartial Relation *of the episode, the King dealt breezily with the spokesman of the Fellows, when they waited upon him at Christ Church:*

The King: What's your name? Aren't you Dr Pudsey?

Dr. Pudsey: Yes, may it please your Majesty.

The King: Did you receive my Letter?

Dr. Pudsey: Yes, Sir, we did.

The King: Then I must tell you and the rest of your Fellows that you have behaved yourselves undutifully to me, and not like gentlemen. You have not paid me common respect. You have always been a stubborn and turbulent college, I have known you to be so these six and twenty years myself. You have affronted me. Know I am King and I will be obeyed. Is this your Church of England loyalty? One would wonder to see so many Church of England men got together in such a thing. Go back and show yourselves good members of the Church of England. Go and admit the Bishop of Oxford, Head, Principal, what do you call it of the college. I mean President of the college. Let them that refuse it look to it. They shall feel the weight of their Sovereign's displeasure. Get you gone home, I say again, go, get you gone, and immediately repair to your Chapel, and elect the Bishop of Oxford, or else you must expect to feel the weight of my hand.

🜲 *Adjourning to their own Chapel, the Fellows decided that, though they would obey the King in anything else, their Statutes forbade them to obey him in this. As they said in a petition to the King, which he promptly rejected:*

Our Founder . . . obligeth us under the pain of perjury, a dreadful anathema, and eternal damnation, not to suffer any of his Statutes to be altered, infringed, or dispensed with, and commands us under the same sacred obligations, not to exercise any Orders or Decree whatsoever, contrary or repugnant to the said Statutes and Oaths, and so we are utterly incapacitated to admit the said Reverend Father in God to be our President.

🜲 *The King left for London, but presently there arrived in Oxford a Royal Commission, escorted by three troops of horse. It comprised the Chief Justice of the King's Bench, the Bishop of Chester, and a Baron of the Exchequer, and on 21 October 1687 the Fellows of Magdalen were summoned to appear before it in their own hall. The Bishop of Chester opened the proceedings, Dr. Aldworth records, with an awful warning:*

Gentlemen: If he, who provokes the King to anger, sins against his own soul, what a complicated mischief is yours, who have done and repeated it in such an ungrateful and indecent manner as you have done, and upon such a trifling occasion. You were the first, and I hope will be the last, who did ever thus undeservedly provoke him. . . . It is neither good nor safe for any sort of men to be wiser than their Governors, nor to dispute the lawful commands of their Superiors. . . . It was neither dutifully nor wisely done of you to drive the King to a necessity of bringing this visitation upon you. And as it must needs grieve every loyal and religious man in the Kingdom to the heart to find men of your liberal education and parts so intractable, and refractory to so gracious a Prince, so it will be very mischievous to you at the Great Day of God's Visitation. Who will then be the greatest losers by your contumacy?

The Commission proceeded to interrogate John Hough, the elected President of Magdalen:

Bishop of Chester: Dr. Hough, will you submit to this Visitation?

President: My Lords, I do declare here in the name of myself and of the greater part of the Fellows, that we submit to the Visitation so far as it is consistent with the Laws of the Land, and the Statutes of the College, and no further. I desire your Lordships that it may be recorded.

Lord Chief Justice: You cannot imagine that we act contrary to the Laws of the Land, and as to the Statutes the King has dispensed with them. Do you think that we come here to act against the Law?

President: It does not become me, my Lords, to say so, but I will be plain with your Lordships. I find that your Commission gives you authority to change and alter the Statutes, and make new ones as you think fit. Now, my Lords, we have an oath not only to observe these Statutes, but to admit of no new ones, nor alterations in these. This must be my behaviour here. I must admit of no alteration from it, and by the Grace of God never will.

Bishop of Chester: Do you submit to the Decree of the Commissioners whereby your Election is declared null?

President: I am possessed of a Freehold according to the Laws of England and the Statutes of the College, having been elected as unanimously, and with as much formality as any one of my Predecessors, who were Presidents of this College, and afterwards admitted by the Bishop of Winchester, our Visitor, as the Statutes of the College require. And therefore I cannot submit to that sentence, because I think that I cannot be deprived of my Freehold but by course of Law in Westminster Hall, or by being in some way incapacitated by the Founder's Statutes.

Bishop of Chester: Will you deliver up the Keys of the President's Office and Lodgings for the use of that Person, whom the King hath appointed your President, as the Statutes require?

President: As the Statutes require, my Lord. My Lord, there neither is, nor can be, a President so long as I live and obey the Statutes of the College, and therefore I do not think fit to give up my Right, the Keys and Lodgings.

Bishop of Chester: Dr. Hough, will you deliver up the Keys, and quiet possessions of the Lodgings to the Person whom his Majesty has appointed President?

President: My Lords, I have neither seen nor heard any thing to induce me to it.

Bishop of Chester: Dr. Hough, I admonish you to depart peacably out of the Lodgings, and to act no longer as President, or pretended President of this College.

The Bishop repeated this injunction three times, and Dr. Hough making no response, they struck his name out of the college books and ordered him to quit the premises, which he shortly did. The next day the Bishop complained bitterly of his reception by the college—'when we came into the Chapel, there was no table, when we went into the Hall, no carpet':

And when we came into Town, as we passed along in the coach through the High Street, I put off my hat to some scholars, that were in a bookseller's shop, and one of them instead of returning the civility, cocks up his hat to show his pretty face. He was one of this House, and I spoke with him this morning, and shall speak with him again before I go out of town, and make him know himself.

Overawed by such bullying, in the following week almost all the Fellows agreed not to oppose the installation of the Bishop of Oxford as President, but they soon regretted this submission. For one thing they were sneered at in Oxford for their weakness—'Here is your Magdalen College conscience!'—and for another the King demanded a more absolute gesture of remorse, 'acknowledging their contempt to his Sacred majesty in Person, imploring his majesty's Pardon, and laying themselves at his Feet'. This was too much:

Bishop of Chester: We do not expect of you to confess any capital crime, only to make some acknowledgement.

George Fulham: My Lord, we were ordered to address ourselves, as having acted in contempt of his majesty's authority, which, my Lord, I look upon as so great a crime that on no account I would be guilty of it. My Lord, we have endeavoured to obey his majesty to the utmost of his power, and seeing your Lordships were pleased to accept our answer of submission on Tuesday, I humbly conceive your Lordships' honour is engaged that nothing further be required of us.

Bishop of Chester: You are a very forward Speaker, and abound in your own sense. Will you obey the Bishop of Oxford as in possession of the Presidency of the college?

Fulham: I cannot because the Bishop hath not lawful possession. He hath not possession in due form of law, nor by proper officers. I am informed that the proper officers to give possession of a freehold is the Sheriff with a *Posse Comitatus*.

Lord Chief Justice: Pray who is the best lawyer, you or I? Your Oxford law is no better than your Oxford divinity. If you have a mind to a *Posse Comitatus* you may have one soon enough.

🎄 *The Fellows, however, would not be intimidated again. The keys of the President's Lodgings being missing, the Commissioners ordered that it should be broken into, so that the Bishop of Oxford could take possession: nobody would do it—a professional locksmith, brought in to do the job, ran away—and in the end the Commissioners had to see to it themselves. Twenty-five Fellows were expelled, but even the King himself, faced with a furious public opinion, recognized that he had gone too far. By the end of 1688 all the Fellows were back in office, Dr. Hough was reinstated as President, and the Bishop of Oxford had been carried out of the President's Lodgings dead.*

Let the King himself have the last word, in itself an epitaph to the divine right of kings which he had tried to restore in Oxford and in England. It is from an autobiographical fragment written in his own hand:

There is no doubt but that the King had done more prudently had he not carried the thing so far, but few Princes are of a temper to receive a baffle patiently in a thing they heartily espouse.

ANTHONY WOOD

🎄 *In 1695 there died in his Oxford lodgings Anthony Wood—Anthony à Wood as he preferred to call himself—who appears so often in the pages of this book, as of all Oxford books. More than anybody else, he expresses the spirit of seventeenth-century Oxford, and though he was a learned and immensely diligent antiquarian, his idiosyncratic, funny, testy, and sometimes malicious personality is expressed best in his day-to-day journal of Oxford life, some of it retrospective, some recorded as it happened. Here is a brief selection from this inexhaustible trove.*

At a stage play in Oxon, at the King's Armes in Holywell, a Cornish man was brought in to wrestle with three Welshmen, one after another, and when he had worsted them all, he calls out, as his part was, 'Have you any more Welshmen?': which words one of Jesus Coll. took in such indignation that he leapt upon the stage and threw the player in earnest.

Dr. Richard Gardiner, a boone companion, ejected from his canonry of Ch. Church, Oxon, by the parliamentarian visitors anno 1648, preached the year following among several ejected loyallists at Magd.

Oxford from the south-west, 1767

The Oxford skyline, 1779, from left to right: New College, All Souls, St Mary's, Christ

parish church in Oxon and dilating himself on Xt's miracle of turning water into wine, said that 'every good fellow could turne wine into water: but who or any other mortall could turne water into wine. This, I say, makes the miracle the greater.'

Thomas Hyll, student of Ch. Church, a great eater, was reported to have eaten up a pound of candles. . . . He was the miller's son of Osney and died as I remember about 1657. He was a good scholar but managed and spent his time so that he comprehended it in these 2 verses:—

'Morn, mend hose, stu. Greeke, breakfaste, Austen, quoque dinner: Afternoon, wa. me., cra. nu., take a cup, quoque supper'

i.e. in the morning, mend his hose or stockings, study Greeke, break his faste, study Austen, then go to dinner; in the afternoone, walk in Ch. Ch. meade, crack nuts, and drink, and then for supper.

July 24, [1658] Thomas Balsar or Baltzar, a Lubecker borne, and the most famous artist for the violin that the world had yet produced, was now in Oxon; and this day Anthony Wood was with him and Mr. Edward Low, lately organist of Ch. Church, at the meeting-house of William Ellis. A.W. did then and there, to his very great astonishment, heare him play on the violin. He then saw him run up his fingers to the end of the finger-board of the violin, and run them back insensibly, and all with alacrity and in very good tune, which he nor any in England saw the like before. . . . He played to the wonder of all the auditory: and exercising his fingers and instrument several wayes to the utmost of his power, Wilson thereupon, the public professor, (the greatest judge of musick that ever was) did, after his humoursome way, stoop downe to Baltzar's feet, to see whether he had a huff [*hoof*] on, that is to say to see whather he was a devil or not, because he acted beyond the parts of man.

Jan 26 [1664] Mr. R[ichard] B[erry] a chapleyn of Xt Ch., one much given to the flesh and a great lover of Eliz. the wife of Funker and daughter of Woods of Bullock's lane, having mind one night in the month of Jan. . . . to vent or coole his passion, sent his servitor (a little boy) to Carfax where shee sold apples to come to his chamber and 12d in apples. The boy forgetting her name went to another huckster and told her that she (who it seems told him that she knew Mr. Berry) must come down to his chamber and bring apples. Well, she comes at a little past 6 at night up to his chamber; who against her comming (supposing her to be Eliz . . .) shut the shuttings of his windows and put out his candle. Ane when shee was come in, he said

'Oh Betty art come? I am glad with all my hart; I have not seene thee a great while': and kissed her and groped her and felt her brests. 'Come, what wilt have to supper? What joynt of meat wilt have?' and the like. 'Come, I have not layd with thee a great while', and soe put his hands under her coates. But shee bid him 'forbear' and told him 'he was mistaken: if her would pay her for her apples, well and good; she would not play such vile actions with him'. 'Who', quoth he 'are you not Eliz. . . .' 'Noe, marry, am I not'. With that he thrust her downstairs and kikt her.

Four *berryes* in the University of Oxford, viz., 1, *black-berry*, that is Dr. Arthur Berry, rector of Exeter Coll., a little black man; 2, *coffey-berry*, that is, Phineas Berry, fellow of Wadham Coll., a great coffey-drinker; 3, *ale-berry*, i.e., Amos Berry, one of the senior fellowes of Corpus Christi College, a great ale-bibber; 4, *goose-berry*, i.e. Richard Berry, chaplaine of Ch. Church, a simple hot-headed coxcombe.

July 17, [1672] Robert Pawling an attorney his servant (Edward Cole sone of Edward Cole of S. Martin's parish) and Marsh his son the butcher stood at the convocation house dore by the altar with gownes on turn'd inside out and the forepart behind; one had a square cap with the hinder part before and the other a round cap. Each of them with a paper on their breast with this written—

> '*For wearing scholars gownes, affronting the proctors, and raising of tumults.*'

By them were two servitors of Allsoules, without their gownes and caps, standing by, who had lent the aforesaid 2 fellowes their gownes: who were severall nights in the Act going before, doing mischeif and beating people, so that being taken by the proctors, this was their punishment, to stand so from 9 to 11 in the morning. The townsmen and attorney laugh at it as unwarrantable.

Mar. 15 [1680] Thomas Hovell that killed John White a servitour of Ball. Coll. was hanged on a gallows against Ball. Coll. gate: died very penitent and hang'd there till 2 or 3 in the afternoone. The next day hanged on a gibbet in chaines on this side Shotover. . . .

Nov. 9 [1695] The King [William III] came into the Theater at half an hour past 10 of the clock . . . afterwards he was desired to descend into the area to tast of a rich banquet then prepared at the charge of the University, or rather an ambigue; but he denied it and went straight out . . . to his coach. Now so it was that the Masters,

Bachelaurs, and Undergraduats being confined to their galleryes, and the women to theirs, there were only some gentlemen and ordinary people and attendants in the Area, who rudely scambled away all the banquet and sweetmeates, all sorts of souse fish (lobsters, crayfish), fruite, etc.—about 50 larg dishes besides very many little or small dishes intermix'd—who swept all away and drank all the wine. The Universitie had employed William Sherwin, the inferior beadle to go to London and provide all rarities that could be; were at great charge; and when all was done, few scholars participated, few gent., and no women.

🐾 *Wood's Last Will and Testament gave precise instructions about his burial:*

Imprimis, I commend my soul into the hands of Almighty God, who first gave it, (professing myself to die in the Communion of the Church of England) and my body to be buried in Merton college church, deeper than ordinary, under, and as close to the wall (just as you enter in at the north on the left hand) as the place will permit, and I desire that there may be some little monument erected over my grave.

🐾 *His wishes were obeyed. By then the Glorious Revolution had occurred, the staunchly Protestant King William and Queen Mary were on the throne, and the Oxford that had been plagued throughout Wood's lifetime by war and furious faction moved into the eighteenth century in a mood of battered escapism.*

Port and Prejudice
1700 – 1800

'Port and Prejudice' is how a contemporary writer characterized eighteenth-century Oxford, and certainly both commodities were plentiful. Nevertheless it was by Oxford standards a tranquil century, and though the drinking was terrific, the idleness disgraceful, and the scholarship often shoddy, it was graced by some great Oxonians and some lovely new Oxford buildings.

THE INNOCENT EYE

Miss Celia Fiennes, age uncertain, travelling for her health, visited Oxford at about the beginning of the century and thought it 'pleasant and Compact':

The high Streete is a very Noble one, soe large and of a Greate Length. In this is ye University Church Called St. Maryes, which is very large and Lofty but Nothing very Curious in it.

There are several good Collegdes I saw most of y*m*. Waddom Hall is but little. . . . In one of the courts of Christ Church is a tower new built for to hang the Mighty Tom, that bell is of a Large size, so great a Weight they were forced to have engines from London to raise it up to the tower. . . . St. John's Colledge had fine gardens and walkes but I did just look into it, so I did into kings, and queens Colledges, and several of the rest I looked into, they are much alike. . . . [*She did not look too hard into King's, which is at Cambridge.*]

In Corpus Christus Colledge w*ch* is but small there I was entertained at supper and eate of their very good bread and beare which is remarkably the best anywhere Oxford bread is.

She liked the domestic arrangements of the Fellows of New College, each with his 'Dineing Rooms, bed Chamber, a studdy and a room for a Servant', and left Oxford with a pleasant glimpse of their academic lives:

Here they may Live very Neatly and well if Sober and here all their Curiosityes they take so much delight in, greens of all sorts, Myrtle, oringe and Lemons and Lorrestine growing in potts of Earth and so moved about from place to plaee and into the aire sometymes.

Through England on a Side-Saddle in the Time of William and Mary, 1888.

THE PEEVISH EYE

Z. C. von Üffenbach, a German scholar, travelling for self-instruction, visited Oxford in 1710 and thought it very second-rate:

In the afternoon, we went to the Bodleian Library, where we were instructed to take the oath; but the Proctor (as he is called), really *Procurator Academiae*, who was to receive it was not there—for an Englishman owing to a general lack of courtesy is seldom up to time.

The Sub-Librarian Crab (an arch-ignoramus who, were it not that this was his living, would have preferred sitting in a tavern to being in the Library) . . . opened the two cabinets which one finds in the first part of [the] cross-corridor at the outset where the contents— mostly playthings and likely to please the ignorant—are always shown. They are for the most part codices, elegantly written and painted or decorated with gold; but Mr. Crab never even mentioned what they are and probably neither knows nor can read them. Of one however he did remark: 'That book is very old—more than eight hundred years'. When I asked him how he knew this, he could reply nothing but: 'It is certain, Dr. Grabe told me so', (i.e. the famous Joh. Ernest Grabius of Königsberg, with whom he considered himself great friends because they have similar sounding names). Thereupon he looked so desperately wise that one could not help laughing.

Afterwards Mr. Crab led us up to the so-called gallery and showed us a poor little room on the right which he called 'the study'. It would perhaps serve as a museum for the Librarian or old Fellows in winter. In here hung some pictures, amongst which were several embroidered in silk. Mr. Crab made a great fuss over them, although I have seen many more beautiful, and even have better ones worked by my own grandmother. . . . After this a great armchair was pointed out to us, as something very special, because it is said to have been made out of the ship in which Captain Drake sailed round the world; also several Chinese staves, bows and arrows, and again a cylinder with some vile figures.

In [Christ Church Cathedral] are various epitaphs which, like the monuments throughout England, are neither very valuable nor well-sculptured, and indeed nowhere, either in London or elsewhere, did we see a memorial approaching in interest those in Lübeck or other places. One cannot but wonder at this, as in most things, the English are so wasteful and expensive.

We saw [Christ Church] hall or dining-room, which is fearfully large and high but otherwise poor and ugly in appearance; it also reeks so strongly of bread and meat that one cannot remain in it, and I should find it impossible to dine and live there. The disgust was increased (for the table was already set), when we looked at the coarse and dirty and loathsome table cloths, square wooden plates and the wooden bowls into which the bones are thrown; this odious custom obtains in all the colleges. The *socii collegiorum* as well as the students or scholars must dine here; but the most important have their meals brought to their rooms at incredibly high cost.

In the centre of [Christ Church quadrangle] is a fountain with a Mercury, which does not play.

At 7 o'clock in the evening the Grassys conducted us to the usual *collegium musicum* held every Wednesday. On this occasion the music was very weak and poor.

The present Professor of Chemistry, Richard Frewyn, troubles himself very little about [the chemistry laboratory]. The operator, Mr. White (who is said to be very debauched) still less. The result is that the furnaces look entirely uncared for . . . and not only are the finest instruments, crucibles and other things belonging to the place almost all of them lying in pieces, but everything is covered in filth. Who could imagine that so fine and costly an undertaking should receive so little attention? Indeed who would believe such a thing of England, which we foreigners hold so much in awe that we believe all subjects, and chemistry in particular, to be passing through a golden age?

Some of Herr von Üffenbach's briefer judgements:

The Physic Garden: wild and overgrown.
New College gardens: very mediocre.
Exeter College: tolerable-looking building in front, at the back old and poor.
Pembroke College: very indifferent and confused building.
Merton College: consists of several ugly old buildings.
Magdalen College: very old and badly built.
Ashmolean Museum: Herr Bürgermeister Reimer in Lüneburg has twice as many specimens.
The return to London: God be thanked.

Oxford in 1710, ed. W. H. and W. J. C. Quarrell, 1928.

THE OXFORD MAN, 1704

Bookwit: How is my Manner? My Mien? Do I move freely? Have I kicked off the trammels of the Gown, or does the Tail on't seem still tuck't under my arm, where my hat is, with a pert Jerk forward, and a little Hitch in my Gate like a Scholastick Beau? This wig, I fear, looks like a Cap. My Sword, does it hang careless? Do I look bold, negligent, and erect . . .? I horribly mistrust myself. I fancy people see I understand Greek. Don't I pore a little in my Visage? Ha'nt I a down bookish Lour, a wise Sadness? I don't look gay enough and unthinking.

Latine: I protest you wrong yourself. You look very brisk and ignorant.

Richard Steele, The Lying Lover, 1704.

KIN AND COUNTRY

🦁 *By the early years of the eighteenth century there were some twenty Oxford colleges. Each had its own character, and most tried to sustain a continuity of style and interest by offering places and scholarships to members of the founder's family, or to students from a particular region. The most resolute of all in preserving a flavour was very properly Jesus College, the Welsh college founded in 1571, whose list of benefactions in 1712 included the following:*

1. Griffith Lloyd, Principal, bequeathed by his will in 1586 lands 'to the finding of some scholar or my own kin for ever, and to no other use'.

2. Lewis Owen founded two scholarships for scholars of Beaumaris School, with preference 'to my kindred if any be found fit'.

3. Owen Wood, Dean of Armagh, founded a fellowship and a scholarship in 1608, with preference 'to some of my name neare me in bloud'.

4. Henry Rowlands, Bishop of Bangor, founded two scholarships in 1616 for North Wales boys, with preference to 'any of my bloud fitte and capable'.

5. Richard Parry, Bishop of St. Asaph, founded a scholarship in 1616 for a poor boy of Ruthin or St. Asaph diocese, with preference to 'one of my kindred and a minister his sonne'. . . .

🦁 *The system worked, and Jesus College was to maintain its Welsh connections ever after.*

LOYALTY AND LEARNING

🦁 *Despite Oxford's unhappy experiences with James II, Jacobite sympathies were still strong in the University. In 1715 a cavalry regiment was sent to the city to root it out, and almost at the same time George II made a gift of books to the University of Cambridge. A famous epigrammatic exchange resulted:*

> The King, observing with judicious eyes,
> The state of both his Universities,
> To one he sent a regiment. For why?
> That learned body wanted loyalty.
> To t'other he sent books, as well discerning
> How much that loyal body wanted learning.
>
> Joseph Trapp (1679–1747).

> The King to Oxford sent his troop of horse,
> For Tories own no argument but force;
> With equal care to Cambridge books he sent,
> For Whigs allow no force but argument.
>
> William Browne (1692–1774).

Sir William Browne was a scholarly and eccentric physician, whose best-known work this is; Trapp was the first Professor of Poetry at Oxford, described by Jonathan Swift as 'a sort of pretender to wit'.

A LITTLE VANITY

About a mile before I reachd Oxford, all the bells toll'd, in different notes; the Clocks of every College answer'd one another; & sounded forth (some in deeper, some a softer tone) that it was eleven at night.

All this was no ill preparation to the life I have led since; among those old walls, venerable Galleries, Stone Portico's, studious walks & solitary Scenes of the University. I wanted nothing but a black Gown and a Salary, to be as meer a Book-worm as any there. I conform'd myself to the College hours, was rolld up in books, lay in one of the most ancient, dusky parts of the University, and was as dead to the world as any Hermite of the desert. If any thing was alive or awake in me, it was a little Vanity. . . . For I found my self receiv'd with a sort of respect, which this idle part of mankind, the Learned, pay to their own Species; who are as considerable here, as the Busy the Gay and the Ambitious are in your World. Indeed I was treated in such a manner, that I could not but sometimes ask myself in my mind, what College I was founder of, or what Library I had built? Methinks I do very ill, to return to the world again, to leave the only place where I make a figure, and from seeing myself seated with dignity on the most conspicuous shelves of a Library, put myself into the abject posture of lying at a Lady's feet in St. *James's* Square.

Alexander Pope, *Letters*, 1717.

TERRAE-FILIUS

🦁 *Terrae-Filius, Son of the Soil, was the name given to the licensed buffoon who, at the degree-giving ceremony called the Act, enlivened the solemnities (originally in St. Mary's church) with his comical, irreverent, and often scurrilous Latin observations on the state of the University. The practice became much degraded, and John Evelyn was shocked by it when he attended the Act in 1669, the first in the new Sheldonian Theatre:*

Terrae Filius entertained the auditory with a tedious, abusive, sarcastical rhapsody, most unbecoming the gravity of the University. . . . It was rather licentious lying and railing than genuine and noble wit.

🦁 *The office was allowed to lapse at the beginning of the eighteenth century, after a series of particularly offensive performances, but the name was appropriated by Nicholas Amhurst, a disaffected former scholar of St. John's College, for a satirical magazine which he started in 1721, and which he meant to perform some of the functions of the old jester. Some examples:*

ON PERJURY

To give a just account of the state of the university of Oxford, I must begin where every *freshman* begins, with *admission* and *matriculation*; for it so happens, that the first thing a young man has to do there, is to prostitute his conscience and enter himself into perjury, at the same time that he enters himself into the university.

If he comes elected from any publick school, as from *Westminster, Winchester*, or *Merchant-Taylors*, to be admitted upon the foundation of any college, he swears to a great volume of statutes, which he never read, and to observe a thousand customs, rights and privileges, which he knows nothing of, and with which, if he did, he could not perhaps honestly comply. . . .

Within fifteen days after his admission into any college, he is obliged to be *matriculated*, or admitted a member of the university; at which time he subscribes the *thirty nine* articles of religion, though often without knowing what he is doing, being order'd to write his name in a book, without mentioning upon what account; for which he pays *ten shillings and sixpence*.

At the same time he takes the oaths of allegiance and supremacy, which he is pretaught to evade, or think null: some have thought themselves sufficiently absolved from them by kissing their *thumbs*, instead of the book; others, in the croud, or by the favour of an *honest* beadle, have not had the book given them at all.

He also swears to another volume of statutes, which he knows no more of than of his private college-statutes, and which contradict one another in many instances, and demand unjust compliances in many others; all which he swallows ignorantly, and in the dark. . . .

What a pack of conjurors were our forefathers!

ABOUT PROFESSORS

There are now founded at OXFORD lectures and professorships of all or most of the arts, sciences, and faculties in the world, with profitable salaries annex'd to them. But it is very merry to observe how preposterously these places are dispos'd of. . . . I have known a profligate *debauchee* chosen professor of moral philosophy; and a fellow, who never look'd upon the *stars soberly* in his life, professor of astronomy; we have had history professors, who never read any thing to qualify them for it, but *Tom Thumb, Jack the gyant-killer* and other suchlike valuable records; we have had likewise numberless professors of *Greek*, *Hebrew*, and *Arabick*, who scarce understood their mother-tongue: and, not long ago, a famous *gamester* and *stock-jobber* was elected professor of *divinity*; so great, it seems, is the analogy between *dusting of cushions*, and *shaking of elbows*; or between *squandering away of estates*, and *saving of souls*!

ADVICE TO FRESHMEN

Leave no stone unturned to insinuate your selves into the favour of the HEAD, and *senior-fellows* of your respective colleges. Whenever you appear before them, conduct yourselves with all specious *humility* and *demureness*; convince them of the great veneration you have for their Persons, by *speaking very low*, and *bowing to the ground at every word*; wherever you meet them, jump out of the way with your *caps* in your hands, and give them the whole street to walk in, be it as broad as it will. Always *seem* afraid to look them in the face, and make them believe that their presence strikes you with a sort of awe and confusion.

Another thing very proper in order to grow the favourites of your HEADS, is first of all to make your selves the favourites of their FOOTMEN. You must often have heard, my lads, of the old proverb, *Love me, and love my dog*; which is not very foreign in this case; for if you expect any favour from the *master*, you must shew great respect to his *servant*.

Have a particular regard how you speak of those *gaudy things* which flutter about *Oxford* in prodigious numbers, in summer time, called TOASTS; take care how you reflect on their parentage, their condition, their Virtue, or their beauty; ever remembring that of the Poet,

Hell has no *Fury* like a *Woman scorn'd*

Especially when they have *spiritual bravoes* on their side, and old *lecherous bully-backs* to revenge their cause on every audacious contemner of *Venus* and her altars.

AN ANECDOTE

A young fellow of *Baliol* college having, upon some discontentment, cut his throat very dangerously, the MASTER of the college sent his *servitor* to the *buttery-book* to *sconce* (that is, *fine*) him *five shillings*, and, says the doctor, tell him that the *next time he cuts his throat, I'll* sconce *him* TEN.

SPENDING TOO MUCH

🦟 *William Pitt the elder, the future Lord Chatham, replies from Trinity College to a severe letter from his father, 1727:*

Hon*ed* Sr,—I rec*d* yr*s* of y*e* 25th in which I find with y*e* utmost concern y*e* dissatisfaction you express at my expences. To pretend to justify, or defend myself in this case would be, I, fear, with reason thought impertinent; tis sufficient to convince me of the extravagance of my expences, that they have met with y*r* disapprobation, but might I have leave to instance an Article or two, perhaps you may not think 'em so wild and boundless, as with all imaginable uneasiness, I can see you do at present. Washing 2*l*. 1s. 0d about 3s 6d per w*k*, of which money half a dozen shirts at 4d. each comes to 2s per week, shoes and stockings 19s 0d. Three pairs of Shoes at 5s each, two pairs of Stockings, one silk, one worcestead, are all that make up this Article, but be it as it will, since S*r*, you judge my expence too great, I must endeavour for y*e* future to lessen it & shall be contented with whatever you please to allow me. . . .

Hon*ed* Sr, yr most Dutifull Son,
W. Pitt.

🦟 *But he left Oxford within a year, because of his gout.*

DR. JOHNSON

Samuel Johnson the lexicographer entered Pembroke College in 1728, but stayed little more than a year, leaving without a degree. Thomas Carlyle described his student days luridly—'What a world of blackest gloom, with sun-gleams and pale tearful moon-gleams and flickerings of a celestial and infernal splendour, was this that now opened for him!' Witnesses nearer the time were more matter-of-fact:

I have heard from some of his contemporaries [*wrote Bishop Thomas Percy to James Boswell*] that he was generally seen lounging at the College gate, with a circle of young students round him, whom he was encouraging with wit, and keeping from their studies, if not spiriting them up to rebellion against the College discipline, which in his maturer years he so much extolled.

Sir J. Hawkins, in his Life of Johnson (1787) *reported that Johnson used to be infuriated by the Pembroke custom of having a servitor knock on every student's door from time to time, to see if they were working, or report their absence to the Master:*

[He] could not endure this intrusion, and would frequently be silent, when the utterance of a word would have insured him from censure; and farther to be revenged for being disturbed when he was profitably employed as perhaps he could be, would join with others of the young men in the college, in hunting, as they called it, the servitor, who was thus diligent in his duty; and this they did with the noise of pots and candlesticks, singing to the tune of Chevy-chace, the words in that old ballad—

> 'To drive the deer with hound and horn', etc.

Oliver Edwards, a contemporary at Pembroke, recalled that Dr. Johnson would never allow his fellow-students to use the word 'prodigious'—'for even then he was delicate in language, and we all feared him'—while Matthew Panting, the Master, reported overhearing Johnson talking to himself one day in his room over the college gateway:

Well, I have a mind to see what is done in other places of learning. I'll go and visit the Universities abroad. I'll go to France and Italy. I'll go to Padua.—And I'll mind my business. For an *Athenian* blockhead is the worst of all blockheads.

William Adams, another contemporary student and later Master of Pembroke, said that Johnson had been 'caressed and loved by all about

*him, was a gay and frolicksome fellow, and passed there the happiest part
of his life'. This Johnson himself sadly disputed:*

Ah, Sir, I was mad and violent. It was bitterness which they mistook
for frolick. I was miserably poor, and I thought to fight my way by
my literature and my wit; so I disregarded all power and authority.

🐾 *He did not, by his own account, work very hard (nor, he says, did anybody
else):*

   Johnson: My tutor [William Jorden of Pembroke] was a very
worthy man, but a heavy man, and I did not profit much by his
instructions. Indeed, I did not attend him much. The first day after
I came to college I waited upon him, and then staid away four. On
the sixth, Mr. Jorden asked me why I had not attended. I answered
I had been sliding in Christ Church meadow. And this I said with as
much *nonchalance* as I am now talking to you. I had no notion that I
was wrong or irreverent to my tutor.
   Boswell: That, Sir, was great fortitude of mind.
   Johnson: No, Sir; stark insensibility.

🐾 *Johnson used to cross the road to the far more opulent college of Christ
Church, to pick up lecture notes from a friend there, until, Boswell says,*

His poverty being so extreme, that his shoes were worn out, and his
feet appeared through them, he saw that this humiliating circumstance
was perceived by the Christ Church men, and he came no more. He
was too proud to accept of money, and somebody having set a pair
of new shoes at his door, he threw them away with indignation. How
must we feel when we read such an anecdote of Samuel Johnson!

🐾 *In later years, though, he remembered his student days with pleasure, and
he often went back to Oxford. His first return visit to Pembroke was in
1754, when he was accompanied by the poet Thomas Warton (whose
account is quoted by Boswell):*

He was highly pleased to find all the College servants which he had
left there still remaining, particularly a very old butler; and expressed
great satisfaction at being recognized by them, and conversed with
them familiarly. He waited on the master, Dr. [John] Ratcliffe, who
received him very coldly. Johnson at least expected that the master
would order a copy of his *Dictionary*, now near publication; but the
master did not choose to talk on the subject, never asked Johnson to
dine, nor even to visit him while he stayed at Oxford. After we had
left the lodgings, Johnson said to me, '*There* lives a man, who lives
by the revenues of literature, and will not move a finger to support

it. If I come to live at Oxford, I shall take up my abode at Trinity.'
We then called on the Reverent Mr. Meeke, one of the fellows, and
of Johnson's standing. Here was a most cordial greeting on both sides.
On leaving him, Johnson said, 'I used to think Meeke had excellent
parts, when we were boys together at the College; but alas!

> 'Lost in a convent's solitary gloom!'

'I remember, at the classical lecture in the Hall, I could not bear
Meeke's superiority, and I tried to sit as far from him as I could,
that I might not hear him construe.'

When he left Mr. Meeke . . . he added 'About the same time of life,
Meeke was left behind at Oxford to feed on a Fellowship, and I went
to London to get my living: now, Sir, see the difference of our
literary characters!'

❧ *He was delighted nevertheless to get an honorary degree from Oxford in
1775 (though the 'Dr.' in 'Dr. Johnson' comes from an earlier Dublin
degree), and during his visits, so Lord Stowell said, delighted in the
protocol of the place:*

[He] prided himself in being . . . accurately academic in all points;
and he wore his gown almost ostentatiously.

❧ *We hear of him dining splendidly, drinking heavily, being heaved through
the gateway of University College, or lost in thought, as Martin Routh,
President of Magdalen, loved to recall, astride the open gutter which ran
down the High Street, surrounded by street-urchins—'none daring to
interrupt the meditations of the great lexicographer'. By his own account
he was often merry:*

I have been in my gown ever since I came here. It was, at my first
coming, quite new and handsome. I have swum thrice, which I had
disused for many years. I have proposed to Vansittart climbing over
the wall, but he has refused me.

❧ *He loved showing other people around, as Hannah More records in her
Memoirs (1834):*

OXFORD, June 13, 1782. Who do you think is my principal *Cicerone*
at Oxford? Only Dr. Johnson! and we so do gallant it about! You
cannot imagine with what delight he showed me every part of his
own College (Pembroke). . . . He would let no one show it me but
himself,—'This was my room; this Shenstone's.' Then after pointing
out all the rooms of the poets who had been of his college, 'In short',
said he, 'we were a nest of singing-birds.—Here we walked, there we
played at cricket.' He ran over with pleasure the history of the juvenile

days he passed there. When we came into the common room, we spied a fine large print of Johnson, framed and hung up that very morning, with this motto: 'And is not Johnson our's, himself a host'.

🦁 *By then he was getting old, and his last visit to Pembroke took place two years later, as Douglas Macleane records in his* Pembroke College (*1900*):

He had expressed a wish to see his old rooms again, but . . . being then unwieldy, asthmatic, and infirm, he was obliged to invoke the aid of the janitor, who lived at the bottom of the narrow stair, to push him up it from behind.

🦁 *Dr. Johnson remained to the end an outspoken, in some eyes an archetypal, Oxford Man:*

He delighted in his own partiality for Oxford, and one day, at my house, entertained five members of the other university with various instances of the superiority of Oxford, enumerating the gigantic names of many men whom it had produced, with apparent triumph. At last I said to him, 'Why there happens to be no less than five Cambridge men in the room now.' 'I did not (said he) think of that till you told me; but the wolf don't count the sheep.'

Hester Lynch Piozzi, Anecdotes of the Late Samuel Johnson, 1786.

JOHNSONUS OXONIENSIS

There is here, Sir, such a progressive emulation. The students are anxious to appear well to their tutors; the tutors are anxious to have their pupils appear well in the college; the colleges are anxious to have their students appear well in the University; and there are excellent rules of discipline in every college. That the rules are sometimes ill observed, may be true; but is nothing against the system. The members of an University may, for a season, be unmindful of their duty. I am arguing for the excellency of the institution.

Lectures were once useful; but now, when all can read, and books are so numerous, lectures are unnecessary. If your attention fails, and you miss a part of a lecture, it is lost; you cannot go back as you do upon a book.

Sir, if a man has a mind to *prance*, he must study at Christ Church and All Souls.

 George III: I hear you have been at Oxford lately. You are fond of going there, are you not?
 Johnson: I am indeed fond of going there sometimes: but I am always glad to come back again.

THE METHODISTS

Methodism was born in Oxford in 1729, when John Wesley of Lincoln College and some friends formed a society, nicknamed the Holy Club, to bring a new enthusiasm and compassion to what they saw as a sluggish spiritual environment. Holding prayer meetings, visiting the prisoners in Oxford Castle, giving religious instruction, they were inevitably accused of priggishness, hypocrisy, and insulting behaviour:

When I happened to be in Oxford in 1742 Mr. [Charles] Wesley, the Methodist, of Christ Church, entertain'd his Audience [at St. Mary's Church] two Hours, and having insulted and abus'd all Degrees, from the highest to the lowest, was in a manner hissed out of the Pulpit by the lads.

> Mr Salmon, *The Foreigner's Companion Through the Universities of Cambridge and Oxford*, 1748.

Here John Wesley, in a letter, recalls the beginnings of the movement:

In November 1729, at which time I came to reside at Oxford, [four of us] agreed to spend three or four evenings in a week together. Our design was to read over the classics, which we had before read in private, on common nights, and on Sunday some book in divinity. In the summer following Mr. M. told me he had called at the gaol, to see a man who was condemned for killing his wife; and that, from the talk he had with one of the debtors, he verily believed it would do much good, if any one would be at the pains of now and then speaking with them. This he so frequently repeated, that on the 24th of August 1730, my brother and I walked with him to the castle. We were so well satisfied with our conversation there, that we agreed to go thither once or twice a week; which we had not done long, before he desired me to go with him to see a poor woman in the town who was sick. In this employment too, when we came to reflect upon it, we believed it would be worth while to spend an hour or two in a week, provided the minister of the parish in which any such person was, were not against it. . . .

Soon after, a gentleman of Merton College, who was one of our little company, which now consisted of five persons, acquainted us that he had been much rallied the day before for being a member of the holy club; and that it was become a common topic of mirth at his college, where they had found out several of our customs, to which we were ourselves utter strangers. . . . About this time there was a meeting . . . of several of the officers and seniors of the college, wherein it was consulted what would be the speediest way to stop the progress of enthusiasm in it. The result we know not, only it was soon publicly reported, that Dr. —— and the censors were going to blow up the Godly Club. This was now our common title; though we were sometimes dignified with that of the Enthusiasts, or the Reforming Club.

The two points, whereunto, by the blessing of God . . . we had before attained, we endeavoured to hold fast: I mean, the doing what good we can, and in order thereto communicating as often as we have opportunity. To these . . . we have added a third, the observing the fasts of the Church; the general neglect of which we can by no means apprehend to be a lawful excuse for neglecting them. And in the resolution to adhere to these and all things else which we are convinced God requires at our hands, we trust we shall persevere, till he calls us to give an account of our stewardship. As for the names of Methodists, Super-erogation-men, and so on, with which some of our neighbours are pleased to compliment us, we do not conceive ourselves to be under any obligation to regard them, much less to take them for arguments.

🦡 *Wesley left Oxford for America in 1734, but Methodism retained some of its impetus (and its reputation) in the University thanks to the fervour of George Whitefield of Pembroke College, who recorded his conversion to the cause in* A Short Account of God's Dealings with the Rev. George Whitefield, *1740:*

I had not been long at the University before I found the benefit of the foundation I had laid in the country for a holy life. I was quickly sollicited to joyn in their excess of riot with several who lay in the same room. God, in answer to prayers before put up, gave me grace to withstand them; and once, in particular, it being cold, my limbs were so benummed by sitting alone in my study, because I would not go out amongst them, that I could scarce sleep all night. But I soon found the benefit of not yielding; for when they perceived they could not prevail, they let me alone as a singular, odd fellow.

I now began to pray and sing psalms thrice every day, besides morning and evening, and to fast every *Friday*, and to receive the Sacrament at a parish church near our college, and at the Castle, where the despised Methodists used to receive once a month. . . . From time to time Mr. *Wesley* permitted me to come unto him, and instructed me as I was able to bear it. By degrees, he introduced me to the rest of his Christian brethren. I now began, like them, to live by rule, and to pick up the very fragments of my time, that not a moment of it might be lost.

By degrees, I began to leave off eating fruits and such like, and gave the money I usually spent in that way to the poor. Afterward, I always chose the worst sort of food, tho' my place furnished me with variety. I fasted twice a week. My apparel was mean. I thought it unbecoming a penitent to have his hair powdered. I wore woollen gloves, a patched gown, and dirty shoes.

It was now suggested to me that *Jesus Christ* was amongst the wild beasts when He was tempted, and that I ought to follow His example; and being willing, as I thought, to imitate *Jesus Christ*, after supper I went into *Christ Church* walk, near our college, and continued in silent prayer under one of the trees for near two hours, sometimes lying flat on my face, sometimes kneeling upon my knees. . . . The night being stormy, it gave me awful thoughts of the day of judgement. I continued I think, until the great bell rung for retirement to the college, not without finding some reluctance in the natural man against staying so long in the cold.

Soon after this, the holy season of *Lent* came on, which our friends kept very strictly, eating no flesh during the six weeks, except on *Saturdays* and *Sundays*. I abstained frequently on *Saturdays* also, and ate nothing on the other days, (except on Sunday) but sage-tea without sugar, and coarse bread. I constantly walked out in the cold mornings till part of one of my hands was quite black. This, with my continued abstinence and inward conflicts, at length so emaciated my body, that, at Passion-week, finding I could scarce creep upstairs, I was obliged to inform my kind tutor of my condition, who immediately send for a physician to me.

This caused no small triumph among the collegians, who began to cry out, 'What is his fasting come to now?' . . .

Thus God dealt with my soul.

In 1768 six Methodists at St. Edmund Hall were accused of failing to honour the Thirty-Nine Articles of the Church of England faith, acceptance of which was compulsory for all members of the University. They were arraigned before the Vice-Chancellor:

James Matthews: Accused that he was brought up to the trade of a weaver—that he had kept a tap-house—confessed—Accused that he is totally ignorant of the Greek and Latin languages; which appeared by his declining all examination—accused of being a reputed methodist by the evidence of Mr. Atkins formerly of Queen's College—that he

maintained the necessity of the sensible impulse of the holy spirit—
that he entered himself of Edmund-Hall, with a design to get into
holy Orders, for which he had offered himself a candidate, tho' he still
continues to be wholly illiterate, and incapable of doing the exercises
of the Hall—proved—That he had frequented illicit conventicles held
in a private house in Oxford—confessed.

Thomas Jones: Accused that he had been brought up to the trade of
a barber, which he had followed very lately—confessed—Had made
a very small proficiency in the Greek and Latin languages—was two
years standing, and still incapable of performing the statutable
exercises of the Hall—that he had been at the meetings at Mrs.
Durbridge's—that he had expounded the scriptures to a mixed congregation at Wheaton-Aston, tho' not in holy Orders, and prayed
extempore. All this he confessed.

Joseph Shipman: Accused that he had been brought up to the trade of a
draper, and that he was totally illiterate; which appeared on his
examination—accused that he had preached or expounded to a mixed
assembly of people, tho' not in Orders, and prayed extempore—all
which he confessed.

Erasmus Middleton: Confesses to have done duty in a chapel of ease
belonging to Cheveley, not being in holy Orders. That he was dis-
carded by his father for being connected with the methodists—That
he had been refused Orders by the Bishop of Hereford—Accused that
he was deficient in learning—that he was attached to Mr. Haweis, who
had boasted that THEY should be able to get him into Orders. That
he holds that faith without works is the sole condition of salvation—
that the immediate impulse of the spirit is to be waited for—that he
had taken frequent occasion to perplex and vex his Tutor.

Benjamin Kay: Confesses that he has been present at the meetings
held in the house of Mrs. Durbridge, where he had heard extempore
prayers frequently offered up by one Hewett a staymaker; that sometimes Mrs. Durbridge has read to them—Accused that he endeavoured
to persuade a young man of Magdalen-College, who was sent into
the country for having been tainted with calvinistical and methodistical
principles, to leave his father—that he holds, that the spirit of God
works irresistibly—that once a child of God always a child of God.

Thomas Grove: Accused that he had preached to a mixed assembly of
people called methodists, not being in Orders, which he confessed,
and likewise that he prayed extempore—that he could not fall down
upon his knees, and worship God in the form of the church of England,
though he thought it a good form.

🕉 *The Vice-Chancellor expelled them all, in effect ending the progress of Methodism in Oxford, and delighting old-school Oxonians. Dr. Johnson, in particular, heartily approved:*

Johnson: Sir, the expulsion of six students from the University of Oxford, who were Methodists and would not desist from publicly praying and exhorting, was extremely just and proper. What have they to do at an University who are not willing to be taught, but will presume to teach? Where is religion to be learnt, but at an University? Sir, they were examined, and found to be mighty ignorant fellows.

Boswell: But was it not hard, Sir, to expel them, for I am told they were good beings.

Johnson: I believe they might be good beings, but they were not fit to be in the University of Oxford. A cow is a very good animal in a field, but we turn her out of a garden.

🕉 *But then he was particularly prejudiced. He was once walking in the garden of New Inn Hall with its principal, Sir Robert Chambers, when Chambers began collecting snails from the ground and tossing them over the wall into the next door premises. Dr. Johnson thought this unneighbourly behaviour:*

Chambers: Sir! my neighbour is a Dissenter.

Johnson: Oh! if so, my dear Chambers, toss away, toss away as hard as you can.

🕉 *And it was the idea of Oxford Methodism that gave rise to one of his most celebrated (but unfortunately apocryphal) retorts:*

Boswell: What would you say, Sir, if you were told that your Dictionary would be superseded by one written by a *Scotch Nonconformist*, living at *Oxford*.

Johnson: Sir, in order to be facetious, it is not necessary to be indecent.

UNDERGRADUATES, *c.* 1730

Having been elected from a public school in the vicinity of Oxford, and brought with me the character of a tolerably good Grecian, I was invited, by a very worthy person now living, to a very sober little party, who amused themselves in the evening with reading Greek and drinking water. Here I continued six months, and we read over Theophrastus, Epictetus, Phalaris's Epistles, and such other Greek authors as are seldom read at school. But I was at length seduced from this mortified symposium to a very different party; a set of jolly, sprightly young fellows, most of them west-country lads; who drank ale, smoked tobacco, punned, and sung bacchanalian catches the whole evening: our 'pious orgies' generally began with,

> Let's be jovial, fill our glasses,
>   Madness 'tis for us to think,
> How the world is rul'd by asses,
>   And the wisest sway'd by chink . . .

Some gentlemen commoners, however, who were my country-men . . . and who considered the above-mentioned as very *low* company, (chiefly on account of the liquor they drank) good-naturedly invited me to their party: they treated me with port-wine and arrack-punch; and now and then, when they had drank so much, as hardly to distinguish wine from water, they would conclude with a bottle or two of claret. They kept late hours, drank their favourite toasts on their knees, and, in short, were what were then called 'bucks of the first head'. This was deemed good company and high life; but it neither suited my taste, my fortune, or my constitution.

There was, besides, a sort of flying squadron of plain, sensible, matter-of-fact men, confined to no club, but associating occasionally with each party. They anxiously enquired after the news of the day, and the politics of the times. They had come to the university in their way to the Temple, or to get a slight smattering of the sciences before they settled in the country. They were a good sort of young people, and perhaps the most rational of the college: but neither with these was I destined to associate. . . .

Richard Graves, *Recollections*, 1788.

THE FRESHMAN'S LETTER TO HIS MOTHER

Oxford, April 13, 1733

Honoured Madam

I have been at Sir Adolphus Oughton's house in London about two weeks, and return'd from thence last Friday to Oxford, where I met with the linen you were so kind as to send me. My sister hints in her letter as if you were displeas'd with my thinking of tea things. I do assure you that I should not have mentioned 'em, only my brother William told me you might probably supply me with any such things

out of your old stock of household furniture; and if you consider 'em
with their consequences, I believe you will find 'em to be the cheapest
and most convenient article we can have in our colledge: for, not to
speak of the great addition they make to the furnishing of our rooms,
we have nothing at all for our own breakfast but what we get ourselves, and I find tea to be at least by one half the cheapest thing I
can get: indeed, as I am no admirer of the common tea that is drank,
I chuse myself to breakfast upon that which is made of herbs, such as
sage, balm, colesfoot and the like: and if I am to make any acquaintance that may be usefull to me in future life, which is the only reason
I am sent to this colledge, I believe I need not tell you that introducing
myself into their company by asking 'em to my rooms to drink a
dish of tea is beyond any comparison the cheapest and most convenient
way I can possibly pretend to. . . .

I am, dear Mamma, with the greatest respect, your very dutiful
and affectionate son

Richard Congreve.

SCHOLAR, 1734

The man who, stretched in Isis' calm retreat,
To books and study gives seven years complete,
See! Strowed with learned dust, his night-cap on,
He walks, an object new beneath the sun!
The boys flock round him, and the people stare;
So stiff, so mute, some statue you would swear,
Stept from its pedestal to take the air!

Alexander Pope, *Horace's Epistles*, 1734.

THOMAS HEARNE

'Pox on't!' said Time to Thomas Hearne,
'Whatever I forget you learn'.

*Oxford's great seventeenth-century antiquarian, Anthony Wood, was
succeeded by Thomas Hearne of St. Edmund Hall (1678–1735), who was
no less diligent, and if anything even more ill-tempered, libellous, and
entertaining. From his collected jottings, which fill eleven volumes, come
these Oxford observations:*

Aug. 13. [1717] Going this Day through Xt Church, I took the
opportunity to view distinctly the Statue just put up in one of the

nitches within the College, by the Dean's Lodgings, of Bp Fell. The Statuary was at work. All People that knew the Bp agree 'tis not like him, he being a thin, grave man, whereas the Statue represents him plump and gay. I told the Statuary that it was unlike, & that he was made too plump. Oh, says he, we must make a handsome Man. Thus this Fellow. Just as if it were to burlesque the Bp, who is put in Episcopal Robes, & yet by the Statue is not represented above 20.

Dec. 6. [1717] The Rector of Exeter-Coll. (Matthew Hole), now very near, if not quite, fourscore Years of Age, courts a young Girl, living at the Turl-Gate, in Oxford, with her Father, named Brickland, and a Seller of Cheney-Ware. She is handsome, but an Ideot.

Feb. 27 (Ashwed.) [1723] It hath been an old Custom in Oxford for the Scholars of all Houses, on Shrovetuesday, to go to dinner at 10 Clock (at wch time the little Bell call'd Pan-cake Bell rings, or, at least, should ring, at St. Marie's) and at four in the Afternoon, and it was always follow'd in Edmund Hall as long as I have been in Oxford, 'till Yesterday, when they went to dinner at 12 and Supper at six, nor were ther any Fritters at Dinner, as ther us'd always to be. When laudable old Customs alter, 'tis a Sign Learning dwindles.

Feb. 22. [1724] Upon the Top of Heddington Hill, by Oxford, on the left hand as we go to Heddington, just at the Brow of the Branch of the Roman Way that falls down upon Marston Lane, is an Elm that is commonly call'd and known by the Name of Jo. Pullen's Tree, it having been planted by the Care of the late Mr. Josiah Pullen of Magdalen Hall, who used to Walk to that place every day, sometimes twice a day, if tolerable Weather, from Magdalen Hall and back again in the Space of half an Hour. This Gent. was a great Walker, & some walks he would call a Mug of two Penny, & others a Mug of Three penny, &c, according to the Difference of the Air of each Place.

🔖 *Part of the trunk of Pullen's elm stood there until 1909, when it was burnt.*

Nov. 15. [1725] About Thursday last Dr. Francis Gastrell, Canon of Christ-Church, of the seventh Stall, and Bishop of Chester, was seiz'd very violently with a Gout in his Head. He was told that if he would take a Bottle of Port Wine, it might, probably, drive it back, but this he absolutely declin'd, saying he had much rather die than drink a whole Bottle of that Wine. Accordingly, he died sometime last Night. . . .

Aug. 16 [1726] Last Sunday Morning, between 8 & 9 Clock, when

they were at Prayers, at Merton College, one Mr. Gardiner, M.A. &
Fellow of that College, cut his own Throat in his Chamber, and died
of the Wound last Night. He was a pretty Man, & was about 25
Years of Age. . . . He hath two Sisters, wch for a good while together
lived at Dr. Gardiner's at All Souls Coll., & used to go to the College
Chappell. One of them was observ'd to be very rampant & wild, &
mighty desirous of Men, so as to be ready to leap out at Window
after them.

April 6. [1727] Last Tuesday (being Easter Tuesday), there being a
Bull baiting at Heddington near Oxford, a Quarrel arose between
some Scholars that were there, & two or three of Heddington, about
a Cat, that the Scholars would have had tied to the Bulls Tayl. The
Scholars being worsted . . . notice was given to other Scholars at
Oxford, whereupon a great Number (some say five hundred, others
about two hundred) of them went immediately with Clubs to Heddington, and committed such strange disorders, as have hardly been
heard of. They broke almost all the windows in the Town, (pulling
down the very window bars) got into Houses, opened Chests, beat and
bruiz 13 several people in an intolerable manner. . . . Heddington
looked very strange after this disaster. Some of the Inhabitants,
upon approach of the Scholars, run away, others hid themselves, the
rest that staid and were found suffered much.

HEARNE'S WHO'S WHO

Dr. Baron of Balliol: A poor snivelling Fellow, and in many respects
a Knave.

Dr. De Laune of St. John's: Imbezzel'd all the Treasury of the
University.

Anthony Wood the antiquarian: Always look'd upon in Oxford as a
most egregious, illiterate, dull Blockhead, and a conceited, impudent
Coxcombe.

Dr. Kennett of St. Edmund Hall: That republican, whiggish, giddy-
headed, and scandalous Divine.

Dr. Tyndal of All Souls: A noted Debauchee & a man of very per-
nicious Principles, sly and cunning.

Dr. Lasher of St. John's: Noted for a silly, Puritannical, prick-ear'd
Whigg, fit to be made a Cuckold of.

Mr. Skelton of Queen's: Being a stout Claretteer kill'd as the Report
goes an Apothecary at Northampton by hard drinking, and to make
his Widdow amends for this like an old Doating Fool married her.

Mr. Wise of Exeter: A certain conceited, muddy-headed Person, a dark, immethodical Prater.

Dr. Bowles of the Bodleian: That most egregious Coxcomb and Rascal, imployed in every pitifull, scoundrell Affair, for wch none is more proper.

Dr. Hall of Pembroke: An admirer of whining, cringing Parasites, and a strenuous Persecutor of truly honest Men as occasion offer'd itself.

Dr. Meare of Brasenose: Never noted for Learning or any thing else.

Dr. Gardiner of All Souls: An indifferent Scholar, a Man of a loose Life, will not stick at any thing to promote his Interest, he being withall, a Man of great Price and Conceit, & of a knavish, tricking disposition, & studying nothing more than to baffle those he does not care for by all the base, treacherous, & malicious Methods he & his Agents can invent.

Dr. Hudson of the Bodleian: Stingy, miserable, selfish Temper, & abominable Love of Money.

Dr. Charlett of University: A Man of a strange Rambling Head. He is justly called by many the stupid Incumberer of the Ground of University College. A very busy Man, very illiterate.

Dr. Lancaster of Queen's: That old hypocritical, ambitious, drunken sot, a Person of a smooth, tricking, trepanning and I know not what principles.

Mr. Wanley of the Bodleian: A very loose, debauched Man, kept Whores, was a very great Sot, & by that means broke to Pieces his otherwise very strong Constitution.

John Vanbrugh the architect: A Blockhead.

G. F. Handel and his company of musicians: A lowsy crew of foreign fidlers.

Hearne's epitaph, in the churchyard of St. Peter-in-the-East, Oxford, was self-composed:

> Here lieth the body of Thomas Hearne AM
> Who studied and preserved Antiquities.

K--G J---S FOR EVER!

The Stuart cause still had friends in Oxford, even after the unsuccessful rising of 1745, as a staunch and indignant Establishment man, the Reverend William Blacow, discovered in 1747.

On Tuesday the 23d day of February 1747, I was in a private Room at Winter's Coffee-House, near the High Street in Oxford, in company with several Gentlemen of the University and an Officer in his Regimental Habit. About seven o'clock in the Evening, a person, belonging to the Coffee-House, came into the Room and told us, There were a number of Gownsmen at the door, shouting k—g J—s for ever, Pr— C—s, and other treasonable Words.

The Rioters, in a short time, returned to Winter's Coffee-House; and then I myself heard them shout aloud G—d bless k—g J—s, Pr— Ch—s, d—n K—g G—e, and other Treasonable and Seditious Expressions.

Being determined to use my utmost endeavors to discover these Rioters, I followed them down the High Street (where I heard them uttering the same Treasons, almost in one continued Shout) and from that Street into St. Mary Hall Lane. And in this Lane, opposite to Oriel College, I saw the Rioters, to the number of about Seven, standing still, and continuing to shout as above.

As I came near the Rioters, one of them, whom I was afterwards informed and believe to be Mr. Whitmore, advanced to me, waving his Cap, and shouting k—g J—s for ever, Pr— Ch—s, G—d bless the Great k—g J—s the Third; and other Treasonable Expressions. Upon which I laid hold of him; and told him, I insisted upon carrying him to the Proctor.

Mr. Whitmore's companions, upon this, came about me; and, at first, desired me to let Mr. Whitmore go. I answered, 'Gentlemen, this is strange Imprudence, let your Party be what it will': and refus'd to let him go. Upon which refusal, some of the Gentlemen pull'd off their cloaths, assaulted, and struck me several times; and endeavour'd to force Mr. Whitmore from me: Which at last they effected.

The Riot still increasing, after Mr. Whitmore had been forced from me, I endeavoured to take refuge in Oriel College: which several Gentlemen, whom I apprehended to belong to that College, strove to prevent; so that tho' I enter'd, it was with great difficulty. Having been, some time within the College, I heard the Rioters, who still continued in the same place, having been join'd by many other persons (as I apprehend, about Forty) continue the same Treasonable Shouts: and one part of the rioters louder than the rest, in crying D—n K—g G—e and all his Assistants, and cursing me in particular. Upon this, stepping to the Gate, I told them, I heard their Treason, and should certainly bring them to justice.

Just upon my appearing at the Gate, came by Mr. Harrison, a Master of Arts, of Corpus Christi College; whom I requested to assist me, in taking proper notice of that Treasonable Riot. But his answer being abusive and insulting, I told him, if he, in the same circumstances, had not acted in the same manner, he must have been perjured.

Immediately after the preceding Conversation with this worthy Clergyman, Mr. Luxmore and Mr. Dawes advanced and laid hold of me. But Mr. Dawes, taking his hand from me, and stripping to fight, said, I am a man who dare not say, God bless k—g James the Third; and tell you, my name is Dawes of St. Mary Hall. I am a man of an independent Fortune, and therefore afraid of no man; or words to that effect. At this instant (seasonably for me) came the Proctor. . . .

I acquainted the Vice-Chancellor circumstantially and truly, with the whole of the Treasonable Riot. The Vice-Chancellor said, He was sorry for what had happen'd; but that nothing could prevent young Fellows getting in liquor; but that they should be severely punished.

It may be proper just to remark here, as to the severe punishment; that it proved, at last, to be no more, than putting off their degrees for one year, and 'an imposition of English to be translated into Latin'.

A Letter to William King, D.D., 1755.

THE OLD FELLOWS

1753: On my way to Hagley I dined at Park Place and lay at Oxford. As I was quite alone, I did not care to see anything; but as soon as it was dark I ventured out, and the moon rose as I was wandering among the colleges, and gave me a charming venerable Gothic scene, which was not lessened by the monkish appearance of the old fellows stealing to their pleasures. . . . Oxford, my passion . . . the whole air of the town charms me.

Horace Walpole (1717–97), *Letters.*

MIND OVER MATTER

They wanted to make an old woman of me, and that I should stuff Latin and Greek at the University. These schemes I cracked like so many vermin as they came before me!

John Hunter, the anatomist (1728–93).

THE HEDONISTS

For all the able men in eighteenth-century Oxford, the overwhelming impression left by the memoirists, satirists, and reporters is one of idle hedonism, pursued in colleges where senior common rooms were rich celibate clubs, and where all too many students had no interest in learning. These miscellaneous extracts, from most decades of the century, are characteristic of many more:

PLEASURES OF THE COFFEE-HOUSE

Besides the libraries of Radcliffe and Bodley and the Colleges, there have been of late years many libraries founded in our coffee-houses for the benefit of such as have neglected or lost their Latin or Greek. . . . As there are here books suited to every taste, so there are liquors adapted to every species of reading. Amorous tales may be perused over Arrack, punch and jellies; insipid odes over Orgeat or Capilaire; politics over coffee; divinity over port; and Defences of bad generals and bad ministers over Whipt Syllabubs. In a word, in these libraries instruction and pleasure go hand in hand; and we may pronounce, in a literal sense, that learning no longer remains a dry pursuit.

Thomas Warton, A Companion to the Guide, c. 1760.

PLEASURES OF THE COMMON-ROOM

These fellowships are pretty things;
We live once more like petty kings,
And dine untax'd, untroubled under
The portrait of our pious founder.

Why did I sell my College life
(He cries) for benefice and wife?
Return, ye days, when endless pleasure
I found in reading or in leisure;

[147]

When calm around the Common Room
I puff'd my daily pipe's perfume;
Rode for a stomach and inspected,
At annual bottlings, corks selected.

Thomas Warton, *The Progress of Discontent*, 1746.

How often did I wish myself transported to the blissful region of the common room fire-side! Delightful retreat, where never female shewed her head since the days of the founder!

Vicesimus Knox, *Essays Moral and Literary*, 1782.

PLEASURES OF DRINK

If on my theme I rightly think,
There are five reasons why men drink,
Good wine, a friend, or being dry,
Or lest we should be, by and by,
Or any other reason why.

From the Latin of Henry Aldrich,
Dean of Christ Church 1689–1710.

Eldon tells a tale of a Doctor of Divinity trying under the influence of some inspiration much stronger than that of the Pierian stream to make his way to Brasenose, through Radcliffe Square. He had reached the library, a rotunda then without railings, and unable to support himself except by keeping one hand on the building, he continued walking round and round, until a friend, coming out of the college, espied the distress of the case, and rescued him from the *orbit* in which he was so unsteadily revolving.

Horace Twiss, *Life of Lord Chancellor Eldon*, 1844.

Talking of drinking wine, Dr. Johnson said, 'I did not leave off wine, because I could not bear it; I have drunk three bottles of port without being the worse for it. University College has witnessed this.'

James Boswell, *Life of Johnson*, 1791.

PLEASURES OF FOOD

Dr. Gardiner, Warden of All Souls Coll., being a perfect Epicurean, minding nothing but eating & drinking, and heaping up Money, and doing Mischief, hath got a new way of stuffing a leg of Mutton roasted. He had lately one stuffed by his own order, with White Herrings out of the Pickle.

Thomas Hearne, *Diary*, 1724.

The Clarendon Building, 1774: then the offices of the University Press

Merton College, 1788

THE HEDONISTS

Dec. 5, 1774—I dined & spent the Afternoon at the Wardens [of New College] . . . we had a most elegant Dinner indeed. The first Course, was Cod & Oysters, Ham & Fowls, boiled Beef, Rabbits smothered with Onions, Harrico of Mutton, Pork Griskins, Veal Collops, New-Coll: Puddings, Mince Pies, Roots &c—The second Course was a very fine rost Turkey, Haunch of Venison, a brace of Woodcocks, some Snipes, Veal Olive, Trifle, Jelly, Blomonge, stewed Pippins, Quinces preserved &c . . . Madeira, Old Hocke, and Port Wines to drink &c. A desert of Fruit after Dinner—we stayed till near 8.

James Woodforde, Diaries, 1774.

GOURMET'S EPITAPH

Here lyes Doctor Sergeant within these cloysters,
Whom if the last trump don't wake, then crye oysters.

On Thomas Sergeant of All Souls.

GOURMAND'S EPITAPH

Here lies Randal Peter,
Of Oriel, the eater,
Whom death at last has eaten:
Thus is the biter bitten.

On Peter Randal of Oriel.

UNDERGRADUATE LIFE

I Rise about nine, get to Breakfast by ten,
Blow a Tune on my Flute, or perhaps make a Pen;
Read a Play 'till eleven, or cork my lac'd Hat;
Then step to my Neighbours, 'till Dinner, to chat.
Dinner over, to *Tom's*, or to *James's* I go,
The News of the Town so impatient to know;
While Law, Locke and Newton, and all the rum Race,
That talk of their Modes, their Ellipses, and Space,
The Seat of the Soul, and new Systems on high,
In Holes, as abstruse as their Mysteries, lie.
From the Coffee-house then I to Tennis away,
And at five I post back to my College to pray.
I sup before eight, and secure from all Duns,
Undauntedly march to the *Mitre* or *Tuns*;
Where in Punch or good Claret my Sorrows I drown,
And toss off a Bowl 'To the best in the Town':
At One in the Morning, I call what's to pay,
Then Home to my College I stagger away.
Thus I tope all the Night, as I trifle all Day.

The Oxford Sausage, edited by Thomas Warton, 1764.

THE EXAMINATION SYSTEM

🦎 *Oxford examinations in the eighteenth century, whether for entry or for degrees, were always oral, and were often as eccentric as they were ineffective, having developed haphazard from the disputations of the Middle Ages, or the quirks of particular dons:*

QUARE

It is . . . well-known to be a custom for the *candidates* either to present their *examiners* with a *piece of gold*, or to give them an handsome *entertainment*, and make them *drunk*; which they commonly do the night before *examination*, and sometimes keep them till morning, and so adjourn, *Cheek by Joul*, from their *drinking room* to the *school*, where they are to be *examined*—*Quare*, whether it would not be very ungrateful of the *examiner* to refuse any candidate a *testimonium*, who has treated him so splendidly over night?

Nicholas Amhurst, Terrae-Filius, 1721.

STRINGS

Every Undergraduate has in his possession certain papers, which have been handed down from generation to generation and are denominated *strings*. . . . These consist of two or three arguments, fairly transcribed in that syllogistical form which is alone admitted. The two disputants having procured a sufficient number of them and learned them by heart, proceed with confidence to the place appointed. From one o'clock till three they must remain seated opposite each other, and if any Doctor should come in, who is appointed to preside over those exercises, they begin to rehearse what they have learned, frequently without the least knowledge of what is meant. . . .

Four times must this farce be performed before the student is qualified for the degree of Bachelor of Arts.

From The Gentleman's Magazine, 1780

🦎 *For an example of syllogistical argument, 'handed down from generation to generation', see page 7.*

'DON'T YOU KNOW THAT?'

The history examination commences. 'What comes in the seventy-ninth chapter of the second book of Herodotus?' A pause. 'Don't you know that? It's about the crocodile. Which jaw does he move?' A pause. 'Don't you know that, sir? Then tell me the three wrong reasons Herodotus gives for the Nile overflowing in summer.' 'I forget them, sir, but I can tell you the right one.' 'I don't want the right one: I want the wrong ones.' Serious looks are passing among the examiners. One more chance is given. 'How many miles is it from the village Agnosté to the village Aneureté?' 'Seven miles.' 'No, sir.' 'Eight miles and a half.' 'Worse again, sir, it is seven miles and a half'; and the examiner sits down with an air of triumph at being able to correct the important difference from a piece of paper in which he has previously copied it out, and which he attempts to hide under his gown. . . .

Francis Trench, A Few Notes from Past Life, 1862.

SURE OF SUCCESS

The masters take a most solemn oath, that they will examine properly and impartially. Dreadful as all this appears, there is always found to be more of appearance in it than reality; for the greatest dunce usually gets his *testimonium* signed with as much ease and credit as the finest genius. . . . If the vice-chancellor and proctors happen to enter the school . . . then a little solemnity is put on, very much to the confusion of the masters, as well as of the boy, who is sitting in the little box opposite to them. As neither the officer, nor anyone else, usually enters the room (for it is reckoned very *ungenteel*), the examiners and the candidates often converse on the last drinking-bout, or on horses, or read the newspaper, or a novel, or divert themselves as well as they can in any manner, till the clock strikes eleven, when all parties descend, and the *testimonium* is signed by the masters. With this *testimonium* in his possession, the candidate is sure of success. The day in which the honour is to be conferred arrives; he appears in the Convocation house, he takes an abundance of oaths, pays a sum of money in fees, and, after kneeling down before the vice-chancellor, and whispering a lie, rises up a Bachelor of Arts.

Vicesimus Knox, Essays Moral and Literary, 1782.

GRADUATED

This is the complete examination in Hebrew and History undergone by John Scott, afterwards Lord Chancellor Eldon, for his Bachelor's degree in 1770, at least by his own account:

Examiner: What is the Hebrew for the place of a skull?
Scott: Golgotha.
Examiner: Who founded University College?
Scott: King Alfred.
Examiner: Very well, sir, you are competent for your degree.

UNPROFITABLE MONTHS

🦁 *Edward Gibbon matriculated at Magdalen College shortly before his fifteenth birthday: after little more than a year, much of it spent away from Oxford, he professed himself a Roman Catholic and was obliged to leave the college. Forty years later, in his* Memoirs (1792), *he recalled the experience in the most celebrated of all Oxford indictments:*

My own introduction to the university of Oxford forms a new aera in my life; and at the distance of forty years I still remember my first emotions of surprise and satisfaction. In my fifteenth year I felt myself suddenly raised from a boy to a man: the persons, whom I respected as my superiors in age and academical rank, entertained me with every mark of attention and civility; and my vanity was flattered by the velvet cap and silk gown, which distinguish a gentleman commoner from a plebeian student. A decent allowance, more money than a school-boy had ever seen, was at my own disposal; and I might command, among the tradesmen of Oxford, an indefinite and dangerous latitude of credit. A key was delivered into my hands, which gave me the free use of a numerous and learned library; my apartment consisted of three elegant and well-furnished rooms in the new building, a stately pile, of Magdalen College; and the adjacent walks, had they been frequented by Plato's disciples, might have been compared to the Attic shade on the banks of the Ilissus. Such was the fair prospect of my entrance (April 3, 1752) into the university of Oxford.

I spent fourteen months at Magdalen College; they proved the fourteen months the most idle and unprofitable of my whole life: the reader will pronounce between the school and the scholar; but I cannot affect to believe that Nature had disqualified me for all literary pursuits. . . . In the discipline of a well-constituted academy, under the guidance of skilful and vigilant professors, I should gradually have risen from translations to originals, from the Latin to the Greek classics, from dead languages to living science: my hours would have been occupied by useful and agreeable studies, the wanderings of fancy would have been restrained, and I should have escaped the temptations of idleness, which finally precipitated my departure from Oxford. . . .

The fellows [of Magdalen] of my time were decent easy men, who supinely enjoyed the gifts of the founder; their days were filled by a series of uniform employments; the chapel and the hall, the coffee house and the common room, till they retired, weary and well satisfied, to a long slumber. From the toll of reading, or thinking, or writing, they had absolved their conscience; and the first shoots of learning and ingenuity withered on the ground, without yielding any fruits to the owners of the public. . . . Dr. —— [his own tutor] well remembered that he had a salary to receive, and only forgot that he had a duty to perform. Instead of guiding the studies, and watching over the behaviour of his disciple, I was never summoned to attend even the ceremony of a lecture; and, excepting one voluntary visit to his rooms, during the eight months of his titular office, the tutor and pupil lived in the same college as strangers to each other.

In all the universities of Europe, excepting our own, the languages and sciences are distributed among a numerous list of effective professors: the students, according to their taste, their calling, and their diligence, apply themselves to the proper masters; and in the annual repetition of public and private lectures, these masters are assiduously employed. Our curiosity may inquire what number of professors has been instituted at Oxford? . . . by whom they are appointed, and what may be the probable chances of merit or capacity? how many are stationed to the three faculties, and how many are left for the liberal arts? what is the form, and what the substance, of their lessons? But all these questions are silenced by one short and singular answer, 'That in the university of Oxford the greater part of the public professors have for these many years given up altogether even the pretence of teaching.'

It might at least be expected, that an ecclesiastical school should inculcate the orthodox principles of religion. But our venerable mother had contrived to unite the opposite extremes of bigotry and indifference: an heretic, or unbeliever, was a monster in her eyes; but she was always, or often, or sometimes, remiss in the spiritual education of her own children. . . . Without a single lecture, either public or private, either christian or protestant, without any academical subscription, without any episcopal confirmation, I was left by the dim light of my catechism to grope my way to chapel and communion table, where I was admitted, without a question, how far, or by what means, I might be qualified to receive the sacrament. Such almost incredible neglect was productive of the worst mischiefs. . . . The blind activity of idleness urged me to advance without armour into the dangerous mazes of controversy; and at the age of sixteen, I bewildered myself in the errors of the church of Rome.

My father was neither a bigot nor a philosopher; but his affection deplored the loss of an only son; and his good sense was astonished at my strange departure from the religion of my country. In the first sally of passion he divulged a secret which prudence might have suppressed, and the gates of Magdalen College were for ever shut against my return.

THE WORKERS

🦅 *Edward Gibbon's celebrated attack on Oxford (page 152) was widely accepted as a just reflection of the state of the eighteenth-century university. There is plenty of evidence though of hard work and scholarship by less eminent Oxonians, and Gibbon's indictment was angrily received by many of his contemporaries. A furious clergyman, for example, James Hurdis, published on a private press* A Word or Two in Vindication of the University of Oxford, *of which this is the most telling passage:*

Were it necessary to attempt a refutation of the charge of indolence as it affects the rest of the University, by calling to mind every studious contemporary of the Historian, the catalogue should begin with the name of Kennicott. But having mentioned this laborious and indefatigable collator, we may rest upon his Atlantaean shoulders the whole weight of our cause, and let him stand alone, an editor of mountainous desert, who lifts his head far above the thunders and luminous coruscations of Gibbon. Let us but compare the work of this writer, *quod humeri fere mille hominium ferre recusarent*, with the smaller labours of Gibbon, and what are they? Let us consider him as the publisher of a Hebrew bible, which gives us at one view all the various readings of nearly *seven hundred* Manuscripts and Editions, preserved in all quarters of the world, and what are the six quartos of Gibbon, and all the researches which attended them? Can they boast of being the consequence of more labour and industry? Can they claim the palm for superior utility? No—while the Colossus Kennicott literally *bestrides the world*, it is the inferior lot of Mr Gibbon to be great only among us *petty men*, who

> Walk under his huge legs, and peep about
> To find ourselves dishonourable graves.

🦅 *John Wesley the divine, who was a tutor at Lincoln College in the 1730s, stayed in college with his pupils all the year—*

I should have thought myself little better than a highwayman if I had not teached them every day in the year but Sundays.

—and Henry Clynes Finton, who entered Christ Church in the last decade of the century, would vehemently have rejected any charge of laziness:

At the time of my leaving Oxford, I possessed the following writers:

Homer	verses	27,000
Pindar		5,560
Aeschylus		8,139
Sophocles		10,341
Aristophanes		15,282
Of Euripedes		3,000

pp 2,310

Thucydides	pp	786
Demosthenes		775
Aeschines		220
Lycurgus		51

pp 1,832

Andocides	81
Lysias (half)	100
Of Plato	500
Of Aristotle	500

pp 1,181

5,323

(I forebear to add Herodotus, Polybii libb. i. ii., *Dionys. Halic. Critica Opera*, because these were works which, though I often inspected, I did not accurately study.)

❧ *Not much time was left for lounging, either, in the four-year course of study suggested by an anonymous Oxford don in* Advice to a Young Student by a Tutor *(1725)—philosophy every morning and evening, classics in the afternoons, divinity on Sundays and Church festivals. These are the books he recommends:*

First Year

Philosophical	Classical
Wingate's *Arithmetic*	Terence
Euclid	*Xenophontis Cyri Institutio*
Wallis's *Logic*	Tully's *Epistles*
Salmon's *Geography*	Phaedrus' *Fables*
Keil's *Trigonometria*	Lucian's *Select Dialogues*
	Theophrastus
	Justin
	Nepos
	Dionysius' *Geography*

Second Year

Harris's *Astronomical Dialogues*	Cambray *On Eloquence*
Keil's *Astronomy*	Vossius' *Rhetoric*

Philosophical	*Classical*
Locke's *Human Understanding*	Tully's *Orations*
Simpson's *Conic Sections*	Isocrates
Milnes' *Sectiones Conicae*	Demosthenes
Keil's *Introduction*	Caesar
Cheyne's *Philosophical Principles*	Sallust
Bartholin's *Physics*	Hesiod
Rohaulti's *Physics*	Theocritus
	Ovid's *Fasti*
	Virgil's *Eclogues*

Third Year

Burnet's *Theory*	Homer's *Iliad*
Whiston's *Theory*	Virgil's *Georgics*
Well's *Chronology*	Virgil's *Aeneid*
Beveridge's *Chronology*	Sophocles
Ethices Compendium	Horace
Puffendorf's *Law of Nature*	Euripedes
Grotius *de Jure Belli*	Juvenal
	Persius

Fourth Year

Hucheson's *Metaphysics*	Thucydides
Newton's *Optics*	Livy
Gregory's *Astronomy*	Diogenes Laertius
	Cicero's *Philosophical Works*

❧ *The student was also advised to read the best English authors, 'such as Temple, Collier,* The Spectator *and the other writings of Addison'.*

❧ *Near the beginning of the century an undergraduate, Benjamin Marshall, described in a long Latin letter what he claimed to be a typical day of his Oxford life.*

	Rise before dawn.
6 a.m.	Public Latin prayers.
	Breakfast.
	A walk with my friends, half an hour.
	Study of the Minor Prophets.
	Study of the poem of Tograeus.
9 a.m.	Study of Philosophy.
10 a.m.	To my Tutor, Mr Pelling, who expounds some portion of Philosophy to me and my friends.

11 a.m.	Luncheon.
	With my friends to coffee-house, where we discuss public affairs.
1 p.m.	Study of the Koran.
4 p.m.	Study of Aristotle's *Rhetoric*.
6 p.m.	Dinner.
	Read Horace's *Odes* or Martial's epigrams, or mix with my friends in a sociable way.
9 p.m.	Public Latin prayers. In the morning we pray for success upon our doings, and in the evening we return thanks for such success as has been secured.

But if this translated abstract sounds a little too diligent to be true, perhaps that is because Benjamin Marshall was addressing his letter to his former headmaster, Dr. John Postlethwayt of St. Paul's.

THE HUTCHINSONIANS

Methodism is quite decayed in Oxford, its cradle. In its stead, there prevails a delightful fantastic system called the sect of the Hutchinsonians. After much enquiry, all I can discover is that their religion consists in driving Hebrew to its fountain head, till they find some word or other in every text of the Old Testament, which may seem figurative of something in the New, or at least of something that may happen, God knows when, in consequence of the New. . . . I could not help smiling at the thought of etymological salvation; and I am sure you will smile when I tell you, that according to their gravest doctors, 'Soap is an excellent type of Jesus Christ. . . .'

Horace Walpole, *Letters*, 1753.

MR. WOODFORDE'S ACCOUNTS

James Woodforde matriculated at New College in 1759. Here are some of his early accounts:

Two Logick Books	0	6	0
A pair of Curling Tongs	0	6	0
A Sack of Coal	0	4	9
2 White Waistcoats	1	16	0
For a Pint of Beer at Abby-Milton	0	0	3
At the Whey House	0	0	0½

Gave away	0	0	0½
For Sweeping my Chimney	0	1	0
At Mother Radford's, for Ale and Oysters	0	0	11
A Betting with Williams at Cards lost	0	0	6
Paid Orthman for 16 Lectures	1	1	0
For tuneing my Harpsichord	0	15	0
Gave Mr. Rice's Workmen to drink	0	0	6
At Mother Yeoman's with Loggin	0	0	4
Gave a Poor Old Woman	0	0	1
Found a Knife in High Street. Value	0	0	4

LITTLE BENTHAM'S FIRST YEAR

Jeremy Bentham the political philosopher entered The Queen's College in 1760. He was twelve years old and small for his age, but a friend of the family reported that he made a good start:

He has stood the stare of the whole University: as his Youth, and the littleness of his Size naturally attract the Eyes of everyone: However all enquiries after him will tend much to his Credit by convincing People that he has multum in parvo.

Here are some glimpses from Bentham's letters home, during his first year at Oxford:

As I was coming home from Dr. Lee's who should I meet in the street but Dr. Burton? he stared at me, and I stared at him for a little while, for I did not know him to be Dr. Burton, tho' I knew his face very well; till at last he spoke to me; 'so little Bentham', (says he) 'how do you do. . . .'

The Dr. seeing that I had pumps on, gave me a long harangue upon the dangers of wearing them in this weather, and he told me I should get cold if I did not get shoes. when I went away he made me a present of some Reflexions on Logick, written by himself.

however you may think me idle, I fancy when you understand how much business I do you will alter your opinion. for what with Logic, Geography, Great Testament, Tully de Oratore and this translation [of Cicero, for his father] I think I shall have pretty well enough to do. at 10 o'clock we go to lecture in logic and as we can never get the bedmaker scarce to come to us till about half an hour after we come out from prayers and we must get by heart some Logic and look it over etc., generally a good deal; we must have our hair dressed and clean ourselves and at half an hour after 12 dine, so that that takes up all the morning almost, besides Geography at 4 o'clock on Thursday and exercise on Saturday morning (though indeed those two days we have no logic lecture) and the classics at night.

That Tooth which had several bits of it broke out and was as I complained to you so extremely sore being very troublesome I with my fingers pulled it out having plucked up a good courage: besides there were two other Teeth one of which had a young one growing by the Side of it; I pulled them both out myself: however my Face swelled: before I pulled the Teeth out one or two of them aked very bad, so that with that and the Swelled face, which succeeded to the aking: I was forced to keep up in my rooms: when Mr Jefferson asked me what was the matter with me, and I told him, he told me aeger was the Latin for idle . . . I told him I thought t'was very hard that I could never be believed by him when I said anything.

Pray to tell Sam [his brother] not to direct me Master, when he writes to

Your dutiful and affectionate Son
J. Bentham

QUITE UNHINGED

James Boswell first visited Oxford in 1763, when he was twenty-two, as the guest of Sir James Macdonald of Christ Church, a Scottish undergraduate so learned that he was already known as 'the Marcellus of the North'. The visit was not a success.

Saturday 23 April. I got to Oxford about six. . . . Sir James received me with much politeness and carried me to sup with Dr. Smith of St. Mary Hall. . . . He had with him a Mr. Pepys, of Devonshire, a Mr. Cornwallis, Mr. Eden of the County of Durham and a Mr. Foote. They were all students and talked of learning too much; and in short were just old young men without vivacity. I grew very melancholy and wearied. At night I had a bed at the Blue Boar Inn. I was unhappy to a very great degree.

Sunday 24 April. I got up in miserable spirits. All my old high ideas of Oxford were gone, and nothing but cloud hung upon me. . . . I now thought that human happiness was quite visionary, and I was very weary of life.

Monday 25 April. I breakfasted with Sir James. Pepys was there. I thought them two very dull and unhappy existences, and was vexed at having deprived myself of the venerable ideas which I had of Oxford. . . . We dined at Dr. Smith's, where I was just as bad as ever.

Tuesday 26 April. Early in the morning we set out. . . . When I got to London I could not view it in the usual light. My ideas were all changed and topsy-turvy. . . . I retained the most gloomy ideas of the University. My mind was really hurt by it. I thought every man I met had a black gown and cap on, and was obliged to be home at a certain hour.

Wednesday 27 April. I found myself quite unhinged.

POOR HAWKINS

At 11. this Night was called out of the Chequer by Webber, to go with him & quell a Riot in George Lane, but when we came it was quiet—however we went to the Swan in George Lane, and unfortunately met with a Gownsman above Stairs carousing with some low-life People—we conducted him to his College—He belongs to University College, is a Scholar there, & his Name is Hawkins, he was terribly frightened & cried almost all the Way to his Coll: & was upon his knees very often in the Street, and bareheaded all the way. He is to appear again to Morrow before Webber.
He was apparently sent down, for he took no degree. His name was Joseph, and he was eighteen.

James Woodforde, Diaries, 1774.

WELCOME TO OXFORD

In 1782 a German pastor, C. P. Moritz, making his way on foot from London to Oxford, fell in with an English clergyman on the road. In his book Travels in England (*'translated by a Lady', 1795), he describes the welcome his companion arranged for him in the city:*

'Now,' said he, as we entered the town, 'I introduce you into Oxford by one of the finest, the longest, and most beautiful streets, not only in this city, but in England, and I may safely add in all Europe.'

The beauty and the magnificence of the street I could not distinguish; but of its length I was perfectly sensible by my fatigue; for we still went on, and still through the longest, the finest and most beautiful street in Europe, which seemed to have no end; nor had I any assurance that I should be able to find a bed for myself in all this famous street. At length my companion stopped to take leave of me, and said he should now go to his college.

'And I,' said I, 'will seat myself for the night on this stone bench and await the morning, as it will be in vain for me, I imagine, to look for shelter in a house at this time of night.'

'Seat yourself on a stone!' said my companion, and shook his head. 'No, No! Come along with me to a neighbouring ale-house, where it is possible they mayn't be gone to bed, and we may yet find company.' We went on a few houses further, and then knocked at a door. It was then nearly twelve. They readily let us in; but how great was my astonishment, when, on being shown into a room on the left, I saw a great number of clergymen, all with their gowns and bands on, sitting round a large table, each with his pot of beer before him. My travelling companion introduced me to them, as a German clergyman, whom he could not sufficiently praise for my correct pronounciation of the Latin, my orthodoxy, and my good walking.

I now saw myself in a moment, as it were, all at once transported into the midst of a company, all apparently very respectable men, but all strangers to me. And it appeared to me extraordinary that I should, thus at midnight, be in Oxford, in a large company of Oxonian clergy, without well knowing how I had got there. Meanwhile, however, I took all the pains in my power to recommend myself to my company, and in the course of conversation, I gave them as good an account as I could of our German universities, neither denying nor concealing that, now and then, we had riots and disturbances. 'Oh, we are very unruly here, too,' said one of the clergymen as he took a hearty draught out of his pot of beer, and knocked on the table with his hand. The conversation now became louder, more general, and a little confused; they enquired after Mr. Bruns, at present professor at Helmstadt, and who was known by many of them.

Among these gentlemen there was one of the name of Clerk, who seemed ambitious to pass for a great wit, which he attempted by starting sundry objections to the Bible. . . . Among other objections to the Scriptures, he started this one to my travelling companion, whose name I now learnt was Maud, that it was said in the Bible that God was a wine-bibber. On this Mr. Maud fell into a violent passion, and maintained that it was utterly impossible that any such passage should be found in the Bible. Another divine, a Mr. Caern referred us to his absent brother, who had already been forty years in the church, and must certainly know something of such a passage if it were in the Bible, but he would venture to lay any wager his brother knew nothing of it.

'Waiter! fetch a Bible!' called out Mr. Clerk, and a great family Bible was immediately brought in, and opened on the table among all the beer jugs.

Mr. Clerk turned over a few leaves, and in the book of Judges, 9th chapter, verse xiii, he read, 'Should I leave my wine, which cheereth God and man?'

Mr. Maud and Mr. Caern, who had before been most violent, now sat as if struck dumb. A silence of some minutes prevailed, when all at once, the spirit of revelation seemed to come on me, and I said, 'Why, gentlemen, you must be sensible that it is but an allegorical expression'; and I added, 'how often in the Bible are kings called gods!'

'Why, yes, to be sure,' said Mr. Maud and Mr. Caern, 'it is an allegorical expression; nothing can be more clear; it is a metaphor, and therefore it is absurd to understand it in a literal sense.' And now they, in their turn, triumphed over poor Clerk, and drank large draughts to my health in strong ale. . . . The conversation now turned on many other different subjects. At last, when morning drew near, Mr. Maud suddenly exclaimed, 'D—n me, I must read prayers this morning at All-Souls!'

Next day, waking with a bad head, Pastor Moritz set out to see the town:

[It] did not, however, appear to me nearly so beautiful and magnificent as Mr. Maud had described it to me during our last night's walk. The colleges are mostly in the Gothic taste, and much over-loaded with ornaments, and built with grey stone; which, perhaps, while it is new, looks pretty well, but it has now the most dingy, dirty, and disgusting appearance that you can possibly imagine. . . .

To me Oxford seemed to have but a dull and gloomy look; and I cannot but wonder how it ever came to be considered so fine a city.

THE LITTLE MONARCH

As a senior fellow, I was a little monarch within the verge of my college [St. John's]. The statutes had required, that persons of the lower degrees should pass before me, nay, stand in the quadrangle whenever I was present, with heads uncovered. From this general obeisance, and from many other circumstances, I had been led to conceive myself a person of great importance. I was so, indeed, in the circumscribed limits of my society. But the misfortune was, that I could not easily free myself from the consciousness of it when no longer a member, and expected a similar degree of deference from all I met, which cannot be paid in the busy world without inconvenience. . . .

Vicesimus Knox, *Essays Moral and Literary*, 1782.

THE ANSWER OF THE ORACLE

In the 1790s a group of undergraduates tried to enliven university life by founding a Society for Scientific and Literary Disquisition, an innocuous association which would debate matters of a studiously uncontroversial character. A deputation was appointed to get its rules approved by Authority:

Dr. [John] Wills was then vice-chancellor. . . . He received the deputation in the most courteous manner, and requested that our laws might be left with him, as much for his own particular and careful examination as for that of other heads of houses or officers whom he might choose to consult. His request was readily and as courteously complied with; and a day was appointed when the answer of the oracle might be obtained. In about a week, according to agreement, the same deputation was received within the library of the vice-chancellor, who, after solemnly returning the column (containing the laws) into the hands of our worthy founder, addressed them pretty nearly in the following words. *'Gentlemen, there does not appear to be anything in these laws subversive of academic discipline, or contrary to the statutes of the university—but* (ah, that ill-omened 'BUT') *as it is impossible to predict how they may operate, and as innovations of this sort, and in these times, may have a tendency which may be as little anticipated as it may be distressing to the framers of such laws, I am compelled, in the exercise of my magisterial authority, as vice-chancellor, to interdict your meeting in the manner proposed.'*

Thomas Dibdin, *Reminiscences of a Literary Life*, 1836.

DEBTS AND REGRETS

Two Oxford graduates, born in the same year, look back in melancholy to their Alma Mater.

Oxford, since late I left thy peaceful shore,
Much I regret thy domes with turrets crowned,
Thy crested walls with twining ivy bound,
Thy Gothic fanes, dim aisles, and cloisters hoar,

And treasured rolls of wisdom's ancient lore;
Nor less thy varying bells, which hourly sound
In pensive chime, or ring in lively round,
Or toll in the slow curfew's solemn roar.
Much, too, thy midnight walks, and musings grave
'Mid silent shades of high-embowering trees,
And much thy sister streams, whose willows wave
In whispering cadence to the evening breeze;
But most those friends, whose much-loved converse gave
Thy gentle charms a tenfold power to please.

Thomas Russell (1762–88), *Oxford Memories.*

I never hear the sound of thy glad bells,
Oxford, and chimes harmonious, but I say,
(Sighing to think how time has worn away)
Some spirit speaks in the sweet tone that swells,
Heard after years of absence, from the vale
Where Cherwell winds. Most true it speaks the tale
Of days departed, and its voice recalls
Hours of delight and hope in the gay tide
Of life, and many friends now scattered wide
By many fates. Peace be within thy walls!
I have scarce heart to visit thee; but yet,
Denied the joys sought in thy shades, denied
Each better hope, since my poor — died,
What I have owed to thee, my heart can ne'er forget.

William Lisle Bowles (1762–1850), *On Revisiting Oxford.*

🎄 *Byron made fun of this mournful poem in* English Bards and Scotch
Reviewers, *mocking Bowles's muse as one which 'lamentably tells |
What merry sounds proceed from Oxford bells'.*

GENTLEMEN-COMMONERS

🎄 *The undergraduate class of Gentlemen-Commoners enjoyed its privileged
heyday in the second half of the eighteenth century:*

The irresistible influx of commercial wealth, continually augmented
by a thousand streams, has succeeded in sapping the deep foundations
of national integrity. Universities may rue the contagion. They were
soon irrevocably infected. In them extraordinary largesses began to
purchase immunities; the indolence of the opulent was sure of absolu-

tion; and the emulation of literature was gradually superseded by the emulation of a profligate extravagance; till a [new] order of pupils appeared; a pert and pampered race, too froward for controul, too headstrong for persuasion, too independent for chastisement; privileged prodigals. These are the *gentlemen-commoners* of Oxford. . . . They are perfectly their own masters, and they take the lead in every disgraceful frolic of juvenile debauchery. They are curiously tricked out in cloth of gold, of silver, and of purple, and feast most sumptuously throughout the year.

From *The Gentleman's Magazine*, 1798.

The two years of my life I look back to as most unprofitably spent were those I passed at Merton. The discipline of the University happened at this particular moment [1763–5] to be so lax, that a Gentleman Commoner was under no restraint, and never called upon to attend either lectures, or chapel, or hall. My tutor, an excellent and worthy man, according to the practice of all tutors at that moment, gave himself no concern about his pupils. I never saw him but during a fortnight, when I took into my head to be taught trigonometry.

James Harris, First Earl of Malmesbury, in a letter, 1800.

MISS BURNEY IN OXFORD

George III and Queen Charlotte visited Oxford in 1786 with a train of princesses and courtiers, among them, as a lady-in-waiting, the writer Fanny Burney (Madame d'Arblay). By then Jacobite feeling was virtually dead in Oxford, and Miss Burney was amused by the efforts of senior academics to do the right thing by the Hanoverians, as she recorded in her journal (Diary and Letters of Madame d'Arblay, *published in 1842*):

The Vice-chancellor and professors begged for the honour of kissing the king's hand. . . . The sight, at times, was very ridiculous. Some of the worthy collegiates, unused to such ceremonies, and unaccustomed to such a presence, the moment they had kissed the king's hand, turned their backs to him, and walked away as in any common room; others, attempting to do better, did still worse, by tottering and stumbling, and falling foul of those behind them; some, ashamed to kneel, took the King's hand straight up to their mouths; others, equally off their guard, plumped down on both knees, and could hardly get up again; and many, in their confusion, fairly arose by pulling His Majesty's hand to raise them.

❧ *As usual on such royal occasions, they had a meal at Christ Church, this time a 'cold collation'.*

It was at the upper end of the hall. I could not see of what it consisted, though it would have been very agreeable, after so much standing and sauntering, to have given my opinion of it in an experimental way. . . . The Duchess of Ancaster and Lady Harcourt stood behind the chairs of the queen and the princess royal. . . . Lord Harcourt stood behind the king's chair; and the vice-chancellor and the head-master of Christchurch, with salvers in their hands, stood near the table and ready to hand, to the three noble waiters, whatever was wanted: while the other reverend doctors and learned professors stood aloof, equally ready to present to the chancellor and the master whatever they were to forward.

❧ *Miss Burney and other 'untitled attendants', meanwhile, starved in a semi-circle at the other end of Christ Church hall:*

It was agreed that we must all be absolutely famished unless we could partake of some refreshment. . . . A whisper was soon buzzed through the semi-circle, of the deplorable state of our appetite apprehensions; and presently it reached the ears of some of the worthy Doctors. Immediately a new whisper was circulated, which made its progress with great vivacity, to offer us whatever we would wish and beg us to name what we chose. Tea, coffee, and chocolate were whispered back. The method of producing, and the means of swallowing them, were much more difficult to settle than the choice of what was acceptable. Major Price and Colonel Fairly, however, seeing a very large table close to the wainscot behind us, desired our refreshments might be privately conveyed there, behind the semi-circle, and that, while all the group backed very near it, one at a time might feed, screened by all the rest from observation. . . .

This plan had speedy success, and the very good doctors soon, by sly degrees and with watchful caution, covered the whole table with tea, coffee, chocolate, cakes, and bread and butter.

❧ *Thomas Babington Macaulay was depressed by the spectacle of the gifted Miss Burney so demeaning herself, as he wrote in his* Essay on Madame D'Arblay *(1843), and wished she had been visiting Oxford on her own account, as 'a woman of true genius':*

She might, indeed, have been forced to travel in a hack chaise, and might not have worn so fine a gown of Chambery gauze as that in which she tottered after the royal party; but with what delight would she have then paced the cloisters of Magdalene, compared the antique gloom of Merton with the splendours of Christ Church, and looked down from the dome of the Radcliffe Camera on the magnificent sea of turrets and battlements below! How gladly would learned men have laid aside for a few hours Pindar's Odes and Aristotle's Ethics, to escort the author of *Cecilia* from college to college! What neat little banquets would she have found set out in their monastic cells! With what eagerness would pictures, medals, and illuminated missals have been brought forth from the most mysterious cabinets for her amusement!

'A GOOD TRICK'

In the morning I had been a shooting; in the evening I invited a party to wine. In the room opposite there lived a man universally laughed at and despised . . . and it unfortunately happened that he had a party on the same day, consisting of servitors and other raffs of every description. . . . My gun was lying on another table in the room . . . and I proposed, as they had closed the casements, and the shutters were on the outside, to fire a volley. It was thought a good trick; and according I went into my bedroom and fired.

This disagreeable episode, disagreeably told, led to the expulsion of the poet Walter Savage Landor from Trinity College in 1794, Landor characteristically refusing to admit that he had fired the gun. Described by Robert Southey, his contemporary at Oxford, as 'a mad Jacobin', and said to have been the first undergraduate to have worn his hair without powder, Landor left behind him at Trinity this rhyme about one of its Fellows, 'Horse' Kett, so called because of his extraordinarily long and horse-like head:

> The head is horseish, but, what yet
> Was never seen in man or beast,
> The rest is human, or at least
> Is Kett.

A LACK OF DREAMS

Robert Southey the poet went up to Balliol in 1792. 'You won't learn anything by my lectures', his tutor told him, 'so if you have any studies of your own, you had better pursue them.' This advice was a failure, and by Southey's own accounts his Oxford years were altogether wasted:

There is no part of my own life which I remember with so little pleasure as that which was passed at the University. . . . I had many causes of disquietude and unhappiness—some imaginary, and some, God knows, real enough. . . . What Greek I took there, I literally left there, and could not help losing; and all I learnt was a little swimming and a little bathing.

With respect to its superiors, Oxford only exhibits waste of wigs and want of wisdom; with respect to the undergraduates, every species of abandoned excess.

Seven years ago I walked through Oxford on a fine summer morning, just after sunrise, while the stage was changing horses: I went under the windows of what had formerly been my own rooms; the majesty of the place was heightened by the perfect silence of the streets, and it had never before appeared to me half so majestic or half so beautiful. But I would rather go a day's journey round than pass through that city again, especially in the day-time, when the streets are full. Other places in which I have been an inhabitant would not make the same impression; there is an enduring sameness in a university like that of the sea and mountains. It is the same in our age that it was in our youth; the same figures fill the streets, and the knowledge that they are not the same persons brings home the sense of change which is of all things the most mournful.

I never remember to have dreamt of Oxford—a sure proof of how little it entered into my moral being.

🦎 *But then eighteenth-century Oxford, with its rational architecture, its boozing and its stagnancy, was hardly the stuff dreams are made of. It was the following century, the graver and grander nineteenth, that brought the lyric poets to Oxford at last, and crowned the ancient University with romance.*

The Country House
1800 – 1850

*'I and my friends', wrote Matthew Arnold the poet,
'lived in the Oxford of our day as in a great country
house.' Early nineteenth-century Oxford did have a
drawing-room flavour. Though there were some first
signs of intellectual revival, and though in the 1830s
and 1840s religious controversy exploded once more,
still the University was marking time—between the
indulgences of the previous century, and the grandeurs
yet to come.*

THE MEANING OF EDUCATION

Reform spluttered fitfully in Oxford at the very start of the nineteenth century, when for the first time a written examination was introduced for Honours degrees. Colleges began to look for Fellows with intellectual qualifications, and professors to take their duties more seriously. It was to be a long time, though, before the University recognized itself as primarily a place of education for young minds, its principal function in mid-Victorian times. Here is the assessment of a shrewd young Scot who spent some time in Oxford in the first years of the century:

There were rules that had in a general way to be obeyed, and there were lectures that must be attended, but as for care to give high aims, provide refining amusements, give a worthy tone to the character of responsible beings, there was none ever even thought of. The very meaning of the word education did not appear to be understood. . . . The only care the Heads appeared to take with regard to the young minds they were supposed to be placed where they were and paid well to help to form, was to keep the persons of the students at the greatest possible distance. They conversed with them never, invited them to their homes never, spoke or thought about them never. A perpetual bowing was their only intercourse; a bow of humble respect acknowledged by one of stiff condescension limited the intercourse of the old heads and the young. . . .

Elizabeth Grant, Memoirs of a Highland Lady, 1797–1830, 1898.

A REMARKABLE PECULIARITY

A professor of the University of Padua, spending some weeks in Oxford at the start of the nineteenth century, was struck by a 'very remarkable peculiarity' of the place: nobody would ever admit ignorance.

Elsewhere I have asked a professor of astronomy some questions regarding anatomy, or botany, and he had the courage and honesty at once frankly to answer, 'I do not know'. But at Oxford it really seemed as if everybody considered himself equally bound to be universal, to know everything, and to be able to give some sort of affirmative answer to every question, however foreign it might be to his ordinary and proper pursuits. There is so much wisdom in answering seasonably, 'I do not know', that in an university which has been celebrated, and

accounted most wise for nine or ten centuries, I thought for the credit
of the place, I ought to get it once, at least, before I went away; so
I tried hard, but I could never attain it. Why was this?

T. J. Hogg, *Life of P. B. Shelley*, 1858.

GOTHIC REMNANTS

Whatever the state of the University, Oxford traditions resiliently survived. Once every hundred years, for instance, the Fellows of All Souls performed their Mallard Ceremony (page 29). In 1801 Reginald Heber, an undergraduate at Brasenose College, watched it from his window and described it in a letter to a schoolfriend at Cambridge:

All Souls is on the opposite side of Ratcliffe square to Brazen Nose,
so that their battlements are to some degree commanded by my garret.
I had thus a full view of the *Lord Mallard* and about forty fellows,
in a kind of procession on the library roof, with immense lighted
torches, which had a singular effect. I know not if their orgies were
overlooked by any uninitiated eyes except my own; but I am sure that
all who had the gift of hearing within half a mile, must have been
awakened by the manner in which they thundered their chorus, 'O
by the blood of King Edward'. I know not whether you have any
similar strange customs in Cambridge, so that, perhaps, such ceremonies . . . will strike you as more absurd than they do an Oxford
man; but I own I am of opinion that these remnants of Gothicism tend
very much to keep us in a sound consistent track; and that one cause
of the declension of the foreign universities, was their compliance,
in such points as these, with the variation of manners.

Two years later Heber won the Newdigate Prize for his poem Palestine,
and shortly afterwards read the manuscript aloud over breakfast to Walter Scott, who was visiting Oxford with his wife. When Scott remarked that the poem, in its verses on the building of Solomon's temple, failed to mention the fact that no tools were used in the construction, Heber produced in a flash an additional couplet soon to become famous:

> No hammer fell, no ponderous axes rung,
> Like some tall palm the mystic fabric sprung.

THE OPIUM-EATER IN OXFORD

Thomas de Quincey, author of Confessions of an Opium-Eater, *entered Worcester College in 1803, but was never at home in Oxford:*

I resolved to spend some cost upon decorating my person. But it always happened that some book, or set of books—that passion being absolutely endless, and inexorable as the grave—stepped between me and my intentions; until one day, upon arranging my toilet hastily before dinner, I suddenly made the discovery that I had no waistcoat . . . which was not torn or otherwise dilapidated; whereupon, buttoning up my coat to the throat, and drawing my gown as close about me as possible, I went into the public 'hall' (so is called in Oxford the public eating-room) with no misgiving. However, I was detected; for a grave man, with a superlatively grave countenance, who happened on that day to sit next me, but whom I did not personally know, addressing his friend sitting opposite, begged to know if he had seen the last Gazette, because he understood that it contained an Order in Council laying an interdict upon the future use of waistcoats. His friend replied, with the same perfect gravity, that it was a great satisfaction to his mind that his Majesty's Government should have issued so sensible an order; which he trusted would be soon followed up by an interdict on breeches, they being still more disagreeable to pay for. This said, without the movement on either side of a single muscle, the two gentlemen passed to other subjects.

De Quincey claimed that during his first two years at Oxford he spoke no more than 100 words, and he ended his career there peculiarly: for having sat for the first part of his final examination, he disappeared permanently before the second.

JACKSON OF CHRIST CHURCH

The Yorkshireman Cyril Jackson (1746–1819) was one of the most forceful Deans of Christ Church, running the College, wrote Reginald Heber, as 'an absolute monarchy of the most ultra-Oriental character'. He had 'a wonderful tact in managing that most unmanageable class of undergraduates, noblemen', and when he walked across the quadrangle even senior tutors removed their caps (only college scouts were excused—'lest the better-looking among them should be mistaken for undergraduates'). Here is the letter he sent to Robert Peel, a former pupil and a future Prime Minister, after Peel's second speech in the House of Commons, in 1810:

My dear Sir,

I learnt by today's post from those on whom I can depend, that on Friday night you surpassed your former self, to use the very expression of one of the letters I have received. I suppose, therefore, you have been reading Homer. I have only one conclusion to draw, and I trust and believe it is your conclusion also.

Work very hard and unremittingly. Work, as I used to say sometimes, like a tiger, or like a dragon, if dragons work more and harder than tigers.

Don't be afraid of killing yourself. Only retain, which is essential, your former temperance and exercise, and your aversion to mere lounge, and then you will have abundant time both for hard work and company, which last is as necessary to your future situation as even the hard work I speak of, and as much is to be got from it.

Be assured that I shall pursue you, as long as I live, with a jealous and watchful eye. Woe be to you if you fail me!

🦎 *Like many Deans of Christ Church, Jackson assumed a natural supremacy among Oxford college heads, as Thomas Dibdin resentfully observed in his* Reminiscences of a Literary Life, *1836:*

In the third year of my residence [at St. John's College] Dr. Michael Marlowe succeeded to the presidentship. . . . On his election, the whole college (dependent and independent members) went *en trein* to do homage at Christ Church. We were received by the dean and chapter, as I supposed it to be, in full costume and formality. Dr. Cyril Jackson (than whom few men could dress themselves in the robes of authority with greater dignity and effect) was the Dean. I thought this a very humbling piece of vassalage.

🦎 *Thomas de Quincey, rather vaguely hoping to enter Christ Church as an undergraduate, called on Jackson in 1803:*

The Dean was sitting in a spacious library or study, elegantly, if not luxuriously, furnished. Footmen, stationed as repeaters, as if at some fashionable rout, gave a momentary importance to my unimportant self, by the thundering tone of their annunciations. All the machinery of aristocratic life seemed indeed to intrench this great Dean's approaches; and I was really surprised that so very great a man should condescend to rise on my entrance. But I soon found that, if the Dean's station and relation to the higher orders had made him lofty, those same relations had given a peculiar suavity to his manners. . . . Dr. Cyril Jackson treated me just as he would have done his equal in station and in age. Coming, at length, to the particular purpose of my visit at this time to himself, he assumed a little more of his official stateliness. He condescended to say that it would have given him pleasure to reckon me amongst his flock; 'But, sir . . . at present I have not a dog-kennel in my college untenanted.' . . .

Just at that moment, the thundering heralds of the Dean's hall announced some man of high rank: the sovereign of Christ Church seemed distressed for a moment; but then, recollecting himself, bowed in a way to indicate I was dismissed. And thus it happened that I did not become a member of Christ Church.

Jackson's brother William was also an eminent clergyman, and when in 1811 the Dean was offered the bishopric of Oxford, this is how he replied:

Nolo episcopari. Try Will.

Will accepted. (He it was who once began a sermon with the words: 'St Paul says in one of his Epistles—and I partly agree with him. . . .')

The Dean, a famous teacher and an astute judge of character, was also a man of decision:

Jackson: Well, boy, what do you know of music?
Candidate for Christ Church choristership: Please, sir, I has no more ear nor a stone, and no more voice nor an ass.
Jackson: Go your ways, boy. You'll make a very good chorister.

'THE DEATH OF A MOUSE'

Dr. Joseph Hoare, who was for forty-four years Principal of Jesus, the Welsh college, remained in office until he was ninety, and stone deaf. In 1802 he inadvertently put the leg of his chair upon a paw of his beloved cat Tom. The cat bit the don, the don died, and the following is the Principal's best-known epitaph:

> Poor Dr. Hoare! He is no more!
>   Bid Cambria's harp-strings mourn.
> The Head of a House died the death of a mouse,
>   And Tom must be hanged in return.

HONOURED BY ACCLAMATION

🦁 *The playwright-politician Richard Brinsley Sheridan, impecunious and out of office, was proposed for an honorary degree at the Encaenia of 1809, but rejected by the votes of two hostile dons. He came to the ceremony as a private guest anyway, to be greeted by a tumultuous cry of acclamation from the undergraduates:*

Sheridan among the Doctors! Sheridan among the Doctors!

🦁 *Order was not restored until Sheridan was ushered to a seat among the honorands, and he described the episode later as one of the supreme moments of his life.*

A POET AT OXFORD

🦁 *The poet Percy Bysshe Shelley entered University College in 1810. He loved much about Oxford, but like many another clever young man despised the prevailing intellectual flavour of the University: and declining to admit the authorship of his anonymous pamphlet* The Necessity of Atheism, *when interrogated by the Master of his college, in 1811 he was sent down. His contemporary T. J. Hogg recalls his career:*

At the commencement of Michaelmas term, that is, at the end of October, in the year 1810, I happened one day to sit next to a freshman at dinner; it was his first appearance in hall. His figure was slight, and his aspect remarkably youthful, even at our table, where all were very young. He seemed thoughtful and absent. He ate little, and had no acquaintance with anyone. . . . The stranger had expressed an enthusiastic admiration for poetical and imaginative works of the German school. I dissented from his criticisms. He upheld the originality of the German writings. I asserted their want of nature.

'What modern literature', said he, 'will you compare to theirs?'

I named the Italian. This roused all his impetuosity. . . . So eager was our dispute, that when the servants came to clear the tables, we were not aware that we had been left alone. I remarked, that it was time to quit the hall, and I invited the stranger to finish the discussion at my rooms. He eagerly assented. He lost the thread of his discourse in the transit, and the whole of his enthusiasm in the cause of Germany; for as soon as he arrived at my rooms, and whilst I was lighting the candles, he said calmly, and to my great surprise, that he was not qualified to maintain such a discussion, for he was alike ignorant of Italian and German, and had only read the works of the Germans in translation, and but little of Italian poetry, even at second hand.

'Then the *oak* is such a blessing,' [Shelley] exclaimed with peculiar fervour, clasping his hands, and repeating often—'the oak is such a blessing!' slowly and in a solemn tone. 'The oak alone goes far towards making this place a paradise. In what other spot in the world, surely in none that I have hitherto visited, can you say confidently, it is perfectly impossible, physically impossible, that I should be disturbed? . . . The bore arrives; the outer door is shut; it is black and solid, and perfectly impenetrable, as is your secret. . . . He may knock; he may call; he may kick, if he will; he may inquire of a neighbour, but he can inform him of nothing; he can only say, the door is shut, and this he knows already. . . . When the bore meets you and says, I called at your house at such a time, you are required to explain your absence, to prove an *alibi* in short, and perhaps to undergo a rigid cross-examination; but if he tells you, "I called at your rooms yesterday at three, and the door was shut", you have only to say, "Did you? was it?" and there the matter ends.'

When one of the innumerable clocks that speak in various notes during the day and the night at Oxford, proclaimed a quarter to seven, [Shelley] said suddenly that he must go to a lecture on mineralogy, and declared enthusiastically that he expected to derive much pleasure and instruction from it. . . . I invited him to return to tea; he gladly assented, promised that he would not be absent long, snatched his cap, hurried out of the room, and I heard his footsteps, as he ran through the silent quadrangle, and afterwards along High-street.

An hour soon elapsed, whilst the table was cleared, and the tea was made, and I again heard the footsteps of one running quickly. My guest suddenly burst into the room, threw down his cap, and as he stood shivering and chafing his hands over the fire, he declared how much he had been disappointed in the lecture. Few persons attended; it was dull and languid, and he was resolved never to go to another. . . .

'What did the man talk about?'

'About stones! about stones!' he answered, with a downcast look and in a melancholy tone, as if about to say something excessively profound. 'About stones!—stones, stones, stones!—nothing but stones! . . .'

'They are very dull people here', Shelley said to me one evening soon after his arrival, with a long-drawn sigh, after musing awhile; 'a little man sent for me this morning, and told me in an almost inaudible whisper that I must read: "you must read", he said many times in his small voice. I answered that I had no objection. He persisted; so, to satisfy him, for he did not appear to believe me, I told him I had some books in my pocket, and I began to take them out. He stared at me, and said that was not exactly what he meant: "you must read *Prometheus Vinctus*, and Demosthenes *de Corona*, and Euclid." Must I read Euclid? I asked sorrowfully. "Yes, certainly; and when you have read the Greek works I have mentioned, you must begin Aristotle's Ethics, and then you may go on to his other treatises. It is of the utmost importance to be well acquainted with Aristotle." This he repeated so often that I was quite tired, and at last I said, Must I care about Aristotle? what if I do not mind Aristotle? I then left him, for he seemed to be in great perplexity.'

It was a fine spring morning on Lady-day, in the year 1811, when I went to Shelley's rooms; he was absent; but before I had collected our books he rushed in. He was terribly agitated. I anxiously inquired what had happened.

'I am expelled', he said, as soon as he had recovered himself a little, 'I am expelled! I was sent for suddenly a few minutes ago; I went to the common room, where I found our master, and two or three of the fellows. The master produced a copy of the little syllabus *The Necessity of Atheism* and asked me if I were the author of it. He spoke to me in a rude, abrupt, and insolent tone. I begged to be informed for what purpose he put the question. No answer was given. . . . 'Do you choose to deny that this is your composition?' the master reiterated in the same rude and angry voice. . . . I told him calmly, but firmly, that I was determined not to answer any questions respecting the publication on the table. He immediately repeated his demand; I persisted in my refusal; and he said furiously, 'Then you are expelled; and I desire you will quit the college early tomorrow morning at the latest'. . . .

He sat on the sofa, repeating, with convulsive vehemence, the words, 'Expelled, expelled!' his head shaking with emotion, and his whole frame quivering.

Having breakfasted together, the next morning, March 26, 1811, we took our places on the outside of a coach, and proceeded to London.

Life of P. B. Shelley, 1858.

DIES IRAE

In Henry Johnson's satire Rhydisel, *1811,* Don Juan Vincentio di Morla *of Seville, sent to Oxford by his Anglophile father, meets the Devil of Oxford, Rhydisel, stuffing a mutilated corpse into a well outside the Anatomy School. After revealing a series of disagreeable truths about the city from several high places, Rhydisel gives the young man a glimpse of Oxford's Last Day.*

In an instant the azure vault of heaven was blackened with clouds, that seemed to thicken and conglomerate till nothing but the devoted city was visible. A solemn stillness pervaded all things for a few minutes, interrupted only by the shout of the spirits of destruction, who were hurrying to the scene, whilst their mingled and distorted voices produced such sounds of wailing and horror, that the heart of Don Juan was appalled.

He cast his eyes below, and saw the terrified inhabitants running in every direction, some naked, some half-dressed; some old gentlemen, letting themselves down from the balconies of the ladies, without staying to put on any thing but their drawers; young and old women in vain beseeching young men to help them to a place of safety. Here a nobleman just going to mount his hunter at the college gate, asked the advice and assistance of his scout; and there a reverend tutor flew for succour to his cook. A girl was seen in a full-bottomed cauliflower-wig, and a woman, who had passed for twenty-six the last thirty years of her life, without any hair at all; whilst a grave doctor was shrouded in a pink petticoat, and a master of arts in his hurry had taken a female peruque. . . . Horses and dogs, men, women and pigs, monkeys, children, and gentlemen-commoners, were all huddled together in one strange scene of confusion. The spirits of the elements approached—and such a tempest of fire and thunder, hail, rain, and wind was poured down upon the city, that all went to wreck in a moment.

CHIMES AT MIDNIGHT

Oxford? Oh! Oxford seems paradise to this. What reckless dogs we were! I had two hundred a-year there, and spent three. No matter, I enjoyed it. I see myself now, in jersey and cap, all of one colour, pulling for very life in the torpid-race, and still hear the cheers from the bank as we bumped Brasenose. And then, when I got a little faster,

how I used to ride across country with Ridout and Stickles. . . .
And how I won that race at Bullingdon on Charlie Symonds' mare. . . .
And then the Commemorations; then the long winter evenings—port
and walnuts, and careless merry talk; and the long summer days on
the Cherwell—lying on cushions in a punt, and reading novels. Ah,
well-a-day! I am a curate now, and here comes my churchwarden
(ah! those long churchwardens we used to smoke!) to complain about
something.

J. C. Thomson, *Almae Matres, by Megathym Splene, B.A.*, 1858.

A GRAND OCCASION

❧ *In 1814 the Prince Regent, supposing the Napoleonic Wars concluded,
announced that he would make a victorious visit to Oxford, accompanied
by the Emperor of Russia, the King of Prussia, Prince Metternich, and
the Prussian military hero Marshal Blücher. The following preparatory
instructions were issued by the Vice-Chancellor:*

That all Members of the University, Undergraduates and Bachelors,
as Masters of Arts, Proctors, Doctors, Heads of Houses, and Noblemen, go out to meet His Royal Highness in their proper academical
habits. . . . The Members of the University will, according to their
respective gowns and seniority, range themselves on each side of
High Street, in lines extending from St. Mary's Church to the further
extremity of Magdalen Bridge, the Seniors being nearest to the
Bridge so as to leave the centre of the Street open for the accommoda-
tion of spectators, between whom and the line of Gownsmen Cavalry
will be stationed. . . . Upon the Prince Regent's approaching Magdalen Bridge, the Chancellor, Vice-Chancellor, Noblemen, Heads
of Houses, Doctors, Proctors, and delegated Masters, will move
forward, when the Chancellor will lay the Bedels' staves at the feet of
His Royal Highness.

LESE-MAJESTY

'Is that Magdalen Tower?' said the Prince Regent as he approached
Oxford. 'Yes, your Royal Highness,' replied his non-flattering
travelling-companion, Mr. Croker, 'that's the tower against which
James II broke his head.'

G. V. Cox, *Recollections of Oxford*, 1868.

❧ *James II had been thwarted in his attempts to foist a Roman Catholic
President on Magdalen College—see page 112.*

The Radcliffe Observatory, 1794: the first Greek revival building in Oxford, based upon the Tower of the Winds at Athens.

AT THE BANQUET:

The room was filled with men of rank and eminence: but among them all, attention was particularly directed to the veteran Blücher, who, sensible of the feeling, rose and addressed the company in his native German; which was immediately and eloquently translated into English by the Prince Regent, omitting only (with that exquisite taste which distinguished him) those parts which were complimentary to himself.

James Ingrams, *Memorials of Oxford*, 1837.

BARBAROUS

At the end of dinner the Prince Regent ordered the buttery book to be brought, and by his command the Dean inserted his name in the list of members of the college, 'Regia celsitudo Georgii Principis Walliae Regentis', Georg*ii* and the other genitives having been, by an afterthought, substituted for the nominative Georg*ius*, as first written. It is barbarous Latinity, whichever reading is right.

Henry L. Thompson, *Christ Church*, 1900.

JOHN HENRY NEWMAN

A rising star of early nineteenth-century Oxford was John Henry Newman, who entered Trinity College in 1817. These extracts from his letters, diaries, and Memoir *trace his career to its first climax as a Fellow of Oxford's most brilliant college of the day, Oriel. (Newman wrote the* Memoir *in the third person, 'not to conceal the hand that penned it, but better to show the simplicity of style in which he desired that all told about himself should be composed'.)*

FIRST DINNER IN HALL

At dinner I was much entertained with the novelty of the thing. Fish, flesh, and fowl—beautiful salmon, haunches of mutton, lamb &c, fine strong beer—served up in old pewter plates and misshapen earthenware jugs. Tell Mama there were gooseberry, raspberry, and apricot pies. And in all this the joint did not go round, but there was such a profusion that scarcely two ate of the same.

KEEPING APART

I am not noticed at all except by being silently stared at. I am glad, not because I wish to be apart from them and ill natured, but because I really do not think I should gain the least advantage from their company. For H. the other day asked me to take a glass of wine with two or three others, and they drank and drank all the time I was there. I was very glad that prayers came half an hour after I came to them, for I am sure I was not entertained with either their drinking or their conversation. They drank while I was there very much, and I believe intended to drink again. They sat down with the avowed determination of each making himself drunk. I really think, if any one should ask me what qualifications were necessary for Trinity College, I should say there was only one, Drink, drink, drink.

TROUBLE WITH THE NEIGHBOURS

Is it gentlemanly conduct to rush into my rooms, and to strut up to the further end of it, and ask me in a laughing tone how I do; and then, after my remaining some time in silent wonder, to run and bolt the door, and say they are hiding from some one?

Then, to tell me they have come to invite me to wine, and, when I answer in the negative, to ask me why, pressing and pressing me to come, and asking me in a gay manner if I do not mean to take a first class, telling me I read too much, and overdo it, and then to turn from me suddenly and to hollow out 'Let him alone, come along', and to throw open the door?

I said such conduct was not the conduct of gentlemen—and ordered them to leave the room. One then said he would knock me down, if I were not too contemptible a fellow. (He was 6 feet 3 or 4 inches high, and stout in proportion. . . .)

FEARS FOR THE COLLEGE WELFARE

Tomorrow is our Gaudy. If there be one time of the year, in which the glory of our College is humbled, and all appearance of goodness fades away, it is on Trinity Monday. O how the Angels must lament over a whole Society throwing off the allegiance and service of their Maker, which they have pledged the day before at His Table, and showing themselves true sons of Belial!

O that the purpose of some may be changed before the time! I know not how to make myself of use. I am intimate with very few. The gaudy has done more harm to the College than the whole year can compensate. An habitual negligence of the awfulness of the Holy Communion is introduced. How can we prosper?

A PRAYER ANSWERED

On Wednesday, the 29th of April I determined to stand for the scholarship in consequence of the advice of my Tutor, who thought I might be likely to attain it. And my heart beat within me, and I was sanguine that I should gain it, and I was fearful that I should be too much set upon it.

I therefore said, 'O God of heaven and earth, Thou hast been pleased of Thine infinite goodness to impart Thy Holy Spirit to me and to enlighten my soul with the knowledge of the truth. Therefore, O Lord, for our Blessed Saviour's sake, hearken to the supplication which I make before Thee. Let me not rely too much upon getting this scholarship . . . grant it not to me, if it is likely to be a snare to me, to turn me away from Thee. . . .'

Thus I prayed, and He was pleased to hear my petition, and yesterday, out of his infinite lovingkindness He gave me the scholarship.

WHY HE FAILED

I went to the University with an active mind, and with no thought but that of hard reading; but when I got there, I had as little tutorial assistance or guidance as it is easy to conceive, and found myself left almost entirely to my own devices . . . I was obliged to teach myself, and a very sorry and unsatisfactory teaching it was.

My greatest effort for the Schools was during the half-year immediately preceding my appearance in them; I was reading classics; still, however, by myself . . . well, for the last 20 weeks I read regularly 12 hours a day; generally, I believe, exactly that number every day, but at all events that average; if I read one day only 9 hours, another I read 15.—So it was; and then I went into the Schools for my examination;—and completely failed. Of course I got a common pass, but of honors nothing. . . .

A WALK WITH A FRIEND

Took a walk with [Edward] Pusey—discoursed on Missionary subjects. I must bear every circumstance in continual remembrance. We went along the Lower London Road, across to Cowley, and coming back just before we arrived at Magdalen Bridge turnpike, he confessed to me—O Almighty Spirit, what words shall I use? My heart is full. How should I be humbled to the dust! what importance I think myself of! my deeds, my abilities, my writings! whereas he is humility itself, and gentleness, and love, and zeal, and self-devotion. Bless him with Thy fullest gifts, and grant me to imitate him.

TRIUMPH

Mr. Newman used to relate the mode in which the announcement of his election to a Fellowship of Oriel College was made to him. The Provost's Butler, to whom it fell by usage to take the news to the fortunate candidates, made his way to Mr. Newman's lodgings in Broad Street, and found him playing the violin. This in itself disconcerted the messenger who did not associate such an accomplishment with a candidateship for the Oriel Common Room; but his perplexity was increased, when, on his delivering what may be supposed to have been his usual form of speech on such occasions, that 'he had, he feared, disagreeable news to announce, viz. that Mr. Newman was elected Fellow of Oriel, and that his immediate presence was required there', the person addressed, thinking that such language savoured of impertinent familiarity, merely answered 'Very well' and went on fiddling. This led the man to ask whether perhaps he had not mistaken the rooms and gone to the wrong person, to which Mr. Newman replied that it was all right. But, as may be imagined, no sooner had the man left, than he flung down his instrument, and dashed down stairs with all speed to Oriel College. And he recollected after fifty years the eloquent faces and eager bows of the tradesmen and others whom he met on his way, who had heard the news, and well understood why he was crossing from St. Mary's to the lane opposite at so extraordinary a pace.

THE VISITORS

Two great romantic poets, three famous essayists, all visited Oxford in the years between 1819 and 1830. None were graduates of the University, and all responded characteristically:

JOHN KEATS

I

The Gothic looks solemn
The plain Doric column
Supports an old Bishop and Crosier;
The mouldering arch
Shaded o'er by a larch
Stands next door to Wilson the Hosier.

II

Vicè—that is, by turns,—
O'er pale faces mourns
The black tasselled trencher and common hat.
The chantry boy sings,
The steeple-bell rings,
And as for the Chancellor—*dominat.*

III

There are plenty of trees,
And plenty of ease,
And plenty of fat deer for Parsons;
And when it is venison,
Short is the benison,—
Then each on a leg or thigh fastens.

Lines rhymed in a letter from Oxford, 1819.

WILLIAM WORDSWORTH

Oxford, May 30, 1820.

Ye sacred Nurseries of blooming youth!
In whose collegiate shelter England's Flowers
Expand, enjoying through their vernal hours
The air of liberty, the light of truth;
Much have ye suffered from Time's gnawing tooth:
Yet, O ye spires of Oxford! domes and towers!
Gardens and groves! your presence over-powers
The soberness of reason; till, in sooth,
Transformed, and rushing on a bold exchange,
I slight my own beloved Cam, to range
Where silver Isis leads my stripling feet;
Pace the long avenue, or glide adown
The stream-like windings of that glorious street—
An eager Novice robed in fluttering gown!

The River Duddon, 1820.

CHARLES LAMB

Here in the heart of learning, under the shadow of the mighty Bodley
. . . I can play the gentleman, enact the student. To such a one as
myself, who has been defrauded in his younger years of the sweet
food of academic institution, nowhere is so pleasant, to while away
a few idle weeks, as at one or other of the Universities. . . . Here I
can take my walks unmolested, and fancy myself of what degree or
standing I please. I seem admitted *ad eundem*. I fetch up past opportunities. I can rise at the chapel-bell, and dream that it rings for *me*.
In moods of humility I can be a Sizar, or a Servitor. When the peacock
vein rises, I strut a Gentleman Commoner. In graver moments, I
proceed Master of Arts. Indeed I do not think I am much unlike that
respectable character. I have seen your dim-eyed vergers, and bed-
makers in spectacles, drop a bow or a curtsy, as I pass, wisely mis-
taking me for something of the sort. I go about in black, which

favours the notion. Only in Christ Church reverend quadrangle, I can be content to pass for nothing short of a Seraphic Doctor.

The walks at these times are so much one's own—the tall trees of Christ's, the groves of Magdalen! The halls deserted, and with open doors inviting one to slip in unperceived, and pay a devoir to some Founder, or noble or royal Benefactress (that should have been ours), whose portrait seems to smile upon their over-looked beadsman, and to adopt me for their own. Then, to take a peep in by the way at the butteries, and sculleries, redolent of antique hospitality: the immense caves of kitchens, kitchen fire-places, cordial recesses; and spits which have cooked for Chaucer! Not the meanest minister among the dishes but is hallowed to me through his imagination, and the Cook goes forth a Manciple.

Antiquity! thou wondrous charm, what art thou? that being nothing, thou art everything!

Essays of Elia, 1822.

WILLIAM HAZLITT

We could pass our lives in Oxford without having or wanting any other idea—that of the place is enough. We imbibe the air of thought; we stand in the presence of learning. We are admitted into the Temple of Fame, we feel that we are in the sanctuary, on holy ground, and 'hold high converse with the mighty dead'. The enlightened and the ignorant are on a level, if they have but faith in the tutelary genius of the place. We may be wise by proxy, and studious by prescription. . . .

Let him then who is fond of indulging in a dreamlike existence go to Oxford, and stay there; let him study this magnificent spectacle, the same under all aspects, with its mental twilight tempering the glare of noon, or mellowing the silver moonlight; let him wander in her sylvan suburbs, or linger in her cloistered halls; but let him not catch the din of scholars or teachers, or dine or sup with them, or speak a word to any of the privileged inhabitants; for if he does, the spell will be broken, the poetry and the religion gone, and the palace of the enchantment will melt from his embrace into thin air!

Sketches of the Principal Picture Galleries, 1824.

WILLIAM COBBETT

Upon beholding the masses of buildings, at Oxford, devoted to what they call 'learning', I could not help reflecting on the drones that they contain and the wasps they send forth! However, malignant as some are, the great and prevalent characteristic is *folly*: emptiness of head; want of talent; and one half of the fellows who are what they call *educated* here, are unfit to be clerks in a grocer's or mercer's shop.—

As I looked up at that they call *University Hall*, I could not help reflecting that what I had written, ever since I left Kensington on the 29th October, would produce more effect, and do more good in the world, than all that had, for a hundred years, been written by all the members of this University, who devour, perhaps, not less than *a million pounds a year*, arising from property, completely at the disposal of the 'Great Council of the nation', and I could not help exclaiming to myself: 'Stand forth, ye big-wigged, ye gloriously feeding Doctors! . . . Stand forth and face me, who have from the pens of my leisure-hours, sent, among your flocks, a hundred thousand sermons in ten months! More than you have all done for the last half-century'— I exclaimed in vain. I dare say (for it was at peep of day) that not a man of them had yet endeavoured to unclose his eyes.

Rural Rides, 1830.

AN AMUSING JOKE

I wrote . . . each succeeding year for the Newdigate [Poetry Prize], which I was never so fortunate as to get; and I think that one reason of my failure, among others, was that I always aimed at the sublime, while my judges properly preferred the simple. One year (1822) when the subject was 'Palmyra', I got rightly served for my tendencies towards the bombastic. I spent some three months in concocting a poem (only, be it remembered, fifty lines) on the merits of the ancient Tadmor of the Desert, in which I threw off what I regarded as a stunning line, and it ran as follows:

> *High o'er the waste of Nature and of Time.*

A wicked wag of my acquaintance came into my room when I was absent, and finding my manuscript, which I had incautiously left on the table, erased the word 'Nature' from the above line, and substituted for it the word 'paper'; so that my pet line reappeared on my return in the following shape:

> *High o'er the waste of paper and of time;*

and was thus converted into a very apt description of the progress and result of my literary labours in this instance.

How strangely things come about in the course of years! The author of this amusing joke, who was as eccentric as he was clever, after-wards became a Catholic and priest, and was murdered some twenty years ago while bathing in the Adriatic, by a party of ruffians. . . .

Frederick Oakeley, in *The Month*, 1865.

SOMEWHAT NOVEL

Robert S. Hawker went to Oxford, 1823, and entered at Pembroke; but his father was only a poor curate, and unable to maintain him at the university. . . . When he retired to Stratton for his long vacation in 1824, his father told him it was impossible for him to send him back. But Robert Hawker had made up his mind that finish his career at college he would. The difficulty was got over in a manner somewhat novel. There lived at Whitstone, near Holsworthy, four Miss I'ans, daughters of Colonel I'ans. They had been left with an annuity of £200 apiece, as well as lands and a handsome place. . . . Directly that Robert Hawker learnt his father's decision, without waiting to put on his hat, he ran from Stratton to Bude, arrived hot and blown at Efford, and proposed to Miss Charlotte I'ans to become his wife. The lady was then aged forty-one, one year older than his mother; she was his godmother, and had taught him his letters. Miss Charlotte I'ans accepted him; and they were married in November, when he was 20.

On Hawker's return to Oxford . . . he took [his wife] there, riding behind him on a pillion.

S. Baring-Gould, *The Vicar of Morwenstow*, 1899.

🐾 *They lived happily ever after, until Mrs. Hawker's death in 1863, and Hawker went on to immortality with the ballad chorus: 'And shall Trelawny die, Here's twenty thousand Cornish men /will see the reason why.'*

OXFORD VICES

🐾 *The old excesses of eighteenth-century Oxford were less apparent by the 1820s, and drinking in particular seemed to be on the decline:*

We do not consider it necessary to remonstrate with our readers on the subject of hard drinking, as we are aware that this vice has of itself very much decayed in the University. The change is fortunate in many respects, and will tend greatly to put a stop to riot and disorder. Oxford youths, when they have had a drop too much, are guilty of very great extravagancies, and have less method in their madness than any set of tipplers that we had ever the pleasure to pledge . . . They sally forth to the street, with whoop and holla, to insult females and bully tradesmen, and break lamps and windows, till they meet with some match, and get a good licking, or are sobered by a sudden rencontre with the Proctor and his myrmidons. Now all this is very boyish, and raffish, and foolish; and we are glad to hear that the decreasing consumption of wine promises increasing security to patent glass, and the peace of the inhabitants.

J. Campbell, *Hints for Oxford*, 1825.

✿ *On the other hand the still compulsorily celibate Fellows continued occasionally to kick over the traces:*

Walked again in Chr. Ch meadow with Mr. Young. He told me he had been in St. John's Gardens, the most beautiful spot in Oxford and had *witnessed* a curious scene about *one o'clock in the day*, namely in a sly corner he surprised one of the very revd. fellows of ——— College *in flagrante delicto* with Miss Brown, eldest daughter of the *Rev. Proctor*!! So much for Oxford morals! He said the man was old enough to be her father, and the girl, a very pretty, fair creature! Oh shame! The old fellow buttoned up his inexpressibles and set off with his *inamorata* to Trinity gardens, where he probably renewed his games.

Frederic Madden, *Diary*, 1825.

✿ *But there was no Proctor called Brown in 1825.*

THE UNION

✿ *The Oxford Union Society, later to be perhaps the most famous debating society in the world, was founded in 1825. Before acquiring its own premises it rented rooms in the High Street, and soon became one of the archetypal Oxford institutions:*

He also attended the Debates, which were then held in the long room behind Wyatt's; and he was particularly charmed with the manner in which vital questions, that (as he learned from the newspapers) had proved stumbling-blocks to the greatest statesmen of the land, were rapidly solved by the embryo statesmen of the Oxford Union. It was quite a sight, in that long picture-room to see the rows of light iron seats densely crowded with young men . . . and to hear how one beardless gentleman would call another beardless gentleman his 'honourable friend', and appeal 'to the sense of the House', and address himself to 'Mr. Speaker'; and how they would all juggle the same tricks of rhetoric as their fathers were doing in certain other debates in a certain other House. And it was curious, too, to mark the points of resemblance between the two Houses . . . and how they went through the same traditional forms, and preserved the same time-honoured ideas, and debated in the fullest houses, with the greatest spirit and the greatest length, on such points as, 'What course is it advisable for this country to take in regard to the government of its Indian possessions, and the imprisonment of Mr. Jones by the Rajah of Humbugpoopoonah?'

Edward Bradley, Verdant Green, by Cuthbert Bede, 1853.

LITTLE PUPPY

On my entrance at Oxford, as a member of Christ Church, I was too foppish a follower of the prevailing fashions to be a reverential observer of academical dress:—in truth, I was an egregious little puppy:—and I was presented to the Vice-Chancellor, to be matriculated, in a grass-green coat, with the furiously be-powder'd pate of an ultra-coxcomb; both of which are proscribed, by the Statutes of the University.

Much courtesy is shown, in the ceremony of matriculation, to the boys who come from Eton and Westminster; insomuch, that they are never examined in respect to their knowledge of the School Classicks; —their competency is consider'd as a matter of course:—but, in subscribing the articles of their matriculation oaths, they sign their *praenomen* in Latin;—I wrote, therefore, GEORGEIUS, thus, alas! inserting a redundant *e*,—and, after a pause, said inquiringly to the Vice-Chancellor,—looking up in his face with perfect *naïveté*,—'pray sir, am I to add *Colmanus?*'

My Terentian father, who stood at my right elbow, blush'd at my ignorance;—the Tutor (a piece of sham marble) did not blush at all,— but gave a Sardonick grin. . . .

The good-natured *Vice* drollingly answer'd me, that,—'the Surnames of certain *profound Authors*, whose comparatively modern works were extant, had been latinized;—but that a Roman termination tack'd to the patronymick of an english gentleman of my age and appearance, would *rather* be a redundant formality.'

There was too much delicacy in the worthy Doctor's satire for my green comprehension,—and I walk'd back, unconscious of it, to my college,—strutting along in the pride of my unstatutable curls and coat, and practically breaking my oath, the moment after I had taken it.

George Coleman, the Younger, Random Records, 1830.

COLLEGE SERVANT, *c.* 1830

Your Scout, it must be own'd, is not an animal remarkable for sloth;—and, when he considers the quantity of work he has to slur over, with small pay, among his multitude of masters, it serves, perhaps, as a salve to his conscience, for his petty larcenies. He undergoes the double toil of Boots at a well-frequented inn, and a Waiter at Vauxhall, in a successful season.—After coat-brushing, shoe-cleaning, and message-running, in the morning, he has, upon an average, half-a-dozen supper parties to attend, in the same night, and at the same hour;—shifting a plate here, drawing a cork there,—running to and fro, from one set of chambers to another,—and almost solving the Irishman's question of 'how can I be in two places at once, *unless I was a bird?*'

George Coleman, the Younger, *Random Records*, 1830.

THE GEOLOGIST

Oxford's first Professor of Geology was Canon William Buckland (1784–1856) whose attempts to reconcile his science with the biblical account of creation created a storm of controversy, and whose unorthodox manner of life was legendary:

THE PROFESSOR'S ROOMS

Here see the wrecks of beasts and fishes
With broken saucers, cups and dishes;
The prae-Adamic systems jumbled,
With Sublapsaria brecchia tumbled,
And post-Noachian bears and flounders,
With heads of crocodiles and founders;
Skins wanting bones, bones wanting skins,
And various blocks to break your shins . . .
The sage amidst the chaos stands
Contemplative, with laden hands,
This, grasping tight his bread and butter,
And that a flint, whilst he doth utter
Strange sentences that seem to say;—
'I see it all as clear as day. . . .'

His eye in a fine frenzy rolling,
He thus around the fragments strolling,
Still entertains a fond illusion
That all the strata's strange confusion

> He shall explain beyond conjecture,
> And clear in the ensuing lecture.
>
> P. B. Duncan, *Picture of the Comforts of a
> Professor's Rooms*, 1821.

CLARIFICATION

> Some doubts were once expressed about the Flood,
> Buckland arose, and all was clear as—mud.
>
> Philip Shuttleworth (1782–1842).

CLINCH

When God made the stones he made the fossils in them.

> John Keble (1792–1866), in reply to Buckland's theories about the Deluge.

MEMORIES

I recall . . . when I was wont to play with Frank Buckland and his brother (Dr. Buckland's children) in their home at the corner of Tom Quad: the entrance hall with its grinning monsters on the low staircase, of whose latent capacity to arise and fall upon me I never quite overcame my doubts; the side-table in the dining-room covered with fossils, 'Paws off' in large letters on a protecting card; the very sideboard candlesticks perched on saurian vertebrae; the queer dishes garnishing the dinner table—horseflesh I remember more than once, crocodile another day, mice baked in batter on a third day—while the guinea-pig under the table inquiringly nibbled at your infantine toes, the bear walked round your chair and rasped your hand with file-like tongue, the jackal's fiendish yell close by came through the open window. . . .

At Palermo, on his wedding tour, [Dr. Buckland] visited St. Rosalia's shrine,

> That grot where olives nod,
> Where, darling of each heart and eye,
> From all the youth of Sicily,
> St. Rosalie retired to God.

It was opened by the priests, and the relics of the saint were shown. He saw they were not Rosalia's: 'They are the bones of a goat', he cried out, 'not of a woman'; and the sanctuary doors were abruptly closed.

Frank used to tell of their visit long afterwards to a foreign cathedral, where was exhibited a martyr's blood—dark spots on the pavement ever fresh and ineradicable. The professor dropped on the pavement and touched the stain with his tongue. 'I can tell you what it is; it is bat's urine!'

> W. Tuckwell, *Reminiscences of Oxford*, 1900.

Dr. Buckland was extremely like Sydney Smith in his staple character; no rival with him in wit, but like him in humour, common sense, and benevolently cheerful doctrine of Divinity. At his breakfast-table I met the leading scientific men of the day . . . Every one was at ease and amused at the breakfast-table,—the menu and service of it usually in themselves interesting. I have always regretted a day of unlucky engagement on which I missed a delicate toast of mice; and remembered, with delight, being waited upon one hot summer morning by two graceful and polite little Carolina lizards, who kept off the flies.

John Ruskin, Praeterita, 1889.

Professor Buckland used to claim that he had eaten his way through the whole of animal creation: at first he thought the mole was the nastiest thing he had ever tasted, but later he decided blue-bottles were worse. At Nuneham, outside Oxford, he was once shown the heart of a king of France, carefully preserved in a snuff-box. 'I have eaten some strange things', said the Professor, 'but never the heart of a king': and so saying, he swallowed it.

BUCKLAND LECTURES: THE ORIGINS OF LIFE

> Man was not then—but Paramoudras were.
> 'Twas silence all and solitude: the Sun,
> If Sun there were, yet rose and set to none,
> Till, fiercer grown the elemental strife,
> Astonished Tadpoles quivered into life,
> Young Ewcrini their quivering tendrils spread,
> And tails of Lizards felt the sprouting head;
> (The specimen I hand about, is rare,
> And very brittle; bless me, Sir, take care!):
> And, high upraised from ocean's inmost caves,
> Protruded Corals broke th'indignant waves.
> These tribes extinct, a nobler race succeeds;
> Now Sea-fowl scream amid the plashing reeds;
> Now Mammoths range where yet in silence deep
> Unborn Ohio's hoarded waters sleep;
> Now, ponderous Whales—
>   (Here, by the way, a tale
> I'll tell of something very like a whale,
> An odd experiment of late I tried,
> Placing a snake and hedgehog side by side;

Awhile the snake his neighbour tried t'assail,
When the sly hedgehog caught him by the tail,
And gravely munched him upwards, joint by joint;—
The story's somewhat shocking, but in point).

Of this enough; on Secondary Rock
Tomorrow, gentlemen, at two o'clock.

Anon, *Specimen of a Geological Lecture by Prof. Buckland*, 1822.

BUCKLAND LECTURES: SELECTIVE CREATION

Buckland has just noticed in his geological lecture the extraordinary
fact, that, among all the hosts of animals which are found and are
proved to have existed prior to 6000 years ago, *not one* is there which
would be at all serviceable to man; *but* that directly you get within
that period, horses, bulls, goats, deer, asses &c are at once discovered.
How strong a presumptive proof from the face of nature of what the
Bible asserts to be the case.

John Henry Newman, in his diary, 1821.

PETRIFICATION

Where shall we our famous Professor inter,
    That in peace he may rest his bones?
If we hew him a rocky sepulchre,
    He'll rise and break the stones,
And examine each stratum that lies around;
For he's quite in his element underground.

But expos'd to the drip of some case-hard'ning spring
    His body let stalactite cover,
And to Oxford the petrified Sage we will bring
    When he is incrusted all over.
There 'mid mammoths and crocodiles, high on a shelf,
He shall stand as a monument rais'd to himself.

Richard Whately (1787–1863).

*In fact when he died in 1856 Buckland was buried in a grave hewn, by
his own wishes, out of the solid rock at Islip, six miles from Oxford.*

FATHER TO THE MAN

I, who had come up to Oxford [in 1832] a mere child of nature, totally
devoid of self-consciousness, to such a degree that I had never thought
of myself as a subject of observation, developed a self-consciousness so
sensitive and watchful, that it came between me and everything I said

or did . . . I did not know where to put my hands, how to look, how to carry myself. I tried in vain to find out by what secret other men moved about so unembarrassed. I remember as if it were yesterday the first time I met the Provost [of Oriel] in the street. When I became aware that he was coming I was seized with such a tremor, that in the thought of how I ought to perform my first act of 'capping', I omitted the ceremony altogether, and passed him in blank confusion. I saw he knew me and smiled, and I tortured myself with conjecture as to what the smile meant—contempt or compassion. A few days afterwards I met him and Mrs. Hawkins again in the back lane; he knew me—he knew us all by sight—and goodnaturedly supposing that on the previous occasion I had not recognized him, he advanced towards me, holding out his hand, 'Good morning, Mr. Pattison'. I was again in a state of nervous collapse, but having prepared myself in imagination for the terrible ordeal, I executed according to the rules 'ad justum intervallum caput aperiendo', but took no notice of the outstretched and ungloved hand proffered me. I remember now the grunt of dissatisfaction which escaped from the Provost as I tore past, discovering my blunder when it was too late to repair it. I think the Provost's aversion for me dated from this gross exhibition of *mal-adresse*; and I am not at all surprised at it.

Mark Pattison, *Memoirs*, 1885.

Some thirty years later Mark Pattison was himself the head of a college, Lincoln, but he seems to have forgotten the agonies of his youth, for the walks upon which he invited undergraduates to accompany him were of a legendary dreadfulness. Andrew Clark in his Lincoln College *(1898) describes the ordeal of one poor student:*

[He] waited at the lodgings at the appointed hour, followed the Rector across the quadrangle, and then, when the two had stepped out through the wicket, essayed a literary opening to the conversation by volunteering, 'The irony of Sophocles is greater than the irony of Euripides'. Pattison seemed lost in thought over the statement, and made no answer till the two turned at Iffley to come back. Then he said, 'Quote'. Quotations not being forthcoming, the return and the parting took place in silence.

Pattison believed in research, not education, as the prime purpose of a university ('science and learning, not keeping a school'), and Lang quotes an eye-witness account of him in his capacity as a reluctant tutor:

He was standing on the hearthrug, with his back to the grate, chatting away, when there was a timid knock on the door, and an undergraduate entered, with a sheet of paper in his hand, a theme or composition of some sort. Pattison beckoned the man to come forward, took the sheet and looked over it, puffing slowly at his cigar. Then he crumpled the pages up in his hand, threw it in the man's face, and pointed to the door. The interview between pupil and tutor was over in a few minutes, without a word said on either side.

❦ *The child is father to the man.*

IMPRESSIONS

❦ *The impressions of two young visitors to Oxford in 1832 were variously recorded. The name of the first is unknown:*

I came through Oxford. . . . There was a young fellow of about five-and-twenty, moustached and smartly dressed, in the coach with me. The coach stopped to dine; and this youth passed half an hour in the midst of that city of palaces. He looked around him with his mouth open, as he re-entered the coach, and all the while that we were driving away past the Ratcliffe Library, the great court of All Souls, Exeter, Lincoln, Trinity, Balliol and St. John's. When we were about a mile on the road he spoke the first words that I had heard him utter. 'That was a pretty town enough. Pray, sir, what is it called?'

Thomas Babington Macaulay (1800–59), in a letter.

The second sight-seer was anything but anonymous:

Thursday, 8th November, 1832. At 10 o'clock we set out for Oxford in a closed carriage and 4 with Lord Abingdon and Lady Charlotte Bertie; the other ladies going in carriages before us. We got out first at the Divinity College, and walked from there to the theatre, which was built by Sir Christopher Wren. The ceiling is painted with allegorical figures. The galleries are ornamented with carving enriched with gold. It was filled to excess. We were most WARMLY and EN-THUSIASTICALLY received. They hurrayed and applauded us immensely for there were all the students there; all in their gowns and caps.

Princess Victoria, the future Queen, in her diary, aged thirteen.

OXFORD AWAKE!

Despite the glimmerings of reform at Oxford, critics still thought the old place incorrigibly reactionary:

Oh! Oxford, Oxford, vainly still endow'd
With wealth and ease to lift thee from the crowd,
With opportunities too richly blest,
Why with slow step still lag behind the rest?
When wilt thou cease, with all thy pompous plan,
To waste on mummies, what was meant for man—
With all thy learning when wilt thou decide
That to be useful it must be applied?

 Lo! on a pile of dusty folio's thron'd,
Her Janus brows with dog-ear'd fools-cap crown'd,
Fenc'd with a footstool, that no step should go
Too rashly near, nor crush her gouty toe,
Obese Tuition sits, and ever drips
An inky slaver from her bloated lips!
Unwholesome vapours round her presence shed,
Dim ev'ry eye, and muddle ev'ry head,
Stunt the young shoots, which smil'd with promise once,
And breathe a deeper dulness on the dunce.

Oxford awake! The land hath borne too long
The senseless jingling of thy drowsy song. . . .

      G. V. Cox, *Black Gowns and Red Coats*, 1834.

THE DUKE

The Duke of Wellington, though neither a scholar nor an Oxford man, was elected Chancellor of the University in 1834. His installation at the Sheldonian Theatre was one of the great Oxford occasions:

The Installation of the good old Duke was a most brilliant vision to me. . . . There was a series of Odes, Latin & English, fired off in his honour from the Rostrums at different times. . . . It was a couplet however in the Newdigate, which produced an explosion of enthusiasm and cheering, such as the old Sheldonian never heard before or since. It ran thus—

And that dark soul a world could scarce subdue
Bent to thy Genius, Chief of Waterloo!

The whole audience rose to their feet, cheered, waved hats, handker-
chiefs, and anything else at hand, for some minutes. The Iron Man
seemed quite non-plussed, and did not know what to do; at last he
rose, gave the old military salute, and raised his trencher. This of
course produced a fresh round of cheers.

The papers which the Duke had to read in Convocation [were]
all in M.S.—not printed. Hence all the blunders and the false quantities which the good old hero made, at which the Undergraduates
laughed long and loudly.

> Letter from a Gentleman-Commoner to his younger brother, 1834.

In [to the Theatre] came the Duke of *Wellington*. I was standing
behind his chair. He had somehow to use the word '*Carolus*'. He turned
round and said—'Quick! is it Carōlus or Carŏlus?' 'Carŏlus, your
Grace.' 'Thank you.'

> George A. Denison, in *Our Memories*, 1890.

When the Duke, at the head of the procession, entered the theatre,
the effect was very striking. I felt most absurdly overcome in seeing
the aged hero and hearing the hearty cheers which greeted him. It was
sad to see how unsteadily he walked, and how entirely his manner was
that of an old man. He had a good deal to read . . . but there was
hardly hearing him, so great was the uproar from the undergraduates
thinking proper to batter in the panels of one of the Gallery doors,
and smash several windows. The Duke tried to effect silence by
shaking his hand, but in vain. . . . On entering, his cap was hind side
before, and on preparing to leave, his splendid robes became so out of
order as to need the assistance of someone to replace them.

> Mrs. Jeune, *Diary*, 1841.

The installation has passed off well. The only mistakes being the
Duke's false quantities, which were treated very indulgently. The
Duke himself, whether it was that his face was lighted up by the
joyfulness of all about him, or whether he looked as he always does,
I don't know; but his face had certainly a much kinder expression
about it than I had expected to see. I should say that, dressed in a Court
dress of black, with his Chancellor's robe on him, he was the most
respectable-looking old gentleman I ever saw. . . . The cheering, by
the by, was more tremendous than anything I ever heard. The theatre
was like a bell going. We all of us, of course, sported the most
correct principles on the occasion. We hissed, groaned, and laughed
at Dissenters and Whigs. . . .

> J. B. Mozley, *Letters*, 1834

🦗 *The Duke himself was much impressed by the reception he got from the undergraduates:*

Let the boys loose in the state in which I saw them, and give them a political object to carry, and they would revolutionize any nation under the sun.

🦗 *Wellington took his Chancellorship very seriously, and was infuriated when there was opposition within the University to his appointment of a Vice-Chancellor:*

Walmer Castle September 29 1844

Dear Sir

Judging from what I see in the Newspapers, and from certain Intimations which I have received, and questions put to me by certain friends of mine, I judge that it is the Intention of a Party in the University, to oppose in the Convocation, the nomination of Dr. Symons to be Vice Chancellor of the University. . . .

I will not submit to any Vote by Convocation on this Subject; which shall convey disapprobation of the nomination made in the usual manner.

I did not solicit the Honour of being elected Chancellor of the University of Oxford. On the Contrary I declined to accept it repeatedly; on the ground of my profession, and Habits of Life; and of my want of Connection with the University at any Time.

My objections were overruled; and I was unanimously elected.

I have zealously served the University for more than ten Years. Others might have served them with more ability; but at least I have given them no trouble; if I have not given them satisfaction.

At all events I cannot serve the University, or any other Authority in the State, if I submit to Disgrace on the part of the Convocation; or to anything like a Declaration of the want of Confidence in me: However advanced I may be in years; it is thought by some that I am still capable of serving the State; and I *will not* submit to an Indignity on the part of any authority whatever; the submission to which would incapacitate me. . . .

The Convocation can have no difficulty in finding a Person to accept the Office of Chancellor; who must suit them better; and whose Services will be more acceptable to them than those of the Person, in whom they will have declared by their vote their want of Confidence.

I have the Honour to be
Dear Sir Your most faithful
and obedient Humble Servant
Wellington

The very Revd. Dr. P. Winter
St. John's College Oxford

🦁 *Convocation submitted, and the Duke's nominee went in. On another level of significance, Wellington was irritated by the innumerable letters he received about trivial University affairs, like one from a Mr. Dyson asking him to recommend a relative of his to be cook at Worcester College:*

The Duke suggests to Mr. Dyson that if he has a female relative unmarried; who is desirous of being married; he might as well desire the Duke to recommend her as a wife!

🦁 *The Duke's last visit to Oxford was in the summer of 1844, when he escorted Prince Wilhelm of Prussia, later Emperor of Germany; an occasion recorded by H. A. Harvey in* Our Memories, *1890:*

I shall never forget how my heart beat when I saw the conqueror of *Napoleon* & 'Chief of Waterloo' . . . with his Academical Robe around him, that of a Nobleman (undress), the gold tuft dangling in his eyes, and himself giving his well-known military salute, two fingers to his cap. . . .

But the most characteristic scene was when the Prussian Prince ascended to the roof of the Radcliffe Library (it was not termed *The Camera* then), and the Duke preferred to stay below and take a seat in the beautiful Hall, not crowded up in those days with desks and book-cases, but spacious and adorned with Statuary. Lord *Westmoreland*, our Ambassador at Berlin, who stood beside him, I heard say to him—'I'm afraid your Grace is tired'. 'Well', said the Duke, 'this sight-seeing is d—d tedious work.'

🦁 *On the night the Duke died, so his housekeeper reported, he took to bed with him a copy of a Royal Commission's report on Oxford, 'with a pencil in it', but even so this conscientious Chancellor's memory was not universally cherished in Oxford:*

I see, in looking at the old account-book, that my total expenses for [my first term at Oriel] were £78. This included my first establishment in furniture, crockery, etc., £2:10s. subscription to the boats, and, alas! £2 they extracted from me for some memorial—I know not what—to the Duke of Wellington.

Mark Pattison, Memoirs, 1885.

A PROPHET AND HIS MOVEMENT

🦁 *In the 1830s and 1840s Oxford was racked by another of its religious controversies, this time self-engendered. Tractarianism, called after the tracts which propagated its arguments, and later dubbed the Oxford*

Movement, aimed at a spiritual revival of the Church of England based upon a recognition of its direct succession from the apostolic church. The Romish implications of this attitude, and the increasingly ritualistic nature of the movement, aroused at once furious hostility and profound, sometimes besotted, loyalty.

The scene of this new Movement was as like as it could be in our modern world to a Greek πόλις, or an Italian self-centred city of the Middle Ages. Oxford stood by itself in its meadows by the rivers, having its relations with all England, but, like its sister at Cambridge, living a life of its own, unlike that of any other spot in England, with its privileged powers, and exemptions from the general law, with its special mode of government and police, its usages and tastes and traditions, and even costume, which the rest of England looked at from the outside, much interested but much puzzled, or knew only by transient visits. And Oxford was as proud and jealous of its own ways as Athens or Florence; and like them it had its quaint fashions of polity; its democratic Convocation and its oligarchy; its social ranks; its discipline, severe in theory and often lax in fact; its self-governed bodies and corporations within itself; its faculties and colleges, like the guilds and 'arts' of Florence; its internal rivalries and discords; its 'sets' and factions. Like these, too, it professed a special recognition of the supremacy of religion; it claimed to be a home of worship and religious training, *Dominus illuminatio mea*, a claim too often falsified in the habit and tempers of life. It was a small sphere, but it was a conspicuous one; for there was much strong and energetic character, brought out by the aims and conditions of University life; and though moving in a separate orbit, the influence of the famous place over the outside England, though imperfectly understood, was recognized and great.

R. W. Church, *The Oxford Movement*, 1891.

At the heart of the new movement stood a prophet:

If such was the general aspect of Oxford society at that time, where was the centre and soul from which so mighty a power emanated? It lay, and had for some years lain, mainly in one man—a man in many ways the most remarkable that England has seen during this century. . . . John Henry Newman. The influence he had gained, apparently without setting himself to seek it, was something altogether unlike anything else in our time. A mysterious veneration had by degrees gathered round him, till now it was almost as though some Ambrose or Augustine of elder ages had reappeared. . . . In Oriel Lane light-hearted undergraduates would drop their voices and whisper, 'There's Newman!' when, head thrust forward, and gaze fixed as though on some vision seen only by himself, with swift, noiseless step he glided by. Awe fell on them for a moment, almost as if it had been some apparition that had passed. . . .

The centre from which his power went forth was the pulpit of St. Mary's, with those wonderful afternoon sermons. Sunday after Sunday, month by month, year by year, they went on, each continuing and deepening the impression the last had made. . . . To call these sermons eloquent would be no word for them; high poems they rather were, as of an inspired singer, or the outpourings of a prophet, rapt yet self-possessed. And the tone of voice in which they were spoken, once you grew accustomed to it, sounded like a fine strain of unearthly music. Through the stillness of that high Gothic building the words fell on the ear like the measured drippings of water in some vast dim cave. . . .

Such was the impression made by that eventful time . . . and the marvellous character of him who was the soul of it.

J. C. Shairp, *Studies in Poetry and Philosophy*, 1872.

Matthew Arnold, the poet, confirmed the mesmeric power of Newman's presence:

Who could resist the charm of that spiritual apparition, gliding in the dim afternoon light through the aisles of St. Mary's, rising into the pulpit, and then, in the most entrancing of voices, breaking the silence with words and thoughts which were a religious music—subtle, sweet, mournful? Happy the man who in that susceptible season of youth hears such voices! They are a possession to him for ever.

Discourses in America, 1885.

The theological battle raged with astonishing force, and dominated University affairs for years. Richard Ford the writer, who had taken his degree at Trinity College twenty years before, returned to Oxford in 1842 after many years in Spain and was shocked by the state of affairs, particularly the influence of Newman's ascetic lieutenant, Canon Edward Pusey of Christ Church:

This Oxford is indeed changed since my time. The youths drink toast and water and fast on Wednesdays and Saturdays. They have somewhat of a priggish, macerated look; der Puseyismus has spread far among the rising generation of fellows of colleges. Pusey, the arch-heretic, has indeed the true Jesuit look. . . . The minds of the young men are perplexed with Puseyism y la Santa Iglesia Catholica y Romana. That evil, and a tremendous habit of smoking cigars, seem to be the *features* of the place, and perplex the tutors and heads of colleges.

❧ *The fury of it all was well expressed in this hymn of hate by the evangelical don H. B. Bulteel (later himself to be the inspiration of a religious movement—Bulteelism):*

They've found a wondrous pilot;
They've found a ready crew:—
O may it ne'er be my lot
To sail with hearts untrue!

There's Newman wise and simple,
How saintly is his smile!
Alas beneath each dimple
Lurk treachery and guile.

There's Pusey's gloomy visage
His down-cast eye and head,
The foremost man of this age
To prove his God his bread . . .

See, see! the Vessel's ready,
Her main-sail woos the breeze,
And all her hands are steady,
Their hearts are all at ease.

Sink, Argo, sink for ever,
A bottom and a shore
Thy keel shall touch—no, never!
Sink, and be found no more!

Amen! we long to see it;
Repeat, ye Saints, Amen!
Ye angels shout, 'so be it!'
Again, again, again!

❧ *In 1844 things came to a head when it was proposed that Newman's* Tract XC, *which was thought particularly subversive, should be publicly condemned by Convocation:*

It was a day in itself sufficiently marked by the violent passions seething within Oxford itself, and aggravated to the highest pitch by the clergy, and laity of all shades and classes who crowded the colleges and inns of Oxford, for the great battle of Armageddon, which was to take place in the Convocation of Oxford that day assembled in the Sheldonian theatre. The agitation penetrated to the very servants and scouts. They stood ranged round the doors of their colleges, waiting for the issue of the writ, filled with the gaudia certaminis. 'Theirs not to reason why.' The excitement of the day was yet more fiercely accentuated by one of the most tremendous snowstorms which had down to that time taken place within the memory of man. . . . Two academics who on the night before had arrived at Swindon, in the hope of finding shelter on that perilous night, found every bed and every corner occupied; yet such was their ardour for the fray that they walked that dismal journey to be in time for the fatal hour. The under-graduates, who ardently participated in the excitement of their seniors, watched the procession, as it passed under their windows, with mingled howls and cheers; and one of them, of more impetuosity than the rest, climbed to the top of the Radcliffe Library, and from that secure position pelted the Vice-Chancellor with a shower of snowballs to testify his detestation of the obnoxious measure.

A. P. Stanley, in the *Edinburgh Review*, 1881.

The proposal of condemnation was defeated by the veto of the proctors, but by then in any case Newman himself had come to recognize that his beliefs were incompatible with membership of the Church of England. He resigned his vicarage of St. Mary's and withdrawing to a semi-monastic community at Littlemore, just outside the city, prepared to enter the Roman Catholic church. On 8 October 1845 he wrote this letter to a number of friends:

I am this night expecting Father Dominic [Barberi], the Passionist, who, from his youth, has been led to have distinct and direct thoughts, first of the countries of the North, then of England. After thirty years' (almost) waiting, he was without his own act sent here. But he has had little to do with conversions. I saw him here for a few minutes on St. John Baptist's day last year. He does not know of my intention; but I mean to ask of him admission into the Fold of Christ. . . .
PS This will not go till all is over. Of course it requires no answer.

The next day the prophet left the fold:

It was a memorable day that 9th of October 1845. The rain came down in torrents, bringing with it the first heavy instalment of autumn's 'sere and yellow' leaves. The wind, like a spent giant, howled forth the expiring notes of its equinoctial fury. The superstitious might have said that the very elements were on the side of Anglicanism—so copiously did they weep, so piteously bemoan, the approaching depar-ture of its great representatives. The bell which swung visibly in the turret of the little gothic church at Littlemore gave that day the usual notice of morning and afternoon prayers; but it came to the ear in that buoyant bouncing tone which is usual in a high wind, and sounded like a knell rather than a summons. The 'monastery' was more than usually sombre and still. Egress and ingress there were none that day; for it had been given out among friends accustomed to visit there, that Mr. Newman 'wished to remain quiet'. One of these friends who resided in the neighbourhood, had been used to attend the evening office in the oratory of the house, but he was forbidden to come 'for two or three days, for reasons which would be explained later'. The 5th of the month passed off without producing any satisfaction to the general curiosity. All that transpired was that a remarkable-looking man, evidently a foreigner, and shabbily dressed in black, had asked his way to Mr. Newman's on the day but one before; and the rumour was that he was a Catholic priest.

Frederick Oakeley, *Historical Notes on the Tractarian Movement*, 1865.

The 'remarkable-looking man' was Father Dominic. Newman, we are told, flung himself at his feet and was received there and then into the Roman Catholic Church. The shock for many of Newman's disciples was terrible. As Anne Mozley, wife of a Tractarian don, wrote:

Such *was* our guide, but he has left us to seek our own path: our champion has deserted us—our watchman, whose cry used to cheer us, is heard no more.

And long afterwards the poet A. G. Butler, looking over Oxford from a stile on Shotover hill, wrote this:

'Nothing remains but Beauty', he said, and wearily sighing,
    Sat upon Shotover stile, gazing on Oxford below,
Minaret-crowned St. Mary's, and Magdalen Tower, and Merton,
    Far-off jewels of light, fringed with a circle of shade,
Set in the shining floods. Oh! not alone in the sunshine
    Fair! yet fairer the faith, glory of men who believed . . .

Ah, what a vision was there! But then a vapour ascending
    Rose over turret and spire, crept over College and Hall,
Death-white, all-enfolding. As when from marshy Maremna,
    Rises a poisonous breath, ghastly—inhale it and die!
'Look! that is me,' he whispered, 'I had it once, I am certain,
    Once I had faith. But now! Now there is mist over all.'

A. G. Butler (1831–1909), from *Oxford from Shotover Stile*.

Newman himself left Oxford for ever in February 1846, as he recorded in his Apologia Pro Vita Sua, *1864:*

Dr. Pusey came up to take leave of me; and I called on Dr. Ogle, one of my very oldest friends, for he was my private tutor when I was an undergraduate. In him I took leave of my first college, Trinity, which was so dear to me, and which held on its foundation so many who have been kind to me both when I was a boy, and all through my Oxford life. Trinity had never been unkind to me. There used to be much snapdragon growing on the walls opposite my freshman's rooms there, and I had for years taken it as the emblem of my own perpetual residence even unto death in my University.

On the morning of the 23rd I left. . . . I have never seen Oxford since, excepting its spires, as they are seen from the railway.

John Henry Newman, *Apologia Pro Vita Sua*, 1866.

With his departure Tractarianism, like a passing whirlwind, left Oxford in peace, though it had its lasting effect upon the Church of England. Newman, presently made a Cardinal of the Roman Catholic Church, looked back on its Oxford origins, it seems, with a certain fatalism:

There are those who, having felt the influence of this ancient school (of Oxford), and being smit with its splendour and its sweetness, ask wistfully if never again it is to be Catholic, or whether at least some footing for Catholicity may not be found there. All honour and merit to the charitable and zealous hearts who so inquire! Nor can we dare to tell what, in time to come, may be the inscrutable purposes of that grace which is ever more comprehensive than human hope and aspiration. But for me, from the day I left its walls, I never, for good or bad, have had anticipation of its future; and never for a moment have I had a wish to see again a place which I have never ceased to love, and where I lived for nearly thirty years.

AN OLD CUSTOM

Just as Fellows of Oxford colleges must still be celibate in the first half of the nineteenth century, so every member of the University must be a member of the Church of England, and subscribe to the Thirty-Nine Articles of its faith:

It would perhaps be amusing to a stranger to hear how the ceremony of reading the thirty-nine articles, previously to being presented to a degree, is performed. The Dean of the College invites the young man to breakfast—a couple of articles are read—then succeeds a *wadding* of cold meat—an interlayer of boiled eggs (if indeed the Dean be so munificent) divides the third and fourth; the doctrine of Predestination requires to be swallowed down with a cup of tea, and the Dean reads the newspaper, while the candidate reads the remainder. Is this an indecent farce, or is it not? If it is, why is it not discontinued? Answer: because it is an old custom.

G. V. Cox, *Black Gowns and Red Coats*, 1834.

THE CHARACTERS

Early nineteenth-century Oxford, the Oxford of celibate dons and life-long Fellowships, was the stage of the Oxford Characters, men who would for the most part be lost to history were it not for their foibles and eccentricities. The historian J. R. Green, writing in The Saturday Review *in 1869, looked back on them with admiring affection:*

They were not as other men are. They had in fact a deep, quiet contempt for other men. Oxford was their world, and beyond Oxford lay only waste wide regions of shallowness and inaccuracy. They were often men of keen humour, of humour keen enough at any rate to see and to mock at the mere pretences of 'the world of progress' around them. Their delight was to take a 'progressive idea' and to roast it over the common-room fire. They had their poetry; for the place itself, and the reverence they felt for it, filled them with a quiet sense of the beautiful; and this refinement and this humour both saved them from bowing before the vulgar gods of the world without. They did not much care for money; they saw their contemporaries struggling for it, and lingered on content with their quiet rooms and four hundred a year. They cared very little for fame . . . although most of them had a great dream-work on hand, of which not a chapter was ever written. What they did care for was strangely blended of the venerable and the ridiculous, for their real love of learning was mingled with a pedantry both of mind and of life, and a feminine rigour over the little observances of society and discipline. Such as they were, however, Young Oxford has no type of existence to show so picturesque, so individual.

The Oxford eccentrics rarely realized how eccentric they were. An anecdote told of many of them is this, applied in this instance to 'Mo' Griffiths of Magdalen and Dr. Macbride, the Principal of Magdalen Hall:

Griffith: There's one change in Oxford, Macbride, that I can't at all make out. Don't you remember, when we were young, there used to be a number of queer old fellows among the Dons—*Quizzes* we used to call them—curious old chaps it made one laugh to see. Now I don't see any such fellows about.

Macbride: Perhaps you and I, Mr. Griffith, may seem to the young men of today the sort of persons of whom you speak.

Griffith: *You and I!* Impossible, impossible!

🦡 *The Oxford originals in fact continued to thrive until the University reforms in the middle of the century—some, indeed, thrived even later. Here is a little gallery of them:*

Doctor [Thomas] Randolph was at that time president of Corpus Christi College. With great learning, and many excellent qualities, he had some singularities, which produced nothing more injurious from his friends than a smile. . . . 'Sir,' said he to me as we came out of chapel one Sunday, 'You *never* attend Thursday prayers.' 'I do *sometimes*, Sir,' I replied. 'I did not see you here last Thursday. And, Sir,' cried the president, rising into anger, 'I will have nobody in my college,' (ejaculating a certain customary guttural noise, something between a cough and the sound of a postman's horn,) 'Sir, I will have nobody in my college that does not attend chapel. I did not see you at chapel last Thursday.' 'Mr. President,' said I with a most profound reverence, 'it was impossible that you should see me, for you were not there yourself.'

Instead of being more exasperated by my answer, the anger of the good old man fell immediately. He recollected and instantly acknowledged, that he had not been in chapel on that day. It was the only Thursday on which he had been absent for three years. Turning to me with great suavity, he invited me to drink tea that evening with him and his daughter.

Richard Lovell Edgeworth, *Memoirs*, 1820.

1815: In this year died Mr. Lloyd, B.C.L., of Wadham, who had been Curator of the Ashmolean Museum for nineteen years. The Museum itself seldom seemed to occupy the time or thought of the Curator; he was a retired, quiet (not to say idle) gentleman, having no pretensions to science or scholarship; seldom, indeed, coming out of his lodgings in Holywell Street, where he amused himself in what his neighbours called 'strumming' on his harp.

1816: In July died Constantine Demetriades, a dirty old Greek (not at all answering to his sounding name), who had hung about the University for some time, professedly as a teacher of Modern Greek, but really as a butt for the young men's jokes, a sponge for their eatables and drinkables, and a recipient of their loose silver. He was never seen without a thick staff, 'to keep the dogs and scouts from the pockets of his old great coat!'

Professor Robertson [Savilian Professor of Astronomy] was married, but had no family; his wife died many years before him. At the Observatory he lived in a very retired manner, being taken care of by an old housekeeper, whom, just before he died, he calmly instructed 'how to treat his corpse, to tie up his chin, to lay him out', &c. &c! . . . I have seen him silently and slowly putting together (secundum artem mathematicam) the particulars of some witty story or anecdote which had been told, and when every one else had forgotten the subject, ejaculating 'Yes, indeed; that's very good, very!'

G. V. Cox, *Recollections of Oxford*, 1868.

[Thomas Short, Senior Tutor of Trinity] lectured in Aristotle's Rhetoric, and after a time he passed by many years the age at which Aristotle says that man's powers are at their best. It became a great enjoyment to various generations of undergraduates to hear him say, when he came to that particular passage, 'In those hot climates, you know, people came to their acme much sooner than with us.'

[E. A. Freeman the historian] was a man of very singular manners . . . He paid no regard at all to what people might think of him, and he was in the habit of repeating poetry to himself as he walked the streets, and occasionally leaping into the air when the poem moved him to enthusiasm.

[H. P. Guillemard of Trinity College] was a genial and kind-hearted man, but the object of some good-natured jests among the scholars. His favourite book was Thucydides. In giving a lecture on it he used Arnold's edition, and had open before him Gollers edition to which to refer. On coming to a hard passage he would turn to the book on the table, and say: 'Goller takes this passage as follows.' Among the scholars there was a witty and sarcastic man, E. T. Turner, who would have it that the supposed 'Goller' was really an English translation, referred to when a difficulty occurred. We all knew that it was not so, but the jest suited a class of merry youngsters, and from that time 'cribs' went among us under the name of 'Gollers'.

Joseph Smith, Bursar of Trinity, was the type of the old Oxford don, solemn and pompous. Once an undergraduate was so irritated by him that he decided to throw a bomb into his room, but Smith was warned not to sit in his study that evening:

On my knocking at his door, he threw it open, brandishing a stout walking-stick in his hand. On seeing me he went back to the table at which he had been sitting, in his front room, with a book before him and his stick by his side. 'I ask no questions,' he said, 'but is it gun-powder?' On my saying that I feared so, he resumed his reading, with the words, majestically pronounced: 'Then I sit here tonight.'

F. Meyrick, *Memories*, 1905.

The wit of Magdalen College in the 1830s was a Fellow named Thomas Newman, a famous mimic and joker:

One day he amused himself by masquerading as a stranger visiting Oxford, and hiring a guide to show him round, which the guide did with the usual illustrative comments. When at last they came to Magdalen, the guide pointed out the Fellows' Common Room. To his surprise and horror Newman bolted into it and was seen no more.

Goldwin Smith, *Reminiscences*, 1910.

Joseph Dornford, who became a Fellow of Oriel in 1816, had already served as a rifleman in the Peninsular War, and for the rest of his life was unrivalled in Oxford when it came to talking about military affairs:

What would be Dornford's position, and duties, and opportunities in the Rifle Brigade, passes my knowledge. He could talk of the war, of the movements and marches, and of the generals, much better than any private soldier could have done, or indeed than most sub-alterns; but of his own particular part he was rather reticent, or had not much to say. The undergraduates, however, were resolved that he must have exchanged shots near enough to be of some purpose, that he had killed his man, or any number of men, and that he might have come to close quarters now and then. They had a story, for which I never heard the authority, that he was once pressed by a gay young partner at a ball to say whether he had killed anybody. . . . Dornford, so it was said, replied that once when in ambush he saw a young officer galloping across the open ground a long way off, evidently carrying orders. He was bound to take a shot, he did, and *felt sure the officer quivered in his saddle.* . . .

Undergraduates have a perverse way of looking at things the wrong side, and they called Dornford 'the Corporal'; indeed some would be ready to swear that he had been actually a non-commissioned officer. It was too true that there was a certain flourish, just an approach to bravado about him, especially when there was gallantry in the question, as there always was when he approached a woman of any description whatever. Young or old, rich or poor, fine lady or the merest peasant, he was always the gentleman to the fair sex. Possibly he had witnessed the havoc done by recruiting sergeants on the hearts of barmaids, and thought he could combine that glory with perfect innocency and high culture. . . .

Wherever he was, indoors or outdoors, walking or riding, he was unmistakably the soldier. Cantering, for that was his usual pace, on a long-legged horse, with his martial cloak flying from his shoulders, beggars, the veriest strangers, addressed him 'noble captain'.

T. Mozley, *Reminiscences*, 1882.

They tell me . . . that it was [the custom of Edward Hawkins, Provost of Oriel from 1828 to 1882] to give one finger to a commoner, the whole hand to a 'Tuft' and that he was somewhat embarrassed when a certain man who went down as Mr. — came back as Lord — of —. An Oriel undergraduate took to preaching in St. Ebb's slums: Hawkins angrily inhibited him. 'But, Sir, if the Lord who commanded me to preach, came suddenly to judgement, what should I do?' 'I,' said Hawkins, 'will take the whole responsibility for that upon myself.' He would grant an exeat during term time only in very special cases. A man begged leave to absent himself in order to bury his uncle. 'You may go,' was the reluctant permission: 'but I wish it had been a nearer relation.'

Dr. Frowd, of Corpus, was a very little man, an irrepressible, un-wearied chatterbox, with a droll interrogative face, a bald shining head, and a fleshy under-lip, which he could push nearly up to his nose. He had been chaplain to Lord Exmouth, and was present at the bombardment of Algiers. As the action thickened he was seized with a comical religious frenzy, dashing round the decks, and diffusing spiritual exhortation among the half-stripped, busy sailors, till the first lieutenant ordered a hencoop to be clapped over him, whence his little head emerging continued its devout cackle, quite regardless of the balls which flew past him and killed eight hundred sailors in our small victorious fleet.

In reading chapters from the Old Testament, he used to pause at a marginal variation, read it to himself half audibly, smile on it auspiciously or knit his brow and shake his head in disapproval. I remember too his preaching in All Saints Church. . . . He climbed up the steep three-decker steps into the high-walled pulpit, and disappeared, till, his hands clinging to the desk and his comical face peering over it, he called down into the reading-desk below, 'Thompson, send up a hassock.'

In the room below him lived Holme, a more advanced Bedlamite even than himself, a pleasant fellow as I remember him in his inter-lunar periods, but who died, I believe, in an asylum. Frowd used to exercise on wet days by placing chairs at intervals round his room and jumping over them. Holme, a practical being, one day fired a pistol at his ceiling while these gymnastics were proceeding, and the bullet whizzed past Frowd, who, less unconcerned than at Algiers, ran downstairs, put his head into the room, and cried, 'Would you, bloody-minded man, would you?' The feeling in the Common Room was said to be regret that the bullet had not been billeted. . . .

Old Dr. Ellerton, Senior Fellow of Magdalen . . . was a picturesquely ugly man; the gargoyle above his window was a portrait, hardly an exaggeration, of his grotesque old face. Years before, when the building was restored and he was College tutor, the undergraduates had bribed the sculptor to fashion there in stone the visage of their old Damoetas; he detected the resemblance, and insisted angrily on alteration. Altered the face was: cheeks and temples hollowed, jaw-lines deepened, similitude for the time effaced. But gradually the unkind invisible chisel of old age worked upon his own octogenarian countenance; his own cheek was hollowed, his own jaw contracted, till the quaint projecting mask became again a likeness even more graphic than before.

William Tuckwell, *Reminiscences of Oxford*, 1900.

Dr. Ellerton, who was very Low Church, could never remember names, so called everybody Mister. The following exchange was reported by a friend:

Friend: Good morning Dr. Ellerton, how are you?
Ellerton: Not very well, Mister.
Friend: Sorry to hear that. What is the matter, sir?
Ellerton: Got an intoning Curate, Mister.

Philip Shuttleworth, Warden of New College from 1822 to 1840, designed an iron-wheeled trolley, based upon the coal tubs in the Durham mines, to circulate the port in the senior common-room, and was the author of this much-admired couplet, from a youthful apostrophe to the goddess of Learning:

Make me, O Sphere-descended Queen,
A Bishop, or at least a Dean.

(She obliged in the end: he became, for the last three years of his life, Bishop of Chichester.)

The Sheldonian Theatre, 1820

THE CHARACTERS

🔖 *Robert Hussey, Professor of Ecclesiastical History from 1842 to 1856, told by a surgeon that his toe must be amputated, allegedly replied:*

Very well, cut it off: but be good enough not to disturb me by talking while you do it, as I have a lecture to prepare.

🔖 *Samuel Smith, Dean of Christ Church from 1824 to 1831, was universally known as 'Presence-of-Mind' Smith. During an undergraduate boating trip his companion had fallen into the river, nearly capsizing the boat with him, and Smith had explained the sequel thus:*

Neither of us could swim, and if I had not with great presence of mind hit him on the head with the boathook we would both have drowned.

THE LITTLE MASTER

🔖 *Richard Jenkyns (1782–1854), for thirty-five years Master of Balliol, not only transformed the academic condition of his college by the institution of open scholarships, but with his curious squat appearance, West Country accent, and often eccentric behaviour, was one of the great personalities of early Victorian Oxford:*

He was a gentleman of the old school, in whom were represented old manners, old traditions, old prejudices, a Tory and a Churchman, high and dry, without much literature, but having a good deal of character. . . . His sermon on the 'Sin that doth so early beset us', by which, as he said in emphatic and almost acrid tones, he meant 'the habit of contracting debts', will never be forgotten by those who heard it. Nor indeed have I ever seen a whole congregation dissolved in laughter for several minutes except on that remarkable occasion. The ridiculousness of the effect was heightened by his old-fashioned pronunciation of certain words, such as 'rayther' 'wounded' (which he pronounced like 'wow' in 'bow-wow'). He was a considerable actor, and would put on severe looks to terrify freshmen, but he was really kind-hearted and indulgent to them.

Benjamin Jowett, in W. G. Ward and the Oxford Movement, 1889.

He was a little man, and he had a little pony, and everyone knows that a little man with a little pony is not likely to escape a great deal of playful criticism on the part of tall men who ride tall horses. Thus, it would sometimes happen that the Master and his pony would fall in with a party of riders returning from the hunt. The riders would trot at a quick pace behind the Master and his pony; and the latter, fretted by the noise behind it, would start off at a gallop, and thus place the Master in the awkward position of seeming to head a party of his own undergraduates on their return from the hunting field.

Frederick Oakeley, in *The Month*, 1866.

🐾 *The Reform Bill of 1832 led to riots between town and gown in Oxford, and in one affray the University Proctors and their supporters were driven by a mob along Cornmarket, into Broad Street, and to the gate of Balliol, where they mustered for a stand. The Master was just sitting down to dinner.*

He said,—'What is all this disturbance outside?'—'Master, it is a great fight—Town and Gown; and they say that Mr. Denison of Oriel is killed.' He said—'Give me my Academicals, and open the door of the house into the street.' The household represented the danger of doing this. The answer was—'Give me my Academicals, and open the door.' The master stood on the doorsteps and had just said to Town,—'My deluded friends' when a heavy stone was pitched into the middle of his body and he fell back into the arms of his servants, crying out, 'Close the door.'

G. A. Denison, in *Our Memories, Shadows of Old Oxford*, 1893.

🐾 *Dr. Jenkyns was very much opposed to the principles of the Oxford Movement, led by John Keble, John Henry Newman, and Edward Pusey in the 1830s, and was especially perturbed when one of his own Balliol Fellows, W. G. Ward, was deprived of his degree by Convocation for his heretical views. When it fell to Ward to read the epistle in the college chapel, the Master got there first:*

Directly the Master saw Mr. Ward advancing to the Epistle side of the table he shot forth from his place and rushed to the Gospel side, and just as Mr. Ward was beginning, commenced in his loudest tones:—'The epistle is read in the first chapter of St. Jude.' Mr. Ward made no further attempt to continue, and the Master, now thoroughly aroused, read *at him* across the communion table. The words of the Epistle were singularly appropriate to the situation, and the Master, with ominous pauses and looks at the irreverent Puseyite, who had sown sedition in the Church and blasphemed the Heads of Houses, read as follows slowly and emphatically: 'For there are certain men crept in unawares' (pause, and look at Mr. Ward) 'who were before of old ordained unto this condemnation' (pause and look), 'ungodly men' (pause and look);—and a little later still more slowly and bitterly he read, 'they speak evil of dignities!'

Wilfrid Ward, *William George Ward and the Oxford Movement*, 1889.

❧ *Among Jenkyns' pupils was an eccentric celebrity of his day, the parodist C. S. Calverley, who is said to have taken part in the following exchange with the Master at a viva voce examination:*

Jenkyns: With what feelings ought we to regard the Decalogue?
Calverley (not being at all sure what the Decalogue is): With feelings of devotion, Master, mingled with awe.
Jenkyns: Quite right, young man, a very proper answer.

❧ *Calverley is also supposed to have displayed the Master to an audience of sightseers by the following formula:*

That is Balliol College. That is the Master's house. And that (*throwing a stone neatly at the study window*) is the Master.

❧ *But the story is told of several undergraduates, several Masters, and several generations of sightseers.*

CATEGORIES OF FAILURE

❧ *A New Art-Teaching how to be Plucked, by Edward Caswall (1835) was an Aristotelian squib on how to fail examinations. Here are some of its categories of certain failure:*

He that goeth to the Ascot races. He that loungeth in Quad. He that readeth many books. He that readeth few books. He that readeth no books. He that readeth novels, for verily pleasant things are novels and entice away the mind exceedingly. He that sporteth many new whips. He that knoweth many pretty girls. He that knoweth one pretty girl. He that hateth Greek. He that being poor sporteth champagne. He that hath gone a second time to a dog fight. He that is a radical albeit his father was a Tory, for such a one thinketh himself clever. He that weareth white kid gloves when shooting. He that breaketh lamps in the street. He that learneth more than two instruments of music. He that eateth too much pudding. He that hath a lock of hair in his desk. He that thinketh he will be plucked. He that thinketh he will not be plucked.

OXFORD ETIQUETTE, 1830s

Let us suppose some eight or nine undergraduates, all strangers to each other, assembled in one room; a person unacquainted with the world might suppose that he could not do better than enter into conversation with his nearest neighbour. But let him do nothing of the kind: he has not been introduced to him. If he speaks to him, he will violate one of the first rules of Oxford etiquette. Let him whistle, if he pleases, and act as if there were no one in the room but himself; but let him not speak, except to his dog, or the waiter, if he is at an inn. If you observe any one inclined to address you, fix your eye upon him proudly, as much as to say, 'I am too good to be spoken to.' Act in the same way on or in a coach: it will impress strangers with a very high notion of your consequence.

If, in a party, a stranger should make a remark of which you do not approve, you must express your sentiments to a friend, loud enough for the offending party to hear, but taking great care not to look towards him; for he would imagine you are addressing him, whereas you are doing nothing of the kind.

It is credibly reported, that four very gentlemanly men of this university once travelled inside from Oxford to Birmingham without exchanging a word. One of them, however, on having his sore toe crushed by his neighbour, gave vent to the exclamation 'Dem!' for which he was mentally considered a great 'cad' by the rest.

From *Hints on Etiquette for the University of Oxford*, 1838.

BURTON *v.* OXFORD

One of Oxford's most consistently irreverent sons was Richard Burton, the explorer and Arabist, who reluctantly entered Trinity College in 1840. He felt he had 'fallen among grocers'. After a brief career of outrageous panache, University life being not at all his style, he contrived to get himself rusticated by illegally riding about in a dog-cart, and never returned. Some years later he described his Oxford days to his comrades-in-arms of the Honourable East India Company's army:

My 'college career' was highly unsatisfactory. I began a 'reading man', worked regularly 12 hours a day, failed in everything—chiefly, I flattered myself, because Latin hexameters and Greek iambics had not entered into the list of my studies—threw up the classics, and returned to the old habits of fencing, boxing and single-stick, handling the 'ribbons', and sketching facetiously, though not wisely, the reverend features and figures of certain half-reformed monks, calling themselves 'fellows'. . . .

At last the Afghan war broke out. . . . I determined to leave Oxford, *coûte que coûte.* The testy old lady, Alma Mater, was easily persuaded to consign, for a time, to 'country nursing' the forward brat who showed not a whit of filial regard for her. So, after two years, I left Trinity, without a 'little-go', in a high dog-cart—a companion in misfortune too-tooing lustily through a 'yard of tin', as the dons started up from their game of bowls to witness the departure of the forbidden vehicle. . . .

So, my friends and fellow-soldiers, I may address you in the words of the witty thief—slightly altered from Gil Blas—'Blessings on the dainty pow of the old dame who turned me out of her house; for had she shown clemency I should now doubtless be a dyspeptic Don, instead of which I have the honour to be a lieutenant, your comrade.'

According to his own account Burton told some members of his family, who did not know what rustication was, that he had been given 'an extra vacation for taking a double-first with the highest honours'. Forty years later he crossed swords with Oxford again when, in preparing his unexpurgated translation of the Thousand And One Nights, *he wanted to have a manuscript sent for his use from the Bodleian in Oxford to the India Office Library in London. The curators refused, as they had refused Charles I and Cromwell, but Burton retaliated in 1885 by dedicating to them Volume VI of the* Supplemental Nights:

To the Curators of the Bodleian Library, Oxford; especially Rev. B. Price and Max Müller.

Gentlemen, I take the liberty of placing your names at the head of this Volume which owes its rarest and raciest passages to your kindly refusing the temporary transfer of the Wortley Montague MS from your pleasant library to the care of Dr Rost, Chief Librarian, India Office. As a sop of 'bigotry and virtue', as a concession to the 'Scribes and Pharisees', I had undertaken, in case the loan was granted, not to translate tales and passages which might expose you, the Curators, to unfriendly comment. But possibly anticipating what injury would thereby accrue to the Volume and what sorrow to my subscribers, you were good enough not to sanction the transfer—indeed you refused it to me twice—and for this step my *clientèle* will be (or ought to be) truly thankful to you.

I am, Gentlemen,
Yours obediently
Richard F. Burton

🦁 *Despite it all, after Burton's death in 1890 his widow Isabel wrote this of the old iconoclast:*

I can testify that at the bottom of his heart he loved Oxford.

MAY DAY, 1840s

🦁 *Since Elizabethan times it had been the custom to sing a hymn on Magdalen Tower at sunrise on May 1st. This is how J. R. Green, the historian, remembered the occasion in his* Oxford Studies, 1901:

We used to spring out of bed, and gather in the grey of dawn on the top of the College tower, where choristers and singing men were already grouped in their surplices. Beneath us, all wrapped in the dim mists of a spring morning, lay the city, the silent reaches of the Cherwell, the great commons of Cowley marsh and Bullingdon now covered with houses, but then a desolate waste. There was a long hush of waiting just before five, and then the first bright point of sunlight gleamed out over the horizon; below, at the base of the tower, a mist of discordant noises from the tin horns of the town boys greeted its appearance, and above, in the stillness, rose the soft pathetic air of the hymn *Te Deum Patrem Colinus*. As it closed the sun was fully up, surplices were thrown off, and with a burst of gay laughter the choristers rushed down the little tower stair and flung themselves on the bell ropes, 'jangling' the bells in rough medieval fashion till the tower shook from side to side. And then, as they were tired, came the ringers; and the 'jangle' died into one of those 'peals', change after change, which used to cast such a spell over my boyhood.

PUSEY OF CHRIST CHURCH

🦁 *The extreme High Churchman Edward Pusey, Canon of Christ Church, was made famous by the Tractarian Movement of the 1830s and 1840s, and became, in the jargon of the day, one of the 'lions of Oxford'. The Rev. W. Tuckwell, in his* Reminiscences of Oxford, 1900, *looks back on the Canon in his heyday:*

In those days he was a Veiled Prophet. . . . As mystagogue, as persecuted, as prophet, he appealed to the romantic, the generous, the receptive natures; no sermons attracted undergraduates as did his. I can see him passing to the pulpit through the crowds which overflowed the shabby, inconvenient, unrestored cathedral, the pale, ascetic, furrowed face, clouded and dusky always as with suggestions of a blunt or half-used razor, the bowed grizzled head, then dropping into the pulpit out of sight until the hymn was over, then the harsh, unmodulated voice, the high-pitched devotional patristicism, the dogmas, obvious or novel, not so much ambassadorial as from a man inhabiting his message; now and then the search-light thrown with startling vividness on the secrets hidden in many a hearer's heart. Some came once from mere curiosity and not again, some felt repulsion, some went away alarmed, impressed, transformed.

Two things impressed me when I first saw Dr. Pusey close: his exceeding slovenliness of person; buttonless boots, necktie limp, unbrushed coat collar, grey hair 'all-to-ruffled'; and the almost artificial sweetness of his smile, contrasting as it did with the sombre gloom of his face when in repose. . . . [He] was looked upon with much alarm in the Berkshire neighbourhood, where an old lady, much respected as 'a deadly one for prophecy', had identified him with one of the three frogs which were to come out of the dragon's mouth.

🎋 *The American Nathaniel Hawthorne treated him frankly as a tourist spectacle:*

Mr. Parker told us that Dr. Pusey. . . . would soon probably make his appearance in the quadrangle, on his way to chapel; so we walked to and fro waiting an opportunity to see him. A gouty old dignitary, in a white surplice, came hobbling along from one extremity of the court; and by-and-by, from the opposite corner, appeared Dr. Pusey, also in a white surplice, and with a lady by his side. We met him, and I stared pretty fixedly at him, as I well might; for he looked on the ground, as if conscious that he would be stared at. . . . He was talking with the lady, and smiled, but not jollily.

English Note-books, 1856.

🎋 *J. B. Mozley, an Oxford contemporary and a comrade in Tractarianism, remembers one of the more endearing Puseyan moments:*

Pusey preached last Sunday. . . . In the course of the sermon there was a piece of friendly advice given to the Heads of Houses, for which they would not be much obliged to him. He had been talking of increase of luxury among the under-graduates, of late years, from which he took occasion to say that those *in station* will do well to live more simply than they did. He dropped his voice at this part, which had the effect of course of giving increased solemnity to the admonition; for there was breathless silence in the church at the time. Pusey however meant the under-graduates not to hear, as he told Newman with the utmost simplicity after. It was to have been a sort of aside from the preacher in the pulpit to the Vice-Chancellor across the way. The Master of Balliol was seen to march out of church afterwards with every air of offended dignity. . . .

🦎 *The Master of Balliol was Richard Jenkyns. See page 213.*

🦎 *Pusey was once in an omnibus when a woman started discussing the ritual excesses of the Tractarian Movement—Dr. Pusey, she had been told, killed a lamb in sacrifice every Friday. Pusey thought it right to intervene:*

My dear madam, I *am* Dr. Pusey, and I do not know *how* to kill a lamb.

🦎 *Though he was a harsh parent and a singularly austere churchman (he walked habitually with his eyes fixed to the ground, in humility), many people found Pusey attractive. When the Swedish singer Jenny Lind visited Oxford in 1848, A. P. Stanley, later Dean of Westminster, wrote this of her:*

Her smile is, with the exception of Dr. Pusey's, the most heavenly I ever beheld.

THE COACHING YEARS

🦎 *Standing as it did in the centre of southern England, Oxford was a great coaching centre, and the everyday affairs of the University were coloured by the arrival and departure of the stage coaches, and the excitements of coaching life:*

It was said in those days that the approach to Oxford by the Henley road was the most beautiful in the world. Soon after passing Littlemore you came in sight of, and did not lose again, the sweet city with its dreaming spires, driven along a road now crowded and obscured with dwellings, open then to cornfields on the right, to unenclosed meadows on the left, with an unbroken view of the long line of towers, rising out of foliage less high and veiling than after sixty more years of growth today. At once, without suburban interval, you entered the finest quarter of the town, rolling under Magdalen Tower, and past the Magdalen elms, then in full unmutilated luxuriance, till the exquisite curves of the High Street opened on you, as you drew up at the Angel, or passed on to the Mitre and the Star. Along that road, or into Oxford by the St. Giles's entrance, lumbered at midnight Pickford's vast waggons with their six musically belled horses; sped stage-coaches all day long—Tantivy, Defiance, Rival, Regulator, Mazeppa, Dart, Magnet, Blenheim, and some thirty more; heaped high with ponderous luggage and with cloaked passengers, thickly hung at Christmas time with turkeys, with pheasants in October; their guards, picked buglers, sending before them as they passed Magdalen Bridge the now forgotten strains of 'Brignall Banks', 'The Troubadour', 'I'd be a Butterfly', 'The Maid of Llangollen', or 'Begone, Dull Care'; on the box their queer old purple-faced, many-caped drivers—Cheeseman, Steevens, Fowles, Charles Homes, Jack Adams, and Black Will.

William Tuckwell, *Reminiscences of Oxford*, 1900.

The good old custom of the Oxford people, particularly the enthusiastic lovers of coaching, was, in the early morn, to see standing ten coaches before the old Angel . . . the coachmen and porters busily loading, and waiting their time to a minute for the Queen's College clock to strike eight, for the nine coaches to make a start for their different destinations, East, West, North, and South. . . . I don't think you could witness in any town in the kingdom the same number of coaches standing before an hotel, and nine out of the ten to start at the same time.

W. Bayzand, *Coaching In and Out of Oxford*, c. 1884.

From the box of the Royal Defiance,
   Jack Adams, who coaches so well,
Dropped me down in that region of Science
   In front of the Mitre Hotel.

'Sure never man's prospects were brighter',
   Cried I, as I dropped from my perch,
'So quickly arrived at the Mitre,
   'I am sure to get on in the Church.'

Anonymous song, c. 1830.

They were in good time for the coach; and the ringing notes of the guard's bugle made them aware of its approach some time before they saw it rattling merrily along in its cloud of dust. What a sight it was when it did come near! The cloud that had enveloped it was discovered to be not dust only, but smoke from the cigars, meerschaums, and short clay pipes of a full complement of gentlemen passengers, scarcely one of whom seemed to have passed his twentieth year. . . . The passengers were not limited to the two-legged ones, there were four-footed ones also. Sporting dogs, fancy dogs, ugly dogs, rat-killing dogs, short-haired dogs, long-haired dogs, dogs like muffs, dogs like mops, dogs of all colours and of all breeds and sizes, appeared thrusting out their black noses from all parts of the coach. Portmanteaus were piled upon the roof; gun-boxes peeped out suspiciously here and there; bundles of sticks, canes, foils, fishing-rods, and whips, appeared strapped together in every direction; while all around about the coach,

'Like a swarth Indian with his belt of beads'

hat-boxes dangled in leathery profusion. The Oxford coach on an occasion like this was a sight to be remembered.

Edward Bradley, *Adventures of Mr. Verdant Green, by Cuthbert Bede*, 1853.

Some University men were coaching enthusiasts themselves:

August 26, 1823. *Rev. C. Atterbury*, senior Student of Ch. Ch. was crushed to death by the upsetting of a Birmingham coach. Riding on the box of one of Costar's coaches had been for years his favourite enjoyment, indeed on that very journey he had, after much wrangling, induced a fellow-traveller to give up to him, as an established claim, the seat of honour and (as it turned out) of death. He used to be called 'Costar's right-hand man', and would frequently go down *the whole line* to see that 'all was right'. He did not take the reins himself, but his hobby was to see the working of a well-appointed coach and to sit behind a fine 'team', skilfully handled. This taste did not vulgarize him, for he was a perfect gentleman.

Mr. Bobart, Esquire Bedel in Law . . . had in early life been for three years a Commoner of University College, but never graduated. . . . Like some few individuals in the University at other and later times, he had acquired a taste for driving; but unlike those individuals, he thought he might as well make his favourite pursuit a source of profit. He therefore invested his patrimony in a four-horse coach, to run between Oxford and London, himself being the *auriga*. He kept up, at least, his acquaintance with Virgil, and was fond of challenging a freshman on his coach-box to '*cap verses*'. Many a youth was led, from this encounter, to argue, 'If Oxford coachmen are such scholars, what must the tutors and the heads be!'

G. V. Cox, *Recollections of Oxford*, 1868.

Bobart was the son of Jacob Bobart the Younger, the learned but unprepossessing professor of page 74. Incapacitated by several accidents from driving his coach, in 1815 he was elected Esquire Bedel to the Vice-Chancellor instead.

The first London coach that I recollect was Bobart's, the Balloon, from the Alfred's Head Inn, High Street, Oxford, adjoining University College. . . . Bobart horsed and drove the coach, and was called 'the classical coachman', being a graduate of the University of Oxford.

Edward Quick, Esq., of New College, four chestnuts and dark coach, with his drab driving-coat built by Mrs. Jones, St. Clement's, to sixteen guineas, with as many capes. Quick always drove very slowly around the town, had a gentlemanly appearance in driving, and was a first-class whip. He stood upwards of six feet high. At other times, you would see him driving a curricle and pair of handsome, thorough-bred chestnuts. Sitting by his side was Master E. Cracknell, his factotum.

W. Bayzand, *Coaching In and Out of Oxford*, c. 1884.

Asked in 1833 why he thought it necessary to return to Oxford, after the long vacation, driving a carriage and four, Frodsham Hodson of Brasenose College replied:

That it should not be said that the first tutor of the first college of the first university of the world entered it with a pair.

Dr. James Norris, President of Corpus Christi from 1843 to 1872, was a keen coaching man, and liked to talk in the slang of the road. This is how he admonished an undergraduate who seemed to be leaving his work for the Schools rather too late:

It is never wise, you know, to *take the paint off your wheels.*

The coaching fancy gradually disappeared with the advent of the railway, (though there was a stage coach service to London until the 1890s—diehards professed to find it more comfortable):

A recent Bampton Lecturer, contrasting the present time with those of his early recollections, congratulated the University on the fact 'that the race of *Jehus* and *Nimrods* had passed away!'

G. V. Cox, *Recollections of Oxford*, 1868.

THE TRAINS COME

For years the University fiercely opposed the building of a railway to Oxford, on the grounds that it would threaten the morals of the student population. When, in 1843, the Great Western Railway was at last permitted to build a branch line to the city, University privileges were carefully written into the enabling Act:

The Vice-Chancellor, the Proctors, and pro-Proctors for the Time being of the University of *Oxford*, and Heads of Colleges and Halls, and the Marshals of the said University . . . shall, at or about the Times of Trains or Carriages upon the said Railway starting or arriving, and at all other reasonable times, have free Access to every Depot or Station . . . and to every Booking Office, Ticket Office or Place for Passengers . . . and shall then and there be entitled to demand and take and have, without any unreasonable Delay . . . such Information as it may be in the Power of any Officer or Servant of the Company to give with reference to any Passenger . . . who shall be a Member of the said University or suspected of being such.

If the said Vice-Chancellor [etc] shall . . . notify to the proper Officers, Book-keeper, or Servant of the said Company that any Person or Persons about to travel in or upon the said Railway is a Member of the University not having taken the Degree of Master of Arts or Bachelor in Civil Law, and require such Officer . . . to decline to take such Member of the University as a Passenger upon the said Railway, the proper Officer shall immediately thereupon, and for the space of 24 hours . . . refuse to convey such Member of the said University . . . notwithstanding such Member may have paid his Fare.

🦎 *The University was not entirely reconciled. Dr. Martin Routh, for instance, the aged President of Magdalen, refused to recognize the existence of the railway. When one undergraduate failed to arrive at the beginning of term, Routh insisted that the roads must be in bad condition, or the coaches full: and when it was cautiously suggested that the youth would be coming by railway, the President was quite annoyed:*

Railway, Sir? *Railway*? I know nothing about *railways*.

🦎 *Almost at once, nevertheless, a new class of tourist began to appear in Oxford, and Abel Hayward's Penny Guide to Oxford was written specifically for them:*

Leaving Paddington, the Royal Castle of Windsor is passed on the left hand, and after a few minutes' stay at Reading, the City of Learning is reached after a pleasant run of 63 miles. . . . The railway traveller, after leaving the station, very soon beholds mysterious and romantic looking edifices, full of antiquity, yet solid, as if designed to last for ever. He walks between buildings castellated, grandly Gothic, or palatial. Instead of the rattle of London, there is thoughtfulness, gravity, and repose; while, in Term, students are seen stepping leisurely along in academic cap and gown, or hourly bells are calling men to prayer. . . . Thus it is that Oxford ever charms the eye, and never disappoints the cultivated mind. A first visit to Oxford is a thing to remember for life.

Presently, too, the railway became an integral part of the Oxford myth, essential to late Victorian Oxford literature, and even entered the nostalgic reminiscences of the alumni:

The Garden Quadrangle at Balliol is where one walks at night and listens to the wind in the trees, and weaves the stars into the web of one's thoughts; where one gazes from the pale inhuman moon to the ruddy light of the windows, and hears broken notes of music and laughter, and the complaining murmur of the railroad in the distance. . . .

Arnold Toynbee, *Remains*, 1884.

OXFORD! PARDONNE-MOI

Oxford! pardonne-moi, je voulais te chanter,
Mais ton auguste Reine, que tout Français admire,
A fait soudain vibrer les cordes de ma lyre,
Et, plein d'un doux transport, je n'ai pu m'arrêter.

C'est là que Robert Peel gagna sa double classe,
Wellington en eût fait tout autant à sa place,
Car il est homme habile, aussi bien que guerrier,
C'est pour cela qu'Oxford l'élut son CHANCELIER.

Comment parler d'Oxford, sans dire quelque chose
De ses bons habitants? Si j'écrivais en prose,
Au lieu d'en nommer trois: Wootten, Wingfield, Thomson,
J'en pourrais citer cent avec même raison.

La mémoire du coeur, cette vertu si rare,
Accumule vos noms que la rime sépare,
Et j'allais t'oublier, vicomte Lewisham,
Si je n'eusse trouvé le Warden de Wadham.

P de Mascarene, *Ode à Oxford*, 1845.

MOONSHINE

There is a magic in the moonlit hour
Which day hath never in his deepest power
Of light and bloom, when bird and bee resound
And new-born flowers imparadise the ground!
And ne'er hath city, since a moon began
To hallow nature for the soul of man,
Steeped in the freshness of her fairy light,
More richly shone, than Oxford shines tonight!
No lines of harshness on her temples frown,
But all in one soft magic melted down;
Sublimer grown, through mellow air they rise,
And seem with vaster swell to awe the skies!
On archèd windows how intensely gleams
The glassy whiteness of reflected beams!
Of mitred founders, in funereal gloom,
Extends, or else in pallid shyness falls
On gothic casements, or collegiate walls.

Robert Montgomery, *Oxford*, 1831.

🎺 *This poem, the* Dictionary of National Biography *tells us, 'elicited much ridicule in Oxford, though not elsewhere'.*

THE OXFORD MAN, 1848

'Perhaps I can read you, Sir, better than you can me. You are an Oxford man by your appearance.'

Charles assented. 'How came you,' he added, 'to suppose I was of Oxford?'

'Not entirely by your looks and manner,' replied the stranger, 'for I saw you jump from the omnibus at Steventon; but with that assistance it was impossible to mistake.'

'I have heard others say the same,' said Charles; 'yet I can't myself make out that an Oxford man should be known from another. It is a fearful thing,' he added with a sigh, 'that we, as it were, exhale ourselves every breath we draw.'

John Henry Newman, *Loss and Gain*, 1848.

PENDENNIS OF BONIFACE

In the mid-Victorian age was born the Oxford novel, a prolific genre. Its classic theme was that of the rake's progress at the University—the corruption and downfall of the innocent freshman, told sometimes in pietism, as an object lesson, sometimes in parody, for fun. The model of this familiar tale was that of Arthur Pendennis, whose career at a thinly disguised Pembroke College was narrated by W. M. Thackeray in Pendennis, 1849.

Considering its size [St. Boniface] has always kept an excellent name in the University. Its *ton* is very good. . . . In the comfortable old wainscoted college hall, and round about Roubilliac's statue of St. Boniface (who stands in an attitude of seraphic benediction over the uncommonly good cheer of the fellows' table) there are portraits of many most eminent Bonifacians. There is the learned Dr. Griddle, who suffered in Henry VIII's time, and Archbishop Bush who roasted him—there is Lord Chief Justice Hicks—the Duke of St. David's, K.G., Chancellor of the University and member of this college—Sprott the poet, of whose fame the college is justly proud—Doctor Blogg, the late master, and friend of Dr. Johnson, who visited him at St. Boniface—and other lawyers, scholars, and divines, whose portaitures look from the walls, or whose coats-of-arms shine in emerald and ruby, gold and azure, in the tall windows of the refectory. . . .

Into this certainly not the least snugly sheltered arbour amongst the groves of Academe, Pen now found his way.

During the first term of Mr. Pen's University life, he attended classical and mathematical lectures with tolerable assiduity; but discovering before very long time that he had little taste or genius for the pursuing of the exact sciences, and being perhaps rather annoyed that one or two very vulgar young men, who did not even use straps to their trousers so as to cover the abominably thick and coarse shoes and stockings which they wore, beat him completely in the lecture room, he gave up his attendance at that course, and announced to his fond parent that he proposed to devote himself exclusively to the cultivation of Greek and Roman Literature.

Mrs. Pendennis was, for her part, quite satisfied that her darling boy should pursue that branch of learning for which he had the greatest inclination; and only besought him not to ruin his health by too much study. . . .

Presently he began too to find that he learned little good in the classical lecture. His fellow-students there were too dull, as in mathematics they were too learned for him. . . . After all, private reading, as he began to perceive, was the only study which was really profitable to a man; and he announced to his mamma that he should read by himself a great deal more, and in public a great deal less.

That excellent woman knew no more about Homer than she did about Algebra, but she was quite contented with Pen's arrangements . . . and felt perfectly confident that her dear boy would get the place which he merited.

Thus young Pen, the only son of an estated country gentleman . . . looked to be a lad of much more consequence than he was really; and was held by the Oxbridge authorities, tradesmen, and undergraduates, as quite a young buck and member of the aristocracy. . . . It must be owned that during his time at the University [he] was rather a dressy man, and loved to array himself in splendour. He and his polite friends would dress themselves out with as much care in order to go and dine at each other's rooms, as other folks would who were going to enslave a mistress. They said he used to wear rings over his kid gloves, which he always denies. . . . That he took perfumed baths is a truth; and he used to say that he took them after meeting certain men of a very low set in hall.

Pen himself never had any accurate notion of the manner in which he spent his money, and plunged himself in much deeper pecuniary difficulties, during his luckless residence at Oxbridge University. . . . His melancholy figure might be seen shirking about the lonely quadrangles in his battered old cap and torn gown, and he who had been the pride of the University but a year before, the man whom all the young ones loved to look at, was now the object of conversation at freshmen's wine parties, and they spoke of him with wonder and awe.

At last came the Degree Examinations. Many a young man of his year whose hob-nailed shoes Pen had derided, and whose face or coat he had caricatured . . . took high places in the honours or passed with decent credit. And where in the list was Pen the superb, Pen the wit and dandy, Pen the poet and orator? Ah, where was Pen the widow's darling and sole pride? Let us hide our heads, and shut up the page. The lists came out; and a dreadful rumour rushed through the University, that Pendennis of Boniface was plucked.

RUSKIN'S OXFORD

For much of the nineteenth century John Ruskin the art critic was a familiar of Oxford. He entered Christ Church in 1836 as a Gentleman-Commoner:

My father did not like the word 'commoner'—all the less, because our relationships in general were not uncommon. Also, though himself satisfying his pride enough in being the head of the sherry trade, he felt and saw in his son powers which had not their full scope in the sherry trade. His ideal of my future,—now entirely formed in conviction of my genius,—was that I should enter at college into the best society, take all the prizes every year, and a double first to finish with; marry Lady Clara Vere de Vere; write poetry as good as Byron's, only pious; preach sermons as good as Bossuet's, only Protestant; be made, at forty, Bishop of Winchester, and at fifty, Primate of England.

Though he did in fact win the Newdigate Prize for poetry in 1839 (the subject was Salsette and Elephanta*), his undergraduate career was generally undistinguished, and was perhaps hampered by the fact that his mother came to live in lodgings in Oxford, to keep him company. As his contemporary Henry Robinson recalled austerely in* London Society, *1887:*

John Ruskin, who gained the Newdigate prize poem . . . was nearly always to be seen with some female relation, which was rather remarkable at that day. . . .

Ruskin was powerfully moved, though, by the services in Christ Church Cathedral, the chapel of Christ Church, which he remembered in Praeterita, *1889:*

In [the Cathedral], written so closely and consecutively with indisputable British history, met every morning a congregation representing the best of what Britain had become—orderly, as the crew of a man-of-war, in the goodly ship of their temple. Every man in his place, according to his rank, age, and learning; every man of sense or heart there recognizing that he was either fulfilling, or being prepared to fill, the gravest duties required of Englishmen. A well-educated foreigner, admitted to that morning service, might have learned and judged more quickly and justly what the country had been, and still had power to be, than by months of stay in court or city. There, in his stall, sat the greatest divine of England—under his commanding niche, her greatest scholar—among the tutors the present Dean Liddell, and a man of curious intellectual power and simple virtue, Osborne Gordon. The group of noblemen gave, in the Marquis of Kildare, Earl of Desart, Earl of Emlyn, and Francis Charteris, now Lord Wemyss—the brightest types of high race and active power. Henry Acland and Charles Newton among the senior undergraduates, and I among the freshmen, showed, if one had known, elements of curious possibilities in coming days. None of us then conscious of any need or chance of change, least of all the stern captain, who, with rounded brow and glittering dark eye, led in his old thunderous Latin the responses of the morning prayer.

🥀 *Ruskin left Christ Church in 1840 because of ill-health (he was given, an honorary double-fourth in 1842), but was soon back to help in the construction of the new Museum (see page 252), and in 1869 was elected the first Slade Professor of Fine Art. In this office, he was an enormous success:*

His lectures testify to the brightness and originality of his mind. . . . No one can appreciate their effect, unless he was so fortunate as to hear them. One saw the strange afflatus coming and going in his eye, his gestures, his voice. The lectures were carefully prepared; but from time to time some key was struck which took his attention from the page, and then came an outburst. In the decorous atmosphere of a University lecture-room the strangest thing befell: and, for example, in a splendid passage on the Psalms of David (in a lecture on Birds) he was reminded of an Anthem by Mendelssohn, lately rendered in one of the College chapels, in which the solemn dignity of the Psalms was lowered by the frivolous prettiness of the music. It was, 'Oh! for the wings', etc., that he had heard with disgust, and he suddenly began to dance and recite, with the strangest flappings of his M.A. gown, and the oddest look on his excited face.

G. W. Kitchin, *John Ruskin at Oxford*, 1904.

🥀 *The subjects of his lectures, like his educational techniques, varied unpredictably. Sometimes he used the Oxford around him to illustrate his principles of art:*

All our colleges—though some of them are simply designed—are yet *richly* built, never pinchingly. Pieces of princely costliness, every here and there, mingle among the simplicities or severities of the student's life. What practical need, for instance, have we at Christ Church of the beautiful fan-vaulting under which we ascend to dine? We might have as easily achieved the eminence of our banquets under a plain vault. What need have the readers in the Bodleian of the ribbed traceries which decorate the external walls? Yet, which of those readers would not think that learning was insulted by their removal? ... In these and also other regarded and pleasant portions of our colleges, we find always a wealthy and worthy completion of all appointed features, which I believe is not without strong, though untraced effect, on the minds of the younger scholars, giving them respect for the branches of learning which those buildings are intended to honour, and increasing, in a certain degree, that sense of the value of delicacy and accuracy which is the first condition of advance in those branches of learning themselves.

Sometimes he adventured into sociology:

It is not therefore, as far as we can judge, yet possible for all men to be gentlemen and scholars. Even under the best training some will remain too selfish to refuse wealth, and some too dull to desire leisure. But many more might be so than are now; nay, perhaps all men in England might one day be so, if England truly desired her supremacy among the nations to be in kindness and in learning. To which good end, it will indeed contribute that we add some practice of the lower arts to our scheme of University education; but the thing which is vitally necessary is, that we should extend the spirit of University education to the practice of the lower arts.

Once, in his attempts to dignify the idea of manual labour, Ruskin persuaded a number of undergraduates to resurface the village road at North Hinksey, west of Oxford, in a voluntary project that became famous as 'the Hinksey Diggings'. He explained his purpose in a letter to his friend Dr. Henry Acland:

Now that country road under the slope of the hill, with its irregular line of trees sheltering yet not darkening it, is capable of being made one of the loveliest things in this English world by only a little tenderness and patience, in easy labour. We can get all stagnant water carried away of course, and we can make the cottages more healthy; and the walk, within little time and slight strength from Oxford, far more beautiful than any College garden can be. So I have got one or two of my men to promise me they will do what work is necessary with their own shoulders. I will send down my own gardener to be at their command, with what under work may here and there be necessary, which they cannot do with pleasure to themselves, and I will meet whatever expenses is needful for cartage and the like.

To one of his students he gave working instructions. Besides resurfacing the rutted road, the volunteers should plant the banks with flowers, create a village green, and stimulate the apparently somewhat stagnant sense of local pride:

When you go a little further you come to a much larger depression; in a space of land about as large as the square before the Duomo of Torcello, but triangular, not square, and with cottages on all sides of it. This space I want filled and turfed over. Which being done, a pretty little piece of grazing ground will be obtained for the geese and the donkeys of the neighbourhood. Without being desirous of expressing too strong a fellow-feeling for those animals, it seems to me wholly desirable that the village green should be kept clean and sweet for them. . . .

You must appoint one among you to be a general guardian of innocent weeds and moss. What shall we call him? You will find out some pretty Latin and dignified name for him if you debate this point. I can't stop to think today, and besides am always doubtful of my crazy Latin. But this office should be charged with the care of the moss on [the cottage] steps, and the recommendation of them also to the care of the cottage inhabitant. Minute prizes, offered to the children of any family, for well-kept door-steps, would I think be a legitimate use of bribes.

🦎 *Many young men responded, if only out of curiosity:*

[An] eloquent and authoritative voice had been raised by our popular prophet, John Ruskin, denouncing the wasteful extravagance of athletics, as so much unproductive expenditure of energy; and he tried to lure us away from the river—I think in 1876—and proposed to us the utilitarian work of road-making in the swamps of Hinksey. As I heard that those who obeyed his call had a good chance of being invited to breakfast by him, and I was most anxious to meet the great man, I went forth and for a long afternoon shovelled away the mud under the prophet's eye. I found the toil more tiring and less attractive than rowing and equally unproductive; for he never asked me to breakfast; and as Troy's wall was taken at the place where a mortal had built it, so—as I heard afterwards—a farmer's cart lost a wheel on our road at the place where my hands had laboured.

I returned to the joyous and uneconomical river, feeling that there were secrets in life that Ruskin did not know.

L. R. Farnell, *An Oxonian Looks Back*, 1934.

🦎 *The work became a national joke, and it was a popular afternoon amusement to stroll out to Hinksey from Oxford and laugh at the diggers.*

The road was terrible, and soon disappeared anyway, but at least the local surveyor could report to the landowner, Edward Harcourt, that the 'young men have done no mischief to speak of'.

'IN EVERY WAY A MARVEL'

A legend of Oxford was the theologian Martin Routh, 'the Venerable Routh', President of Magdalen from 1791 to 1854, rector of Tilehurst in Berkshire, who wore a wig into the 1850s, who remembered seeing Dr. Johnson lost in thought astride a High Street drain, and whose 'phantom authority' at Magdalen remained unchallenged until he died, still in office, in his hundredth year:

He was in every way a marvel. Spared to fulfil a century of years of honourable life, he enjoyed the use of his remarkable faculties to the very last. His memory was unimpaired; his 'eye was not dim'. More than that, he retained unabated till his death his relish for those studies of which he had announced the first-fruits for publication in 1788. Was there ever before an instance of an author whose earliest and whose latest works were 70 years apart? . . . Everything about him was interesting—was marvellous: his costume, his learning, his wisdom, his wit, his *wig*.

His way was, after giving his cap to his servant, to say grace himself:— before meat,—'For what we are about to receive, the LORD be praised!' Very peculiar was the emphasis with which on such occasions he would pronounce the Holy Name, giving breadth to the 'o' till it sounded as if the *word* 'awe' as well as the sentiment was to be found in it; rolling forth the 'r' in the manner which was characteristic of him; and pronouncing the last words with a most sonorous enunciation. His manner at such times was to extend his hands towards the viands on the table.

The President wanted (or thought he wanted) no assistance in finding his books; and to the last would mount his library-steps in quest of the occupants of the loftier shelves. Very curious he looked, by the way, perched up at that unusual altitude, apparently as engrossed in what he found as if he had been reclining in his chair. . . . Once (it was in February 1847) a very big book, which he had pulled out unaided, proved 'too many' for him, and grazed his shin. . . . The injury might have proved dangerous, and it did occasion the President serious inconvenience for a long time. A friend (I think it was Dr. Ogilvie) called to condole. The old man, after describing the accident minutely, added very gravely in a confidential voice, 'A *worthless* volume, sir! a *worthless* volume!' *This* it evidently was which weighed upon his spirits.

J. W. Burgon, Lives of Twelve Good Men, 1888.

It was as a *spectacle* that he excited popular interest; to see him shuffle into Chapel from his lodgings a Sunday crowd assembled. The wig, with trencher cap insecurely poised above it, the long cassock, ample gown, shorts and buckled shoes; the bent form, pale venerable face, enormous pendent eyebrows, generic to antique portraits in Bodleian gallery or College Halls, were here to be seen alive—

> Some statue you would swear,
> Stepped from its pedestal to take the air.

Mrs. Routh was as noticeable as her husband. She was born in the year of his election to the Presidency, 1791; so that between 'her dear man', as she called him, and herself—'that crathy old woman', as *he* occasionally called *her*—were nearly forty years. But she had become rapidly and prematurely old: with strongly marked features, a large moustache, and a profusion of grey hair, she paraded the streets, a spectral figure, in a little chaise drawn by a donkey and attended by a hunchback lad named Cox. 'Woman', her husband would say to her, when from the luncheon table he saw Cox leading the donkey carriage round, 'Woman, the ass is at the door.'

W. Tuckwell, Reminiscences of Oxford, 1900.

His introit and exit at chapel were very peculiar, owing to his gliding, sweeping *motion*, I can hardly call it *gait*; for he *moved along* (as the heathen deities were said to move) without seeming to divaricate or take alternate steps. This effect was of course produced by his long gown and cassock, and his peculiar movement. His gestures during the service were remarkable, his hands being much in motion, and often crossed upon his breast. His seat or pew being large and roomy, he was wont to move about in it during service, generally joining aloud in the responses, but without any relation to the right tone.

G. V. Cox, Recollections of Oxford, 1868.

Sometimes he used strong language. One day one of the Fellows met his butler in the middle of the day and asked him how the President was. He replied, 'He has called me a *fool*, a *thief* and a *liar* already this morning; so I think the old gentleman is pretty well.'

The last time [he] ever went up to Convocation, he walked up the High St. accompanied by Mr. *Bulley*, then Senior Tutor and afterwards President. On their way up the street they met two chimney-sweeps with their sacks and brushes, who had never seen such an apparition before as the old President with his big wig, cassock and enormous shoe-buckles. The sweeps promptly backed into the gutter and pulled their locks. 'How do you do, gentlemen?' said Dr. *Routh* bowing to them, and turning to *Bulley*,—'Who are those gentlemen, *Bulley*?' 'They are sweeps, sir', said *Bulley*. 'Eh? Ah! Humph!' said the President and passed on.

John Fisher, in *Our Memories*, 1890.

His deafness, increased by his wig, combined with his old-fashioned respect for rank, once led to a funny incident. A Gentleman-Commoner, son of a Baronet, having been beyond measure lawless, was being reprimanded by the tutors. The President, who had been looking the other way, hearing the loud sound of voices, turned round, saw a Baronet's son on the opposite table, and taking it for granted that the Tutors were paying him compliments, chimed in with: 'I am very happy, Mr. Blank, to hear what the tutors say of you. Pray tell Sir Charles with my compliments that you are a credit to the College.'

He was never seen but in full canonicals of the fashion of the last century. Somebody bet that he would show Routh without his canonicals, and thought to win the bet by crying 'fire', of which Routh was horribly afraid, at the dead of night under his window. Routh at once appeared, in a great fright—but in full canonicals.

Goldwin Smith, *Reminiscences*, 1910.

Though I can laugh now at the indolence and uselessness of the collegiate life of my boy-days, my boyish imagination was over-powered by the solemn services, the white-robed choir, the long train of divines and fellows, and the president—moving like some mysterious dream of the past among the punier creatures of the present. . . . We boys used to stand overawed as the old man passed by, the keen eyes looking out of the white, drawn face, and feel as if we were looking on some one from another world.

J. R. Green (1837–83), in a letter.

It was only 48 hours before Dr. Routh died that his powers began to fail. . . . They tried to get him upstairs to bed, but he struggled with the banisters as with an imaginary enemy.

A. J. C. Hare, *The Story of My Life*, 1896.

[Routh] died . . . so Blagrave, his brother-in-law and man of business admitted, through chagrin at the fall of Russian Securities at the time of the Crimean war—a very respectable way of breaking one's heart, according to Mr. Dombey. . . .

W. Tuckwell, *Reminiscences of Oxford*, 1900.

The whole College, with all its train of past generations that survived, followed the old President to the grave. The majestic music and solemn wailings of the choir seemed to mourn over some great edifice that had fallen, and left a vast void, which looked quite strange and unaccountable. . . .

T. Mozley, Letters, 1873.

ROUTH VERBATIM

Beware, sir, of acquiring the habit of reading catalogues; you will never get any good from it, and it will consume much of your time.

Don: Master! something appalling has happened. One of the Fellows has killed himself.

Routh: Pray do not tell me who, Sir. Allow me to guess.

Dear Master, Something very sad has occurred in your absence. The great plane tree in front of your Lodgings has been blown down by the wind.

Dear Sir, Put it up again.

Aged Don: We are now getting very old, and I hope we are prepared for the change which shortly awaits us.

Routh, aged ninety: Yes, Sir, I am prepared: but I read in the newspaper this morning, Sir, the death of a Dissenting Minister who lived to be 104. I should wish, Sir, for the Church to beat Dissent.

J. W. Burgon: Mr. President, give me leave to ask you a question I have sometimes asked of aged persons, but never of any so aged or so learned as yourself. Every studious man, in the course of a long and thoughtful life, has had occasion to experience the special value of some one axiom or precept. Would you mind giving me the benefit of such a word of advice?

Routh, after thought: I think, sir, since you come for the advice of an old man, sir, you will find it a very good practice *always to verify your references, sir!*

❧ *He used to preach learned sermons to the rustic congregation in his Berkshire parish. This is how he began one:*

I know, my friends, that you may object to the remarks I am about to make by quoting the arguments of St. Irenaeus. . . .

❧ *He was very fond of his dogs. When told that the Fellows of the College had resolved to enforce a rule against keeping dogs in college, Dr. Routh replied:*

Then, sir, I suppose I must call mine—cats.

He was opposed to University reform, and had a stock answer for its advocates:

Wait, Sir, till I am gone.

His last words were spoken to his housekeeper, Mrs. Druce, who had suggested changing the dressing on his leg, to make him more comfortable:

Don't trouble yourself.

GAISFORD OF CHRIST CHURCH

Thomas Gaisford, Dean of Christ Church from 1831 to 1855, once ended a sermon in the Cathedral with the following advice:

Nor can I do better, in conclusion, than impress upon you the study of Greek literature, which not only elevates above the vulgar herd, but leads not infrequently to positions of considerable emolument.

John Ruskin, in Praeterita, *1889, remembered him supervising a college examination in Christ Church hall:*

Scornful at once, and vindictive, thunderous always, more sullen and threatening as the day went on, he stalked with baleful emanation of Gorgonian cold from dais to door, and door to dais, of the majestic torture chamber. . . . Though venerable to me, from the first, in his evident honesty, self-respect, and real power of a rough kind, [the Dean] was yet in his general aspect too much like the sign of the Red Pig which I afterwards saw set up in pudding raisins with black currants for eyes by an imaginative grocer in Chartres fair; and in the total bodily and ghostly presence of him was to me only a rotundly progressive terror, or sternly enthroned and niched Anathema.

EXCHANGES

Message from the Dean of Oriel: The Dean of Oriel presents his compliments to the Dean of Christ Church, and wishes to know what time the examination will be.

Reply from Dr. Gaisford: Alexander the Great presents his compliments to Alexander the Coppersmith, and begs to inform him that he knows nothing whatever about it.

Undergraduate: I have doubts about the 39 Articles [of the Anglican Church].
Gaisford: How much do you weigh, sir?
Undergraduate: About 10 stone, I should think, sir.
Gaisford: And how tall are you to half an inch?
Undergraduate: I really don't know to half an inch.

Gaisford: And how old are you to the hour?

Undergraduate: I really don't know that either, sir.

Gaisford: Yet you walk about saying 'I am 20 years old, I weigh 10 stone, and am five feet eight inches high'. Go, sign the articles: it will be a long time before you find anything that you can have *no* doubts about.

THE STYLE OF THE PLACE

※ *Even to the end of the 1840s, Oxford was in many ways recognizably an eighteenth-century University; witness these extracts from a memoir by T. E. Kebbel, writing in* The National Review *in 1887:*

A Don in my day was only partially associated in the undergraduate mind with the ideas of education and learning. Each college was then a close, powerful and wealthy corporation, doing what it liked with its own, repelling interference from without, and, perhaps it is hardly too much to say, a little University in itself. The members of this corporation, as long as they remained unmarried, and unbeneficed, held their fellowships for life, and were practically irremovable. A fellowship was a freehold; and the tenant of it was simply in the position of a small landed proprietor, rich in the possession of an income sufficient for all his wants. . . .

In Oxford and Cambridge alone were found these ancient immemorial nests of life-long leisure, the occupants of which succeeded each other like rooks in a rookery, where the tall elms tell of centuries of undisturbed repose and inviolate prescription.

No man was ever seen in the streets of Oxford after lunch without being dressed as he would have been in Pall Mall. Tail coats were sometimes worn in those days in the morning, and the fast men still wore cutaways. But the correct thing for the quiet gentlemanly undergraduate was a black frock-coat, and tall hat, with the neates of gloves and boots, and in this costume he went out for his country walk, the admired of all beholders, as he passed through Hinksey or Headington. . . . The whole body of Oxford men were, in many respects like one gigantic common-room: all members of a highly-exclusive society; all members of the Church, and, with some very few exceptions, which did not in the slightest degree affect the tone or manners of the place, all gentlemen.

※ *It was not to last. Half-way through the nineteenth century, the eighteenth was dismissed from Oxford University, and the style changed.*

The Power House
1850–1914

The second half of the nineteenth century, with the first decade of the twentieth, was Oxford's grandest period. Vigorously reformed, intellectually awake, socially cohesive, the University was a conscious instrument of Victorian national greatness. Its confidence bordered upon arrogance, until like the nation itself it was awoken from its splendid island dream by the world reality of war.

REFORM!

🐾 *The end of the absorbing but debilitating Tractarian Movement, which had obsessed Oxford through two decades, cleared the air for a fresh start in the 1850s, and almost at once, in and out of Oxford, there were demands for a modernization of the University.*

If any Oxford man had gone to sleep in 1846 and had woke up again in 1850 he would have found himself in a totally new world. In 1846 we were in Old Tory Oxford; not somnolent because it was as fiercely debating, as in the days of Henry IV, its eternal Church question. There were Tory majorities in all the colleges; there was the unquestioning satisfaction in the tutorial system, *i.e.* one man teaching everybody everything; the same belief that all knowledge was shut up between the covers of four Greek and four Latin books; the same humdrum questions asked in the examination; and the same arts of evasive reply. In 1850 all this was suddenly changed as by the wand of a magician. The dead majorities of head and seniors, which had sat like lead upon the energies of young tutors, had melted away. Theology was totally banished from Common Rooms, and even from private conversation. Very free opinions on all subjects were rife; there was a prevailing dissatisfaction with our boasted tutorial system. A restless fever of change had spread through the colleges—the wonder-working phrase, University reform, had been uttered, and that in the House of Commons. The sounds seemed to breathe new life into us.

Mark Pattison, *Memoirs*, 1885.

🐾 *In April, 1850, the Commons appointed a Royal Commission, headed by the Bishop of Norwich, to investigate the state of the University and recommend changes. Oxford traditionalists, always sensitive to outside interference in University affairs, gave a prickly response. When the Commissioners approached the Visitors of the various colleges, enlisting their support, five of them did not even bother to reply, while the Bishop of Exeter, Visitor of Exeter College, drew an angry analogy with 'the fatal attempt by James II to subject the colleges to his unhallowed control' (page 112):*

I cannot see without the deepest concern and astonishment, the name of our present Gracious Sovereign used by Her advisers to 'authorize and empower' . . . an inquisition which no precedent could justify. . . . It is under the solemn conviction that your Lordship and the other eminent persons who have consented to act on the Commission, have no right whatever 'to call before you' any Members of the College of which I am Visitor, or 'to call for and examine' all such Books, Documents, Papers and Records as you' may 'judge likely to afford you' any 'information' concerning that chartered body 'on the subject of this Commission', that I shall require the Rector, Fellows and other Members, to weigh well all the injunctions of their statutes before feeling themselves at liberty to testify any deference to your authority. Especially I shall enjoin them, under the sacred obligation of their oaths, to beware how they permit themselves to answer any inquiries, or to accept any directions or interference whatsoever, which may trench upon that visitatorial authority, which their statutes, under the known law of the land, have entrusted solely to the Bishop of this See.

🐾 *Many of the dons approached gave evasive or ambiguous answers, like this one from P. B. Duncan, Keeper of the University Archives:*

There is nothing in which I should feel more pride and delight in doing than in giving any useful information for the improvement of my beloved University, were it in my power to suggest anything worthy of your attention for its advantage.

🐾 *A few were franker, like James Garbett, the Professor of Poetry:*

I cannot aid in an object which I condemn, and an inquisition against which I protest. No respect for the abilities, attainments, and position of the members of your body can remove from thoughtful men the apprehension that formidable innovations, and in our opinion disastrous changes, are contemplated under the present Commission. . . . We crave peace and you give us chaos. If all objections on the score of expediency were removed, I should oppose the present Commission as illegal and unconstitutional *in its whole spirit and purpose* if not in the letter, and in an age of professed and in many points real liberalism and improvement, a despotic stretch of antiquated prerogatives. It recalls the worst times, and the worst precedents. *Absit omen!*

🐾 *Old Oxford found much sympathy for these attitudes. Thomas de Quincey, for instance, hardly one of Oxford's most characteristic sons (page 173), was moved to a new loyalty:*

Oxford, ancient Mother! hoary with ancestral honours, time-honoured, and haply it may be time-shattered power . . . Of thy vast riches I took not a shilling, though living among multitudes who owed to thee their daily bread. Not the less I owe thee justice; for that is a universal debt. And at this moment when I see thee called to thy audit by unjust and malicious accusers—men with the hearts of inquisitors and the purposes of robbers—I feel towards thee something of filial reverence and duty.

Autobiography, 1853.

🔏 *And the American Ralph Waldo Emerson suspected that the true meaning of Oxford stood above Royal Commissions or parliamentary inquiries:*

Genius exists there . . . but it will not answer a call of a committee of the House of Commons. England is the land of mixture and surprise,—and when you have settled it that the universities are moribund, out comes a poetic influence from the heart of Oxford, to mould the opinions of cities, to build their houses as simply as birds their nests, to give veracity to art, and charm mankind, as an appeal to moral order always must.

English Traits, 1856.

🔏 *In Oxford it was feared that the Commission would lead to a betrayal of English academic methods to the currently fashionable German style of scholarship—a trend satirized by C. L. Dodgson (Lewis Carroll) in* The Vision of the Three Ts:

Now-a-days, all that is good comes from the German. Ask our men of science: they will tell you that any German book must needs surpass an English one. Aye, and even an English book, worth naught in its native dress, shall become, when rendered into German, a valuable contribution to science. . . . No learned man doth now talk, or even so much as cough, save only in German. The time has been, I doubt not, when an honest English 'Hem!' was held enough, both to clear the voice and rouse the attention of the company, but now-a-days no man of science, that setteth any store by his good name, will cough otherwise than thus, Ach! Euch! Auch!

🔏 *Another satirical response was H. L. Mansel's epic irony* Phrontisterion, *concerned particularly with German definitions of the Infinite, which included the following Chorus of Philosophical Professors:*

> With deep intuition and mystic rite
> We worship the Absolute-Infinite,
> The Universe-Ego, the Plenary-Void,
> The Subject-Object Identified,
> The great Nothing-Something, the Being-Thought,
> That mouldest the mass of Chaotic Nought,

Whose beginning unended and end begun
Is the One that is All, and the All that is One.
   Hail Light with Darkness joined!
   Thou Potent Impotence!
   Thou Quantitative Point
   Of All Indifference!
Great Non-Existence, passing into Being,
Thou two-fold Pole of the Electric One,
Thou Lawless Law, thou Seer all Unseeing,
Thou Process, ever doing, never done!
   Thou Positive Negation!
   Negative Affirmation!
Thou great Totality of every thing,
That never is, but ever doth become,
   Thee do we sing,
   The Pantheist's King,
With ceaseless bug, bug, bug, and endless hum, hum, hum.

   Professors we,
   From over the sea,
From the land where Professors in plenty be;
And we thrive and flourish, as well we may,
In the land that produced one Kant with a K
   And many Cants with a C.

🔊 *The Commission proceeded nonetheless, and interviewed many witnesses. Some of the opinions expressed were obscurantist, like those of Edward Pusey of Christ Church (page 218). The 'problem and especial work of a university', he thought, was*

. . . not how to advance science, not how to make discoveries, not to form new schools of mental philosophy, nor to invent new modes of analysis; not to produce new works in Medicine, Jurisprudence, or even Theology; but to form minds religiously, morally, intellectually, which shall discharge aright whatever duties God, in His Providence, shall appoint to them. . . . It would be absurd to make a distinction between theological and non-theological professors. All the sciences move like planets round the sun of God's truth, and if they left their course, they would soon be hurled back into chaos.

🔊 *The Royal Commission's report, notwithstanding, led to an Act of Parliament, 1854, which really did revolutionize the University of Oxford, widening the system of study, weakening the autonomous pride of the colleges, doing away with the more flagrant class distinctions, enhancing*

The Martyrs' Memorial, 1850

The University Museum, 1860

the importance of the faculty Professors and modernizing the University's government. Archbishop Laud's Statutes (page 78) were largely superseded at last: before long college Fellows were allowed to marry, and the University Tests, obliging every Oxford man to be a member of the Church of England, were abolished—

> At many a Hall and College,
> By many a traitorous stroke,
> The Tree of Christian Knowledge
> Falls like the forest oak.

🦡 *The reforms produced some new kinds of Oxford character, and did away with some of the old:*

The Oxford Professor . . . was quite a young man, of advanced opinions on all subjects, religious, social, and political. He was clever, extremely well-informed, so far as books can make a man knowing, but unable to profit even by that limited experience of life from a restless vanity and overflowing conceit, which prevented him from ever observing and thinking of anything but himself. He was gifted with a great command of words, which took the form of endless exposition, varied by sarcasm and passages of ornate jargon. He was the last person one would have expected to recognize in an Oxford Professor; but we live in times of transition.

Benjamin Disraeli, Lothair, 1870.

> The days of port and peace are gone,
> I am a modern Oxford Don;
> No more I haunt the candle's gloom,
> The cosy chairs of Common-Room;
> No more the senior man discourses,
> Of wine, of women fair and horses,
> Tells with cracked voice and mellow pride
> Old stories of the covert-side,
> Or with sad sigh dim forms recalls
> Whose grey slabs line the cloister walls.
> Student and hunter both are fled,
> All their age is 'lapped in lead'.

Anon., in the Oxford Magazine, 1894.

🦡 *But for better or for worse, the reforms of the 1850s enabled the University to enter the grandest period of its history with a new and more virile image.*

STUDENT CAVALRY

There is no part of our social life so entirely novel, and so well worth exhibiting to a foreigner, as a 'Meet' near Oxford, where in scarlet and in black, in hats and in velvet caps, in top-boots and black-jacks, on twenty pound hacks and two hundred guinea hunters, finest specimens of Young England are to be seen. There is not such raw material for cavalry in any other city in Europe.

Samuel Sidney, *Rides on Railways*, 1851.

VERDANT GREEN

In 1853, cheerfully ignoring all Commissions and Reforms, came Verdant Green *by 'Cuthbert Bede', the most lastingly popular of Oxford novels. Its real author was the Revd. Edward Bradley, described by the* Dictionary of National Biography *as being of 'a somewhat ancient Worcestershire and clerical family', who was not an Oxford man at all, but had spent a year in the city after taking his degree at Durham. The Oxford background was drawn in faithful pastiche, all the same, and several well-known Oxford figures were clearly recognizable, notably Frederic Plumptre, the Master of University College (compare page 250). In our extracts we accompany the ingenuous Mr. Green in his first experiences of Oxford:*

They soon found a guide, one of those wonderful people to which show-places give birth, and of whom Oxford can boast a very goodly average; and under this gentleman's guidance Mr. Verdant Green made his first acquaintance with the fair outside of his Alma Mater.

The short, thick stick of the guide served to direct attention to the various objects he enumerated in his rapid career. 'This here's Christ Church College,' he said, as he trotted them down St. Aldate's, 'built by Card'nal Hoolsy four underd feet long and the famous Tom Tower as tolls wun underd and wun hevery night that being the number of stoodents on the foundation'; and thus the guide went on, perfectly independent of the artificial trammels of punctuation, and not particular whether his hearers understood him or not: that was not his business. And as it was that gentleman's boast that he 'could do the alls, collidges, and principal hedifices in a nour and a naff', it could not be expected but that Mr. Green should take back to Warwickshire otherwise than a slightly confused impression of Oxford.

Instead of the stern, imposing-looking personage that Mr. Verdant Green had expected to see in the ruler among dons, and the terror of offending undergraduates, the master of Brazenface was a mild-looking old gentleman, with an inoffensive amiability of expression and a shy, retiring manner that seemed to intimate that he was more alarmed at the strangers than they had need to be at him. Dr. Portman seemed to be quite a part of his college, for he had passed the greatest portion of his life there. He had graduated there, he had taken Scholarships there, he had even gained a prize-poem there; he had been elected a Fellow there, he had become a tutor there, he had been Proctor and College Dean there; there, during the long vacations, he had written his celebrated 'Disquisition on the Greek Particles', afterwards published in eight octavo volumes; and finally, there he had been elected Master of his college, in which office, honoured and respected, he appeared likely to end his days. He was unmarried; perhaps he had never found time to think of a wife; perhaps he had never had the courage to propose for one; perhaps he had met with early crosses and disappointments, and had shrined in his heart a fair image that should never be displaced. Who knows? for dons are mortals, and have been undergraduates once.

The little hair he had was of a silvery white, although his eyebrows retained their black hue; and to judge from the fine fresh-coloured features and the dark eyes that were now nervously twinkling upon Mr. Green, Dr. Portman must, in his more youthful days, have had an ample share of good looks. He was dressed in an old-fashioned reverend suit of black, with knee-breeches and gaiters, and a massive watch-seal dangling from under his waistcoat, and was deep in the study of his favourite particles. He received our hero and his father both nervously and graciously, and bade them be seated.

'I shall al-ways,' he said, in monosyllabic tones, as though he were reading out of a child's primer,—'I shall al-ways be glad to see any of the young friends of my old col-lege friend Lar-kyns; and I do re-joice to be a-ble to serve you, Mis-ter Green; and I hope your son, Mis-ter-Mis-ter Vir-Vir-gin-ius'—

'Verdant, Dr. Portman,' interrupted Mr. Green, suggestively, 'Verdant.'

'I am sorry, Mis-ter Green,' said Dr. Portman, 'that my en-gage-ments will pre-vent me from ask-ing you and Mis-ter Virg- Verdant, to dine with me today; but I do hope that the next time you come to Ox-ford I shall be more for-tu-nate.'

Old John, the Common-room man, who had heard this speech made to hundreds of 'governors' through many generations of freshmen, could not repress a few pantomimic asides, that were suggestive of anything but full credence in his master's words. But Mr. Green was delighted with Dr. Portman's affability, and perceiving that the interview was at an end, made his congé, and left the Master of Brazenface to his Greek particles.

In the ante-chamber of the Convocation House, the edifying and imposing spectacle of Matriculation was enacted. In the first place, Mr. Verdant Green took divers oaths, and sincerely promised and swore that he would be faithful and bear true allegiance to her Majesty Queen Victoria. He also professed (very much to his own astonishment) that he did 'from his heart abhor, detest, and anjure, as impious and heretical that damnable doctrine and position, that princes excommunicated or deprived by the pope, or any authority of the see of Rome, may be deposed or murdered by their subjects, or any other whatsoever.' And having almost lost his breath at this novel 'position', Mr. Verdant Green could only gasp his declaration, 'that no foreign prince, person, prelate, state, or potentate, hath, or ought to have, any jurisdiction, power, superiority, pre-eminence, or authority, ecclesiastical or spiritual, within this realm.' When he had sufficiently recovered his presence of mind, Mr. Verdant Green inserted his name in the University books as 'Generosi filius natu maximus'; and then signed his name to the Thirty-nine Articles,— though he did not endanger his matriculation, as Theodore Hook did, by professing his readiness to sign forty if they wished it! Then the Vice-Chancellor concluded the performance by presenting to the three freshmen (in the most liberal manner) three brown-looking volumes, with these words: 'Scitote vos in Matriculam Universitatis hodie relatos esse, sub hac conditione, nempe ut omnia Statuta hoc libro comprehensa pro virili observetis.' And the ceremony was at an end, and Mr. Verdant Green was a matriculated member of the University of Oxford. He was far too nervous,—from the weakening effect of the popes and the excommunicate princes, and their murderous subjects,—to be able to translate and understand what the Vice-Chancellor had said to him, but he thought his present to be particularly kind; and he found it a copy of the University Statutes, which he determined forthwith to read and obey.

PRE-RAPHAELITES

🦁 *Two leading members of the Pre-Raphaelite Brotherhood entered Exeter College in the same year, 1853. William Morris and Edward Burne-Jones became such close friends that by their own account they habitually spoke to 'but three or four other men' in the whole college. Burne-Jones reported home enthusiastically:*

Oxford is a glorious place; godlike! at night I have walked round the colleges under the full moon, and thought it would be heaven to live and die here. The Dons are so terribly majestic, and the men are men, in spirit as in name—they seem overflowing with generosity and good-nature. . . . I wonder how the examiners ever have the heart to pluck such men.

❧ *Morris, described by his tutor as 'a rather rough unpolished youth, who exhibited no especial literary tastes or capacity', was less receptive:*

Oxford in those days still kept a great deal of its earlier loveliness: and the memory of its grey streets as they were then has been an abiding influence and pleasure in my life, and would be greater still if I could only forget what they are now; a matter of far more importance than the so-called learning of the place could have been to me in any case, but which, as it was, no one tried to teach men, and I did not try to learn. Since then the guardians of this beauty & romance so fertile of education, though professedly engaged in the higher education (as the futile system of compromises which they follow is nicknamed), have ignored it utterly, have made its preservation give way to the pressure of commercial exigencies, & are determined apparently to destroy it altogether. There is another pleasure for the world gone down the wind. . . .

From *The Aims of Art*, 1886.

❧ *To the end of his life he used the word 'don' as a synonym for narrowness and pedantry:*

To [a don] Oxford was the place where knowledge was to be acquired and honours gained. To [Morris] it was, in his own striking words, 'a vision of grey-roofed houses and a long winding street, and the sound of many bells'. When Morris found that the internal life did not accord with its external aspects . . . he had not the heart to look for the virtues which it did possess.

W. K. Stride, *Exeter College*, 1900.

❧ *Nevertheless only four years later both young men, encouraged by Dante Gabriel Rossetti, enthusiastically volunteered to decorate with murals the new debating hall of the Oxford Union. Their theme was the Arthurian legend, and they were joined by Val Prinsep and other friends, and watched sometimes by John Ruskin and Algernon Swinburne the poet. Supplied with endless free soda-water by the Union, they had a splendid time, as Prinsep recorded:*

What fun we had at the Union! What jokes! What roars of laughter!

🦁 *Ruskin, watching from the outside, foresaw snags to the venture—*

'The fact is, they're all the least bit crazy and it is difficult to manage them'

—and he was right, for the exuberant artists had failed to prepare the walls properly for their pictures. At first the murals looked lovely. Coventry Patmore, the writer, who saw them soon after their completion, marvelled at their brilliant colours:

. . . so brilliant as to make the walls look like the margin of an illuminated manuscript.

🦁 *But they very soon faded, Lancelot, Tristram, and Guinevere crumbled into the distemper, and despite repeated attempts to restore them, they remained virtually invisible ever after.*

LETTERS TO MAMA

🦁 *Mr. Augustus Hare reports to his mother upon his arrival at University College, Oxford, in 1853:*

It is from my own rooms, 'No. 2, Kitchen Staircase', that I write to my mother—in a room long and narrow, with yellow beams across the ceiling, and a tall window at one end admitting dingy light, with a view of straight gravel-walks, and beds of cabbages and rhubarb in the Master's kitchen garden. Here, for £32. 16s. 6d. I have been forced to become the owner of the last proprietor's furniture—curtains which drip with dirt, a bed with a ragged counterpane, a bleared mirror in a gilt frame, and some ugly mahogany chairs and tables. 'Your rooms might be worse, but your servant could not,' said Mr. Hedley when he brought me here. . . . How shy I have just felt in Hall, sitting through a dinner with a whole set of men I did not know and who never spoke to me.

🦁 *Mr. Hare tells his mother about his Matriculation:*

The Master [Frederic Plumptre] asked what books I had ever done, and took down the names on paper. Then he chose Herodotus. I knew with that old man a mistake would be fatal, and I did not make it. Then he asked me a number of odd questions—all the principal rivers in France and Spain, the towns they pass through, and the points where they enter the sea; all the prophecies in the Old Testament in their order relating to the coming of Christ; all the relationships of Abraham and all the places he lived in. These things fortunately

I happened to know. Then the Master arose and solemnly made a little speech—'You have not read so many books, Mr. Hare, not nearly so many books as are generally required, but in consideration of the satisfactory way in which you have passed your general examination, and in which you have answered my questions, you will be allowed to matriculate, and this, I hope, will lead you,' &c. &c. But for me the moral lesson at the end is lost in the essential, and the hitherto cold countenance of Mr. Hedley now smiles pleasantly.

Then a great book is brought out, and I am instructed to write— 'Augustus Joannes Cuthbertus Hare, Armigeri filius'. Then there is a pause. The Master and Dean consult how 'born at Rome' is to be written. The Dean suggests, the Master does not approve; the Dean suggests again, the Master is irritated; the Dean consults a great folio volume, and I am told to write 'de urbe Roma civitate Italiae'. When this is done, Mr. Hedley stands up, the Master looks vacant. I bow, and we go out.

Mr. Hare tells his mother about a strange fellow-student:

I breakfasted the other day at Wadham with a most extraordinary man called R., whose arms and legs all straggle away from his body, and who holds up his hands like a kangaroo. His oddities are a great amusement to his friends, who nevertheless esteem him. One day a man said to him, 'How do you do, R.?' and he answered, 'Quite well, thank you.' Imagine the man's astonishment at receiving next day a note—'Dear Sir, I am sorry to tell you that I have been acting a deceptive part. When I told you yesterday that I was quite well, I had really a headache: this has been upon my conscience ever since.' The man was extremely amused, and showed the letter to a friend, who, knowing R.'s frailties, said to him, 'Oh R., how could you act so wrongly as to call Mr. Burton "Dear Sir"—thereby giving him the impression that you liked him, when you know that you dislike him extremely?' So poor R. was sadly distressed, and a few days later Mr. Burton received the following:—'Burton, I am sorry to trouble you again, but I have been shown that, under the mask of friendship, I have been for the second time deceiving you: by calling you dear sir, I may have led you to suppose I liked you, which I never did and never can do. I am, Burton, yours &c.!'

Mr. Hare recalls in later years the Master of his college:

It would be impossible to discover a more perfect old 'gentleman' than Dr. Plumptre, though he was often laughed at. When he was inquiring into any fault, he would begin with, 'Now pray take care what you say, because whatever you say I shall believe'. He had an old-fashioned veneration for rank, and let Lord Egremont off lectures two days in the week that he might hunt—'it was so suitable'.

THE CLANG OF WAR

A.D. 1854 The opening year loomed heavily with the approaching Crimean horrors, in the war against Russia. If it be asked, 'What had Oxford in particular to do with this, or this with Oxford?' the answer is at hand; Oxford did, in fact, very soon begin to feel its effect, first in the reduced number of Matriculations, and soon after in the departure of many of its actual members for the seat of war. When it is considered that for nearly forty years there had been no field of the kind opened to our English youths, it was no wonder that the excitement and clang of war should be responded to by young men of spirit and enterprise. And so they went, full of health and ardour and cheerful hopes, to waste that strength, exhaust that ardour, and quench those hopes in the protracted miseries of a siege, and to die (as many of them did) in the trenches of Sebastopol. But though perhaps not one of these deserters from our peaceful camp would have changed the battle cry for 'cedant arma togae', yet 'dulcem moriens reminiscitur (Oxford)' was perchance his latest thought.

G. V. Cox, *Recollections of Oxford*, 1868.

ALL IN THE MUSEUM

Even before the great university reforms of 1854, lively minds in Oxford were pressing for the extension of scientific studies, long neglected and often sneered at by classicists and theologians, and in 1855 work began on a new University Museum, intended to be a fulcrum for scientific life in Oxford.

It has taken some centuries from the epoch of Roger Bacon, followed here by Boyle, Harvey, Linacre and Sydenham, besides nearly 200 years of unbroken publication of the Royal Society's Transactions, to persuade this great English University to engraft, as a substantial part of the education of her youth, any knowledge of the great material design of which the Supreme Master-Worker has made us a constituent part. 'The study of mankind', indeed, was 'Man'; but in Oxford it was Man viewed apart from all those external circumstances and conditions by which his probation on earth was made by his Maker possible, and through whose agency, for good and evil, his life here, and preparation for life hereafter, were ordained.

Seeing them, all these things, many here in Oxford, not so much by concert, as by that strange unanimity which comes to some subjects in the fulness of their time, felt as by an instinct, that they might not rest until means for rightly studying what is vouchsafed for man to know of this universe were accorded to the youth committed to their care, and to themselves. From such causes, and from so deep convictions, has arisen the Oxford Museum.

Henry W. Acland, *The Oxford Museum*, 1893.

A leading activist in the project was the young sage John Ruskin, who saw to it that the Museum was built in the highest Venetian Gothic style—'no other architecture, I felt in an instant, could have thus adapted itself to a new and strange office'—and that its chemical laboratory was based upon the Abbot's Kitchen at Glastonbury. Ruskin wanted the building to illustrate his theories about Art and Labour, too, and the workmen on the site found themselves treated with an unaccustomed solicitude.

During this summer, and indeed during the whole long period while the Museum was being built, it was very gratifying to see the good effects of the thoughtful, good feeling of the University authorities, who caused a temporary mess-room and a reading-room to be constructed, close to the site of the Museum, for the use and comfort of the workmen there employed. This building was opened by the Vice-Chancellor, &c. with a dinner for the workmen; books &c. were purchased for their leisure hours, a respectable female was established there, with a fire for cooking, and morning prayers undertaken to be kept up by the Vice-Chancellor and other clergymen.

G. V. Cox, *Recollections of Oxford*, 1868.

The architect, Benjamin Woodward of Dublin, bravely entered into the spirit of the enterprise, and brought over from Ireland a whole family of stone-carvers, the O'Sheas, whom he encouraged to express, in the ornamental masonry, their own vision of the Cosmos. This led to difficulties:

O'Shea rushed into my house one afternoon and—in a state of wild excitement—related as follows.

'The Master of the University', cried he, 'found me on my scaffold just now.' 'What are you at?' says he. 'Monkeys', says I. 'Come down directly', says he; 'you shall not destroy the property of the University.' 'I work as Mr. Woodward orders me.' 'Come down directly', says he; 'come down'.

'What shall I do?' said O'Shea to me. 'I don't know; Mr. Woodward told you monkeys, the Master tells you no monkeys. I don't know what you are to do.' He instantly rushed out as he came, without another word.

The next day I went to see what had happened. O'Shea was hammering furiously at the window. 'What are you at?' said I. 'Cats', says he. 'The Master came along, and says, "You are doing monkeys when I told you not."' 'Today it's cats', says I. The Master was terrified and went away. . . .

It did not however so end; Shea was dismissed. I went to wish him goodbye with mixed and perplexed feelings. I found [him] on a single ladder in the porch, wielding heavy blows such as one imagines the genius of Michael Angelo might have struck when he was first blocking out the design of some immortal work. 'What are you doing, Shea? I thought you were gone, and Mr. Woodward has given no design for the long moulding in the hard green stone.'

Striking on still, Shea shouted,

> 'Parrhots and Owwls!
> Parrhots and Owwls!
> Members of Convocation!'

There they were, blocked out alternately.

What could I do? 'Well', I said, meditatively, 'Shea, you must knock their heads off.'

'Never', says he.

'Directly', said I.

Their heads went. Their bodies, not yet evolved, remain. . . .

Henry W. Acland, *The Oxford Museum*, 1893.

🐾 *Ruskin was not altogether pleased with the look of the building when it was completed in 1859, but then he viewed it chiefly as a symbol of higher things:*

Now therefore, and now only, it seems to me, the University has become complete in her function as a teacher of the youth of the nation, to which every hour gives wider authority over distant lands; and from which every rood of extended dominion demands new, various, and variously applicable knowledge of the laws which govern the constitution of the globe. . . .

🐾 *The judgement of Alfred Tennyson, though, delivered as he walked by the Museum one day in 1860, was tauter:*

Perfectly indecent.

🐾 *Among the Museum's best-loved possessions was to be the skeletonic Dodo which had first come to Oxford with the Tradescant collection in 1683* (*page 110—'Dodar, from the Island Mauritius'*), *and which Hilaire Belloc would immortalize in his* Cautionary Verses:

> The voice which used to squawk and squeak
>   Is now for ever dumb—
> Yet you may see his bones and beak
>   All in the Mu-se-um.

AN INQUIRY

Lord Tennyson was given an honorary Doctorate of Civil Law in 1855. He entered the Sheldonian with his hair, as usual, flowing in poetical dishevelment over his shoulders, and was greeted by a cry from an undergraduate in the gallery:

> Did your mother call you early, dear?

Arthur Baker's monumental Concordance to Tennyson's Poetical and Dramatic Works, *1914, records no use of the word 'Oxford'.*

MORAL CRACKERS

In 1857 somebody produced a book of mottoes to be put into Oxford crackers, intended to 'impart to the mind of youth, in its festive moments of relaxation, lessons of prudence and morality'. Here are some of its jingles:

> Always wear your cap and gown,
> Little Freshman, in the town;
> When a walk you're bent upon,
> You may put your beaver on.

> In town or in country, wherever you go,
> If you meet with a Don, you should cap him, you know,
> For the statutes of Oxford determine 'tis right
> That her scholars should be oh! so very polite.

> He who would the Dons delight
> Hard must study day and night,
> Never play at Cards or Pool, or
> He will find them growing cooler.

> If to sporting you're inclined,
> Guns are all forbidden, mind:

> Should you doubt it, please to look
> At that statute in the book,
> Which in every freshman's hand is,
> 'De bombardis non gestandis'.

🎜 *The Statute mentioned in the last motto, a survivor from Archbishop Laud's seventeenth-century code, forbade not only guns, but also bows and arrows.*

MATTHEW ARNOLD

🎜 *The Oxford poet par excellence was Matthew Arnold, scholar of Balliol (1841), Newdigate prize-winner (1843), Fellow of Oriel (1845), Professor of Poetry (1857). Here are the three best-known lyric pieces ever written about Oxford, two in poetry, one in prose:*

'I DREAM!'

> Screen'd is this nook o'er the high, half-reap'd field,
>   And here till sun-down, shepherd! will I be.
>   Through the thick corn the scarlet poppies peep,
>   And round green roots and yellowing stalks I see
>   Pale pink convolvulus in tendrils creep;
>   And air-swept lindens yield
>   Their scent, and rustle down their perfumed showers
>   Of bloom on the bent grass where I am laid,
>   And bower me from the August sun with shade;
>   And the eye travels down to Oxford's towers.

> And near me on the grass lies Glanvil's book—
>   Come, let me read the oft-told tale again!
>   The story of the Oxford scholar poor,
>   Of pregnant parts and quick inventive brain,
>   Who, tired of knocking at preferment's door,
>   One summer-morn forsook
>   His friends, and went to learn the gipsy-lore,
>   And roam'd the world with that wild brotherhood,
>   And came, as most men deem'd, to little good,
>   But came to Oxford and his friends no more.

> But once, years after, in the country-lanes,
>   Two scholars, whom at college erst he knew,
>   Met him, and of his way of life enquired;
>   Whereat he answer'd, that the gipsy-crew,

His mates, had arts to rule as they desired
   The workings of men's brains,
And they can bind them to what thoughts they will.
   'And I', he said, 'the secret of their art,
   When fully learn'd, will to the world impart;
But it needs heaven-sent moments for this skill'.

This said, he left them, and return'd no more.
   But rumours hung about the country-side,
   That the lost Scholar long was seen to stray,
   Seen by rare glimpses, pensive and tongue-tied,
   In hat of antique shape, and cloak of grey,
   The same the gipsies wore.
Shepherds had met him on the Hurst in spring;
   At some lone alehouse in the Berkshire moors,
   On the warm ingle-bench, the smock-frock'd boors
Had found him seated at their entering.

But, 'mid their drink and clatter, he would fly.
   And I myself seem half to know thy looks,
   And put the shepherds, wanderer! on thy trace;
And boys who in lone wheatfields scare the rooks
   I ask if thou has't passed their quiet place;
   Or in my boat I lie
Moor'd to the cool bank in the summer-heats,
   'Mid wide grass meadows which the sunshine fills,
   And watched the warm, green-muffled Cumner hills,
And wonder if thou haunt'st their shy retreats.

For most, I know, thou lov'st retired ground!
   Thee at the ferry Oxford riders blithe,
   Returning home on summer nights, have met
Crossing the stripling Thames at Bab-lock-hythe,
   Trailing in the cool stream thy fingers wet,
   As the punt's rope chops round;
And leaning backward in a pensive dream,
   And fostering in thy lap a heap of flowers
   Pluck'd in shy fields and distant Wychwood bowers,
And thine eyes resting on the moonlit stream.

And once, in winter, on the causeway chill
   Where home through flooded fields foot-travellers go
   Have I not pass'd thee on the wooden bridge,

Wrapt in thy cloak and battling with the snow,
    Thy face tow'rd Hinksey and its wintry ridge?
        And thou hast climb'd the hill,
And gain'd the white brow of the Cumnor range;
    Turn'd once to watch, while thick the snow-flakes fall,
    The line of festal light in Christ Church hall—
Then sought thy straw in some sequester'd grange.

But what—I dream! Two hundred years are flown
    Since first thy story ran through Oxford halls,
        And the grave Glanvil did the tale inscribe
That thou wert wander'd from the studious walls
    To learn strange arts, and join a gipsy-tribe;
        And thou from earth art gone
Long since, and in some quiet churchyard laid—
    Some country-nook, where o'er thy unknown grave
    Tall grasses and white flowering nettles wave,
Under a dark, red-fruited yew-tree's shade. . . .

From *The Scholar-Gipsy*, 1853.

🦌 *For the origin of the story, see page* 98.

THE LAST ENCHANTMENTS

Beautiful city! so venerable, so lovely, so unravaged by the fierce intellectual life of our century, so serene!

   'There are our young barbarians all at play!'

And yet, steeped in sentiment as she lies, spreading her gardens to the moonlight, and whispering from her towers the last enchantments of the Middle Age, who will deny that Oxford, by her ineffable charm, keeps ever calling us nearer to the true goal of all of us, to the ideal, to perfection,—to beauty in a word, which is only truth seen from another side. . . . Adorable dreamer, whose heart has been so romantic! who hast given thyself so prodigally, given thyself to sides and heroes not mine, only never to the Philistines! home of lost causes, and forsaken beliefs, and unpopular names, and impossible loyalties!

From *Essays in Criticism*, 1865.

NOTHING KEEPS THE SAME

How changed is here each spot man makes or fills!
    In the two Hinkseys nothing keeps the same;
        The village street its haunted mansion lacks,
    And from the sign is gone Sibylla's name,

And from the roofs the twisted chimney-stacks—
    Are ye too changed, ye hills?
See, 'tis no foot of unfamiliar men
    To-night from Oxford up your pathway strays!
    Here came I often, often, in old days—
Thyrsis and I; we still had Thyrsis then.

Runs it not here, the track by Childsworth Farm,
    Past the high wood, to where the elm-tree crowns
    The hill behind whose ridge the sunset flames?
    The signal-elm, that looks on Ilsley Downs,
      The Vale, the three lone weirs, the youthful Thames?—
      This winter-eve is warm,
    Humid the air! leafless, yet soft as spring,
      The tender purple spray on copse and briers!
      And that sweet city with her dreaming spires,
She needs not June for beauty's heightening.

Lovely all times she lies, lovely tonight!—
    Only, methinks, some loss of habit's power
    Befalls me wandering through this upland dim;
    Once pass'd I blindfold here, at any hour;
      Now seldom come I, since I came with him.
      That single elm-tree bright
    Against the west—I miss it! is it gone?
      We prized it dearly; while it stood, we said,
      Our friend, the Gipsy-Scholar, was not dead;
While the tree lived, he in these fields lived on.

But hush, the upland hath a sudden loss
    Of quiet!—Look, adown the dusk hill-side,
    A troop of Oxford hunters going home,
    As in old days, jovial and talking, ride!
      From hunting with the Berkshire hounds they come.
      Quick! let me fly, and cross
    Into yon farther field!—'Tis done; and see,
      Befalls me wandering through this upland dim,
      The orange and pale violet evening sky,
Bare on its lonely ridge, the Tree! the Tree!

              From *Thyrsis*, 1867.

APES *v.* ANGELS

🐾 *Science found many sceptics in mid-Victorian Oxford—'Physics without God', thought Edward Pusey of Christ Church, 'would be but a dull inquiry into certain meaningless phenomena'—and in 1860 occurred a famous confrontation between Thomas Huxley the scientist and 'Soapy Sam' Wilberforce, the fundamentalist Bishop of Oxford, concerning the theory of evolution. It happened at a meeting of the British Association in the brand new University Museum, and it began innocuously enough with an exchange of views from the floor:*

Mr. Gresley, an old Oxford don, pointed out that in human nature at least orderly development was not the necessary rule: Homer was the greatest of poets, but he lived 3000 years ago, and [mankind] has not produced his like.

Admiral Fitzroy was present, and said he had often expostulated with his old comrade of the *Beagle* for entertaining views which were contradictory to the First Chapter of Genesis.

Sir John Lubbock declared that many of the arguments by which the permanence of species was supported came to nothing, and instanced some wheat which was said to have come off an Egyptian mummy, and was sent to him to prove that wheat had not changed since the time of the Pharaohs; but which proved to be made of French chocolate.

W. H. Fremantle, *Charles Darwin, his Life Told, &c.*, 1892.

🐾 *The climax approached as the Bishop of Oxford rose to his feet and addressed a mocking refutation of the evolutionary doctrine directly at Huxley, one of its most eminent proponents:*

The Bishop spoke for full half an hour with inimitable spirit, emptiness and unfairness. In a light, scoffing tone, florid and fluent, he assured us that there was nothing in the idea of evolution; rock-pigeons were what rock-pigeons had always been. Then, turning to his antagonist with a smiling insolence, he begged to know, was it through his grandfather or his grandmother that he claimed his descent from a monkey?

On this Mr. Huxley slowly and deliberately arose. A slight, tall figure, stern and pale, very quiet and very grave, he stood before us and spoke those tremendous words—words which no one seems sure of now, nor, I think, could remember just after they were spoken, for their meaning took away our breath, though it left us in no doubt as to what it was. He was not ashamed to have a monkey for his ancestor; but he would be ashamed to be connected with a man who used great gifts to obscure the truth. No one doubted his meaning, and the effect was tremendous. One lady fainted and had to be carried out; I, for one, jumped out of my seat.

'Reminiscences of a Grandmother', in *Macmillan's Magazine*, 1898.

This celebrated exchange was succeeded by a moment of farce:

From the back of the platform emerged a clerical gentleman, asking for a blackboard. It was produced, and amid dead silence he chalked two crosses at its opposite corners, and stood pointing to them as if admiring his achievement. We gazed at him, and he at us, but nothing came of it, till suddenly the absurdity of the situation seemed to strike the whole assembly simultaneously. . . . Again and again the laughter pealed, as purposeless laughter is wont to do; under it the artist and his blackboard were gently persuaded to the rear, and we saw him no more. He was supposed to be an Irish parson, scientifically minded; but what his hieratics meant or what he wished to say remains inscrutable. . . .

William Tuckwell, *Reminiscences of Oxford*, 1900.

The evolutionists were generally thought to have won the day, though most of the ladies in the audience seem to have supported the Bishop, but four years later Benjamin Disraeli, in a speech at the Sheldonian, offered a last sally on behalf of 'Soapy Sam':

What is the question now placed before society with a glib assurance the most astounding? The question is this: Is man an ape or an angel? I, my Lord, I am on the side of the angels.

AT PLAY

In the 1860s Oxford undergraduates devised the sport of chair-tilting, in which participants propelled themselves backwards in arm-chairs from opposite sides of a quadrangle, meeting violently in the middle.

This tilting mania was, in my opinion, the swiftest and most absorbing mental epidemic that has ever overrun a city. Men really seemed to think and talk of nothing but their tournaments, and every object was referred to one single standard.

One man would ask another, 'What do you think of Smith, of Ch. Ch.?' to whom he had just been introduced, and would probably get for an answer, 'Oh! he's not much good; Jones on my staircase would knock him over like ninepins.' Or perhaps the opinion might be more favourable—'Smith of Ch. Ch.? First-rate! Saw him cut over half-a-dozen of our men without giving them a chance.' Or perhaps the opinion might run thus:—'What do I think of Smith? Why, I think he has the best chair in the 'Varsity.'

All such mental fevers, like those of the body, *must* run their course.

'An Old Oxonian', in *The Boys' Own Volume*, 1863.

THROUGH THE LOOKING-GLASS

Lewis Carroll, the author of Alice in Wonderland, *was really the Revd. C. L. Dodgson (1832–98), for twenty-six years a mathematics tutor at Christ Church: it is said that when Queen Victoria told him how much she had enjoyed* Alice *he delightedly sent her a copy of its predecessor,* A Syllabus of Plane Algebraical Geometry. *Throughout his career the fantastic in him overlapped the academic, as these extracts show:*

AN OXFORD TUTORIAL

It is the most important point, you know, that the tutor should be dignified and at a distance from the pupil, and that the pupil should be as much as possible *degraded*.

Otherwise, you know, they are not humble enough.

So I sit at the further end of the room; outside the door (which is shut) sits the scout; outside the outer door (also shut) sits the sub-scout; half-way downstairs sits the sub-sub-scout; and down in the yard sits the *pupil*.

The questions are shouted from one to another, and the answers come back in the same way—it is rather confusing till you are well used to it. The lecture goes something like this:

Tutor: What is twice three?
Scout: What's a rice-tree?
Sub-scout: When is ice free?
Sub-sub-scout: What's a nice fee?
Pupil (*timidly*): Half a guinea!
Sub-sub-scout: Can't forge any!
Sub-scout: Ho for Jinny!
Scout: Don't be a ninny! . . .

And so the lecture proceeds. Such is life.

From a letter.

To find the value of a given Examiner.

Example: A. takes in ten books in the Final Examination and gets a 3rd class; B. takes in the examiners, and gets a 2nd. Find the value of the Examiners in terms of books.

To continue a given series.

Example: A. and B., who are respectively addicted to Fours and Fives, occupy the same set of rooms, which is always at Sixes and Sevens. Find the probable amount of reading done by A. and B. while the Eights are on.

From *Notes by an Oxford Chiel*, 1874.

A SYLLOGISTICAL PROBLEM

1 Babies are illogical;
2 Nobody is despised who can manage a crocodile;
3 Illogical persons are despised.
Univ. 'persons'; a = able to manage a crocodile;
             b = babies;
             c = despised;
             d = logical.

Answer: Babies cannot manage crocodiles.

From *Symbolic Logic*, 1896.

THE TRANSATLANTIC VIEW

Two eminent Americans publish, in the same year, their impressions of Oxford University.

THE QUALITY OF THE EDUCATION

The logical English train a scholar as they train an engineer. Oxford is a Greek factory, as Wilton mills weave carpets and Sheffield grinds steel. They know the use of a tutor, as they know the use of a horse; and they draw the greatest benefit out of both. The reading men are kept by hard walking, hard riding and measured eating and drinking at the top of their condition, and two days before the examination do no work, but lounge, ride or run, to be fresh on the college doomsday. The effect of this drill is the radical knowledge of Greek and Latin, and of mathematics, and the solidity and taste of English criticism. Whatever luck there may be in this or that award, an Eton captain can write Latin longs and shorts, can turn the Court-Guide into hexameters, and it is certain that a Senior Classic can quote correctly from the *Corpus Poetarum*, and is critically learned in all the humanities. . . . Oxford sends out yearly twenty or thirty very able men, and three or four hundred well-educated men.

The diet and rough exercise secure a certain amount of old Norse power. A fop will fight, and, in exigent circumstances, will play the manly part. In seeing these youths, I believed I saw already an advantage in vigour and colour and general habit, over their contemporaries in the American colleges. No doubt much of the power and brilliancy of the reading men is merely constitutional or hygienic. With a hardier habit and resolute gymnastics, with five miles more walking, or five ounces less eating, or with a saddle and gallop of twenty miles a day, with skating and rowing matches, the American would arrive at as robust an exegesis, and as cheery and hilarious a tone.

Ralph Waldo Emerson, *English Traits*, 1856.

THE QUALITY OF THE STONE

How ancient is the aspect of these college quadrangles! So gnawed by time as they are, so crumbly, so blackened, and so grey where they are not black—so quaintly shaped, too, with here a line of battlement and there a row of gables; and here a turret, with probably a winding stair inside; and lattice windows, with stone mullions, and little panes of glass set in lead; and the cloisters, with a long arcade looking upon the green or pebbled enclosure. The quality of the stone has a great deal to do with the apparent antiquity. It is a stone found in the neighbourhood of Oxford, and very soon begins to crumble and decay superficially, when exposed to the weather; so that twenty years do the work of a hundred, so far as appearances go. If you strike one of the old walls with a stick, a portion of it comes powdering down. The effect of this decay is very picturesque. . . .

Nathaniel Hawthorne, *English Note-Books*, 1856.

THE CHOICE

We won't have Evans at any price,
And as for Price, O 'eavens.

Anon., on the candidates for the Mastership of Pembroke College, 1864.

🦎 *Evan Evans won, to be succeeded at his death by Bartholomew Price.*

🦥 *Reconciled at last to the presence of the railway, the University was up in arms again in 1865 when the Great Western Railway proposed to establish a carriage works in the city. Industry, it was argued, would wreck the tone of Oxford:*

Everybody knows what evils are apt to result when a University is placed in the midst of a great city, and the students are allowed unrestrained access to the population. Everybody knows the character of the students of Paris. If such a state of things, or anything approaching to it, were to arise in Oxford, I do not know who would wish to share the responsibility of bringing the youth of England away from their English homes.

Goldwin Smith, in a letter to the Daily News, 1865.

🦥 *The Reverend Richard Greswell, one of the few academics to support the carriage works proposal, argued that it would give local young people a better kind of employment than Oxford presently offered. At the University Press, he pointed out in a letter to* The Times, *the Bible presses were worked by children aged twelve to sixteen, from six in the morning to six at night, with an hour and a half for breakfast, one for dinner, and a half day on Saturdays:*

There is absolutely nothing to be learnt; the work required to be done, is done as perfectly at the end of their first day's service, as at the end of their fourth year, when . . . they are turned adrift by their Alma Mater, the University of Oxford (for an Alma Mater she is, or rather, or truly speaking, she *ought to be*, to these her youngest children), merely because, when they have reached the full age of 16, they cannot any longer be expected to be contented with the wages of a child.

🦥 *The University was unmoved by the plea, as was* The Times *itself, which thought the carriage works would 'disfigure an ancient and beautiful city and damage the interests of a national University', and the Great Western Railway set up the factory at Swindon instead.*

INJUNCTION, 1860s

The Meadow Keepers and Constables are hereby instructed to prevent the entrance into the Meadow of all beggars, all persons in ragged or very dirty clothes, persons of improper character or who are not decent in appearance and behaviour; and to prevent indecent, rude, or disorderly conduct of every description.

From a notice at the entrance to Christ Church Meadow.

ON THE RIVER, NINETEENTH CENTURY

Rowing on Oxford's two rivers, the Thames (or Isis) and the Cherwell, had always been a University pleasure. In the nineteenth century it was elevated into a fetish, as the annual college rowing championship, Eights Week, became at once the principal athletic and the happiest social occasion of the year.

CHEAP AND EXCELLENT

In good summer weather the river affords to the sturdy rower an excellent and yet a cheap amusement. As exercise, there can be nothing in the world better, and we specially advise every reading character to let no fine evening pass without having a good stout pull. The sport too is not to be despised. There are worse things than cutting down swiftly, amid light breezes and pleasant sunshine, to Sandford; quaffing a cup of Mrs. Davies's Anno Domini, accompanied with a few refreshing whiffs of fragrant Virginia, and then away home again, pulling like Trojans, bumping all opponents, hailed with cheers of victory from . . . both sides of the river.

J. Campbell, *Hints for Oxford*, 1823.

FEELING AT HOME

My father, of course, took me 'on the water'—his own favourite amusement. We were sculled down to Iffley, and he enjoyed paying the overcharge, 'eighteenpence each gentleman as went in the boat, and two shillings the man'; being overcharged made him feel that he was in Oxford. . . .

Mark Pattison, *Memoirs*, 1885.

'ALICE' IS BORN

All in the golden afternoon,
  Full leisurely we glide;
For both our oars, with little skill,
  By little arms are plied.
While little hands make vain pretence
  Our wanderings to guide.

Ah, cruel Three! In such an hour,
　Beneath such dreamy weather,
To beg a tale of breath too weak
　To stir the tiniest feather?
Yet what can one poor voice avail
　Against three tongues together?

Thus grew the tale of Wonderland:
　Thus slowly, one by one,
Its quaint events were hammered out—
　And now the tale is done,
And home we steer, a merry crew,
　Beneath the setting sun.

C. L. Dodgson (Lewis Carroll), *Alice in Wonderland*, 1865.

BUMPED!

The shouts come all in a heap over the water. 'Now, St. Ambrose, six strokes more.' 'Now, Exeter, you're gaining; pick her up.' The water rushes by, still eddying from the strokes of the boat ahead. Tom fancies now he can hear their oars and the workings of their rudder, and the voice of their coxswain. In another moment both boats are in the Gut, and a perfect storm of shouts reaches them from the crowd, as it rushes madly off to the left to the footbridge, amidst which, 'Oh, well steered, well steered, St. Ambrose!' is the prevailing cry. Then Miller, motionless as a statue till now, lifts his right hand and whirls the tassel round his head: 'Give it her now, boys; six strokes and we are into them.' Old Jervis lays down that great broad back, and lashes his oar through the water with the might of a giant, the crew catch him up in another stroke, the tight new boat answers to the spurt, and Tom feels a little shock behind him and then a grating sound, as Miller shouts, 'Unship oars, Two and Three,' and the nose of the St. Ambrose boat glides quietly up the side of the Exeter till it touches their stroke oar.

'Take care what you're coming to.' It is the coxswain of the bumped boat who speaks.

Dear readers of the gentler sex! you, I know, will pardon the enthusiasm which stirs our pulses, now in sober middle age, as we call up again the memories of this the most exciting sport of our boyhood (for we were but boys then, after all. . . .)

Thomas Hughes, *Tom Brown at Oxford*, 1861.

After a champagne dinner, each college lights a bonfire—sometimes, when materials fail, furniture and doors are burnt. . . . At about 10 o'clock the sky was illuminated with red lights, which cast on the old walls reflections as of fire, while the breeze every now and then brought the echo of the chorus of songs, which were bawled out like the songs soldiers sing on the evening of the capture of a town.

> J. Bardoux, tr. W. R. Barker, *Memories of Oxford*, 1899.

PRIORITIES

It is characteristic of the authorities at Oxford that they should consider a month too little for the preparation of a boat-race, and grudge three weeks to the rehearsals of one of Shakespeare's plays.

> George Bernard Shaw, in the *Saturday Review*, 1898.

VANDALISM

When in 1870 a group of Christ Church undergraduates removed a number of valuable statues from the college library and made a bonfire around them in Peckwater Quadrangle, they were rebuked by a severe editorial in The Times:

Astonishment mingled at first with incredulity, is the feeling with which the story that comes from Oxford has been received by the public, and especially by University men. The practical jokes of Undergraduates are sufficiently notorious, and have at times verged on sacrilege and misdemeanour, but this exceeds anything that lingers in the memory of the oldest inhabitant. It must go forth to the world that the most brutal and senseless act of Vandalism that has disgraced our time has been committed by members of the great Foundation of Christ Church, young men belonging to the higher classes of England, brought up in the midst of the most refined civilization, and receiving the most costly education that the country can provide. . . . Truth is stranger than fiction, even on a subject which has so much exercised the invention of novelists as University life.

C. L. Dodgson (Lewis Carroll), a don at Christ Church, stood up for his young men in a letter to the Observer:

The punishment awarded to the offenders was deliberated on and determined on by the Governing Body, consisting of the Dean, the Canons, and some 20 Senior Students. . . . When all was over, we had the satisfaction of seeing ourselves roundly abused in the papers on both sides, and charged with having been too lenient, and also with having been too severe. The truth is that Christ Church stands convicted of two unpardonable crimes—being great, and having a name. Permit me, as one who has lived here for thirty years and has taught for five-and-twenty, to say that, if the writer of your leading article has had an equal amount of experience in any similar place of education, and has found a set of young men more gentlemanly, more orderly, and more pleasant in every way to deal with, than I have found here, I cannot but think him an exceptionally favoured mortal.

As for the Dean, Henry Liddell, he was not at all surprised by the affair:

Young men of large fortune have little to fear from such penalties as we can impose. . . . The late Lord Lyttelton, who turned out a very steady, useful man, was the first who painted the Dean's Door. The late Lord Derby is believed to have been the ringleader of a party who pulled down the figure which still gives name to the fountain in the Great Quadrangle. The attack in my garden last summer . . . was led by two noble Lords, one of whom had never been a member of any University, the other did not belong to us but had graduated with honour from a College of high repute in the University and actually held, as he still holds, the position of a Lord of Her Majesty's Treasury.

Can it be a matter of surprise that, when such things receive such countenance, there should be individuals in each successive generation of wealthy undergraduates who think it a noble pastime to imitate and improve upon the freaks of their predecessors?

The statues survived anyway, though ever afterwards they were coated in limewash to hide the blackening of the bonfire.

KILVERT'S OXFORD

Francis Kilvert the diarist, who graduated from Wadham College in the early 1860s, nostalgically revisits his old University a decade later:

1874: THE SPIRIT OF THE DREAM

Rose early, missed the New College Matins at 7.30 by being a minute too late. Walked round the gardens in the green light of the great lime cloister and wandered round to Wadham gardens. All was usual, the copper beech still spread a purple gloom in the corner, the three glorious limes swept their luxuriant foliage flat upon the sward, the great poplars towered like a steeple, the laburnum showered its golden rain by the quiet cloisters and the wisteria still hung its blue flower clusters upon the garden wall. The fabric of the college was unchanged, the grey chapel walls still rose fair and peaceful from the green turf. But all else was altered, a change had come over the spirit of the dream. The familiar friendly faces had all vanished, some were dead and some were out in the world and all had gone away. Strange faces and cold eyes came out of the doorways and passed and repassed the porter's lodge. One or two of the College servants remembered my face still, almost all had forgotten my name. 'The place thereof shall know it no more.' I felt like a spirit revisiting the scenes of its earthly existence and finding itself strange, unfamiliar, unwanted.

1876: SHORTENING THE FOCUS

After dinner we went down to the river and saw the Boat races very well from the Queen's Barge. In [the] Meadows we overtook 'David' Laing, now Fellow of Corpus, and we came upon him again on board the Barge. David was in an odd excitable defiant mood and whilst walking backwards like a 'peacock in his pride' and declaring that he would rather be a drunkard than a teetotaller, because there would be some pleasure and satisfaction out of drink and drunkenness, he was very like to have got enough to drink and to have put his paradox to the test for he suddenly staggered as if he were really intoxicated, over-balanced himself, and nearly fell into the river. Then David suddenly became hospitable and invited us to breakfast on Saturday, but shortening his notice of invitation like a telescope he gradually brought us nearer to his view and heart and at last it was settled that we should breakfast with our old college friend in his rooms at Corpus tomorrow.

1876: THE OLD SCENE

. . . The old scene passed before my eyes like a familiar dream, the moving crowd upon the banks, the barges loaded with ladies and their squires, the movement of small boats, canoes and skiffs darting about the river, punts crossing with their standing freights of men huddled together, then the first gun booming from Iffley, people looking at their watches, the minute gun five minutes later and last the report which started the boats and told us they were off. Then the suspense, the listening, the straining of the eyes, the first movement in the distant crowd now seen to be running, the crowd pouring over the Long Bridges, the far away shouting rising into a roar as the first boat came round the point with the light flashing upon the pinion-like motion of the rising and falling oars, the river now alive with boats, the strain and the final struggle, the plash of oars, the mad uproar, the frantic shouting as the boats pass the flag scatheless, then the slow procession following, the victors rowing proudly in amongst plaudits from the barges and the shore while the vanquished come humbly behind.

Kilvert's Diary, edited by William Plomer, 1938.

'BRANCHY BETWEEN TOWERS'

Poetry at Oxford. It is a happy thing that there is no royal road to poetry. The world should know by this time that one cannot reach Parnassus except by flying thither. Yet from time to time more men go up and either perish in its gullies fluttering *excelsior* flags or else come down again with full folios and blank countenances. Yet the old fallacy keeps its ground. Every age has its false alarms.

So wrote Gerard Manley Hopkins as an undergraduate at Balliol in 1864. Oxford nevertheless helped him on his way to the sacred mount. Converted under the influence of the Oxford Movement to Roman Catholicism, he later returned to the city as a Catholic priest, and he wrote three sonnets to Oxford, this one in 1865:

> Thus, I come underneath this chapel-side,
> So that the mason's levels, courses, all
> The vigorous horizontals, each way fall
> In bows above my head, as falsified
> By visual compulsion, till I hide
> The steep-up roof at last behind the small
> Eclipsing parapet; yet above the wall
> The sumptuous ridge-crest leave to poise and ride.
> None besides me this bye-ways beauty try.
> Or if they try it, I am happier then:
> The shapen flags and drillèd holes of sky
> Just seen, may be to many unknown men
> The one peculiar of their pleasured eyes,
> And I have only set the same to pen.

Much more often quoted, though, is his celebration, written in 1879, of Duns Scotus's Oxford:

Towery city and branchy between towers;
Cuckoo-echoing, bell-swarmèd, lark-charmèd, rook-racked, river-
    rounded;

The dapple-eared lily below thee; that country and town did
Once encounter in, here coped and poised powers;

Thou hast a base and brickish skirt there, sours
That neighbour-nature thy grey beauty is grounded
Best in; graceless growth, thou hast confounded
Rural rural keeping—folk, flocks and flowers.

Yet ah! this air I gather and I release
He lived on; these weeds and waters, these walls are what
He haunted who of all men most sways my spirits to peace;

Of realty the rarest-veinèd unraveller; a not
Rivalled insight, be rival Italy or Greece;
Who fired France for Mary without spot.

THE JOWLER

*The cherubic Benjamin Jowett (1817–93) became Master of Balliol in
1870, and dedicating himself to the production of a cultivated and
responsible élite—'inoculating England with Balliol'—made the college a
unique factor in British public life. Scholarly, courageous, often intimi-
dating and sometimes cruel, for many years the passionate pen-friend of
Florence Nightingale, 'The Jowler' saw Oxford essentially as a place of
education rather than research, but as his pupil Augustus Hare phrased it, he
was 'at once the terror and the admiration of those he wished to be kind to':*

During the whole four years I was up I scarcely passed a term without
being invited to a solitary meal with him, either breakfast or dinner.
The time passed in almost complete silence. Now and again I used to
venture an embarrassed remark, but as likely as not the reply would
be, 'You wouldn't have said that if you'd stopped to think', and after a
silence more glacial still he dismissed me with a brief 'Good morning'.

Then there were those formidable occasions when, six together,
you read essays to him on philosophical subjects he had set, and again
silence was his weapon. You read your essay feeling a warm glow at
the eloquence of its closing passage. Sometimes he rewarded you with
a brief 'Good essay' or 'Fair essay', but there were other occasions
when he looked at you for an interminable minute and then slowly
shifted his gaze to your neighbour and said 'Next essay, please.'

J. A. Spender, in *Oxford*, 1936.

When the essay was over, Jowett made tea, or drank a glass of wine with me—far more often we had tea of the uncomfortable college sort, lukewarm, out of a large metal pot, in big clumsy cups. Conversation did not flow. Occasionally the subject of the essay led to some remarks from Jowett, but rarely. More often there was spasmodic talking about things in general—Jowett never suggesting a topic—I blunderingly starting one hare after another—meeting silence or a quenching utterance—feeling myself indescribably stupid, and utterly beneath my own high level, but quitting the beloved presence with no diminution of an almost fanatical respect. Obscurely, but vividly, I felt my soul grow by his contact as it had never grown before. That was enough, and more than enough.

J. A. Symonds (1807–71).

I was profoundly grateful to Mr. Jowett, but being constantly asked to breakfast alone with him was a terrible ordeal. Sometimes he never spoke at all, and would only walk round the room looking at me with unperceiving, absent eyes as I ate my bread and butter, in a way that, for a very nervous boy, was utterly terrifying. Walking with this kind and silent friend was even worse: he scarcely ever spoke, and if, in my shyness, I said something at one milestone, he would make no response at all till we reached the next, when he would say abruptly, 'Your last observation was singularly commonplace.'

A. J. C. Hare, *The Story of My Life*, 1896.

The youth faltered in his easy speech; he paused for comment and none came; he began again, and was the more pert perhaps for the effort to stifle his growing uneasiness. At last conscious that his language was less and less effective, he stopped short. Still the little gentleman by the fireplace seemed to be rapt in contemplation of his own little square-toed shoe and the little piece of bluish knitted sock which was visible between the shoe and the black trouser. Still he stood sideways and gave small chance of reading his soft, enigmatic countenance. Stephen perceiving with impatience that his chief might be considering his case, or a difficult passage in the *Phaedrus*, or the price of vegetables as supplied to the College kitchen, found this characteristic silence intolerable. He was obliged to speak, and he spoke with unconcealed irritation. 'Anyway,' he said sharply, 'I'm doing no good here.' The little gentleman did not even shift his shoe from the fender. In a clear, passionless, high tone he said, 'You will do no good anywhere.'

It was like the chant of a little rosy choir-boy; but it stung the youth to fury. He made for the door with his teeth clenched. But unluckily for him a quick temptation to further speech seized him.

With his hand on the handle he turned; the clenched teeth parted, and with concentrated bitterness he spoke. 'If', said he, 'I were going to be a duke or the Ireland scholar, you would take some interest in my—my career.' Clear and high came the answer, brief and clear, 'Yes!'

Julian Sturgis, Stephen Calinari, 1901.

Jowett, who introduced Hegelian philosophy to Oxford, and whose religious views were formidably outspoken, made many enemies and was by no means universally admired:

There was no clinch in his mind. He would have doubted and kept other people doubting for ever. Whatever was advanced, his first impulse was always to deny.

Goldwin Smith, Reminiscences, 1910.

I never could have fully appreciated a man who was both a Platonist and no sportsman. It is told of him that he wouldn't believe that trout lay with their heads up stream!

Hely Hutchinson Almond, in a letter, 1894.

I have always admired Mr. Jowett's wonderful reticence and refinement coupled with sternness and *swift, decided action* when needful, in cases where moral corruption called for drastic measures. At the same time he never seemed to give any man up as hopeless, or beyond the reach of sympathy and help. It was different as regards unhappy or vicious women. Here his somewhat defective *experience* was a disadvantage.

Josephine Butler (1828–1906), in a letter.

> If only the good were clever,
>   If only the clever were good,
> The world would be better than ever
>   We thought that it possibly could.
> But, alas, it is seldom or never
>   That either behave as they should;
> For the good are so harsh to the clever,
>   The clever so rude to the good.

Elizabeth Wordsworth, Poems and Plays, 1931.

JOWETT'S REMARKS

On being asked what his lady-love, Florence Nightingale, was like:

Violent, very violent.

On observing the approaching denouement of a dubious after-dinner story at his table:

Shall we adjourn this conversation to the drawing-room?

On religion, to a young lady:

You must believe in God, my dear, despite what the clergymen say.

On a suggestion that the Newdigate prize poem should be written in blank verse:

You can't get reason out of young men, so you might as well get rhyme.

On 'Soapy Sam' Wilberforce, Bishop of Oxford:

Samuel of Oxford is not unpleasing if you will resign yourself to being semi-humbugged by a semi-humbug.

On seeing evidence of a rag, 'even to the extent of broken windows':

Ah, the mind of the college is still vigorous: it has been expressing itself.

On Cardinal Newman:

His conscience had been taken out, and the Church put in its place.

On truth:

Very often troublesome, but the world cannot get on without it.

On truth again:

What is Truth against an *esprit de corps*?

On his alleged snobbery:

Anyone who tries to get hold of young men of rank or wealth must expect to be accused of snobbishness, but one must remember how important it is to influence towards good those who are going to have an influence over hundreds of thousands of other lives.

On failure:

I have a general prejudice against all persons who do not succeed in the world.

On life:

Never retreat. Never explain. Get it done and let them howl.

Don: A priest is more important than a judge. A judge can only say 'You be hanged', but a priest can say 'You be damned'.
Jowett: Yes, but if a judge says 'You be hanged', you *are* hanged.

The President of Corpus Christi College: Master, I must congratulate you on the appearance of your new volume of Plato. May I send you a few suggestions?

Jowett: Please don't.

Jowett (*on being read a new poem by the Poet Laureate*): I think I wouldn't publish that, if I were you, Tennyson.

Tennyson: Well if it comes to that, Master, the sherry you gave us at luncheon was beastly.

FRANK BUCKLAND

🦥 *Francis* (*Frank*) *Buckland the naturalist* (1826–80), *was the son of Professor William Buckland, Oxford's first Professor of Geology* (*page 191*), *and his peculiar upbringing in the family home at Christ Church made him almost as eccentric as his father:*

Besides the stuffed creatures which shared the hall with the rocking-horse, there were cages full of snakes, and of green frogs, in the dining-room, where the sideboard groaned under successive layers of fossils, and the candles stood on ichthyosauri's vertebrae. Guinea-pigs were often running over the table; and occasionally the pony, having trotted down the steps from the garden, would push open the dining-room door, and career round the table, with three laughing children on his back, and then, marching through the front door, and down the steps, would continue his course around Tom Quad.

In the stable yard and large wood-house were the fox, rabbits, guinea-pigs and ferrets, hawks and owls, the magpie and jackdaw, besides dogs, cats, and poultry, and in the garden was the tortoise (on whose back the children would stand to try its strength), and toads immured in various pots, to test the truth of their supposed life in rock-cells.

G. C. Bompas, *Life of Frank Buckland*, 1885.

🦥 *Buckland himself vividly remembered an episode with a turtle:*

A live turtle was sent down from London, to be dressed for the banquet in Christ Church Hall [in honour of the Duke of Wellington, 1834]. My father [Professor Buckland] tied a long rope around the turtle's fin, and let him have a swim in 'Mercury', the ornamental water in the middle of the Christ Church 'Quad', while I held the string. I recollect, too, that my father made me stand on the back of the turtle while he held me on (I was then a little fellow), and I had a ride for a few yards as it swam round and round the pond. As a

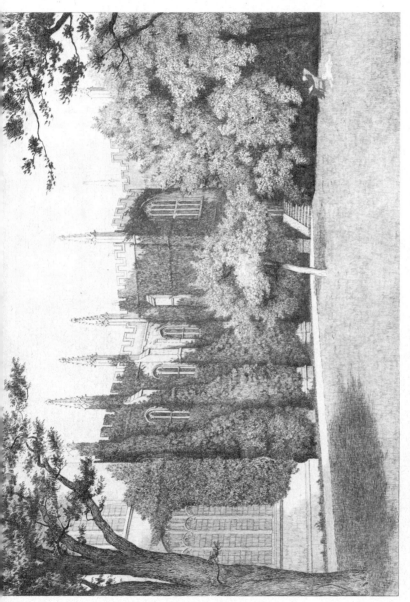

The Divinity School mostly obscured by ivy; above are Duke Humphrey's Library and the Selden End of the Bodleian Library, 1886

treat I was allowed to assist the cook to cut off the turtle's head in the college kitchen. The head, after it was separated, nipped the finger of one of the kitchen boys who was opening the beast's mouth.

🦊 *As an undergraduate Buckland continued the tradition, as William Tuckwell recalled in his* Reminiscences of Oxford, *1900:*

Few men can now recall those unique breakfasts in Frank's room in the corner of Fell's Buildings; the host, in blue pea-jacket and German student's cap, blowing blasts out of a tremendous wooden horn; the various pets who made it difficult to speak or move; the marmots, and the dove, and the monkey, and the chameleon, and the snakes, and the guinea-pigs; the after-breakfast visits to the eagle, or the jackal, or the pariah dog, or Tiglath-pileser the bear, in the little yard outside. 'Why Tiglath-pileser?' . . . Thus it was. On a certain morning in May the bear escaped from Buckland's yard, and found his way into the chapel, at the moment when a student was reading the first lesson, 2 Kings xvi, and had reached the point at which King Ahaz was on his way to meet Tiglath-pileser, King of Assyria, at Damascus. . . . The bear made straight for the Lectern, its occupant fled to his place, and the half-uttered name on his lips was transferred to the intruder.

🦊 *Other episodes in Buckland's incorrigible student career were remembered by his contemporaries in Bompas's biography:*

One day I met Frank just outside Tom Gate. His trowsers' pockets were swollen out to an enormous size; they were full of slow-worms in damp moss. Frank explained to me, that this combination of warmth and moisture was good for the slow-worms, and that they enjoyed it. They were certainly very lively, poking their heads out incessantly, while he repressed them with the palms of his hands.

I was in chapel on that Sunday morning when [his] eagle came in at the eight o'clock service. The cloister door had been left open, and the bird found its way into the church, while the Te Deum was being sung, and advanced with its wings nearly spread out. Two or three men left their places to deal with it; Dean Gaisford looked unspeakable things.

He came down to me one day for the purpose of telling me what he had for dinner the day before—namely, panther chops! He was a great friend of the curator of the then existing Surrey Zoological Gardens. From him Frank heard one day that the panther was dead. 'I wrote up at once', he said, 'to tell him to send me down some chops. It had, however, been buried a couple of days, but I got them to dig it up and send me some. *It was not very good.*'

One evening when I was devoting an hour to coaching him up for his littlego, I took care to tuck up my legs, in Turkish fashion, on the sofa for fear of a casual bite from the jackal which was wandering about the room. After a time I heard the animal munching up something under the sofa, and was relieved that he should have found something to occupy him. When our work was finished, I told Buckland that the jackal had found something to eat under the sofa. 'My poor guinea-pigs!' he exclaimed. . . .

🦊 *Buckland's love for, and ease with, animals lasted all his life, and his last words were reputed to be these:*

I suppose I shall see many strange creatures there.

WILDE OF MAGDALEN

🦊 *Oscar Wilde entered Magdalen College in 1874, having already taken a degree at Trinity College, Dublin. His Oxford career, which included winning the Newdigate Poetry Prize, was one of insouciant and sometimes arrogant ease.*

While lying in bed on Tuesday morning with Swinburne (a copy of) was woken up by the Clerk of the Schools to know why I did not come up. I thought I was not in till Thursday. About one o'clock I *nipped* up and was ploughed immediately in Divinity and then got a delightful *viva voce*, first in the Odyssey, where we discussed epic poetry in general, *dogs*, and women. Then in Aeschylus where we talked of Shakespeare, Walt Whitman and the *Poetics*. He had a long discussion about my essay on Poetry in the Aristotle paper and altogether was delightful. Of course I knew I had got a First, so swaggered horribly.

From a letter, July 1876.

A story is told of a viva voce examination, in which one of the subjects was the New Testament in Greek. My father [Wilde] . . . had not even troubled to look at it, and the examiner, suspecting this and being anxious to teach my father a lesson, told him to turn to Chapter 27 of the Acts of the Apostles and to start translating. This chapter is probably the most difficult in the whole of the New Testament, being the description of St. Paul's shipwreck on his way to Italy; it contains a number of obscure nautical terms which no one could be expected to know unless they had studied them. My father translated it perfectly, and when the foiled examiner told him that he had done enough, he replied: 'Please may I go on? I want to see what happened. . . .'

Vyvyan Holland, *Son of Oscar Wilde*, 1954.

It was impossible to overlook him in any company of College men. He might be disliked by some as a poseur and as being conceited and affected, but he was a brilliant talker and said clever things and could, if he tried, make himself a pleasant companion to the ordinary under-graduate, and he had a reputation of being something out of the common. So, speaking of my own time, I would say that the College was proud of him. . . . It may have been different afterwards.

W. W. Ward, quoted in Vyvyan Holland's *Son of Oscar Wilde*, 1954.

'Let's go and rag Wilde and break some of that furniture he's so proud of.' No sooner said than done. Three or four inebriated in-truders burst into their victim's room, the others followed up the stairs as spectators of the game. To the astonishment of the beholders, number one returned into their midst propelled by a hefty boot thrust down the stairs; the next received a punch in the wind that doubled him up on top of his companion below; a third form was lifted up bodily from the floor and hurled on to the heads of the spectators. Then came Wilde triumphant, carrying the biggest of the gang like a baby in his arms.

Frank Benson, *My Memoirs*, 1930

Wilde was rusticated for a term in 1877 for returning to college a month late after a journey to Greece—'sent down', he liked to say, 'for being the first undergraduate to visit Olympia'. But he loved Oxford—'the most beautiful thing in England . . . nowhere else are life and art so exquisitely blended'—and expressed his affection touchingly in The Burden of Itys:

Ah! the brown bird has ceased: one exquisite trill
  About the sombre woodland seems to cling
Dying in music, else the air is still,
  So still that one might hear the bat's small wing
Wander and wheel above the pines, or tell
Each tiny dew-drop dripping from the bluebell's brimming cell.

And far away across the lengthening wold,
  Across the willowy flats and thickets brown,
Magdalen's tall tower tipped with tremulous gold
  Marks the long High Street of the little town,
And warns me to return: I must not wait.
Hark! 'tis the curfew booming from the bell in Christ Church gate.

Wilde was disappointed not to be offered a Magdalen Fellowship, but when asked once what he proposed to be when he had taken his degree, replied:

God knows; I won't be an Oxford don, anyhow. I'll be a poet, a writer, a dramatist. Somehow or other I'll be famous, and if not famous, I'll be notorious.

THE CRAMMER, *c.* 1880

🦁 *The freelance academic coach, supplementing college teaching, was at his most prosperous in the last years of the nineteenth century, when hundreds of not very bright undergraduates needed his help to get through their examinations:*

There was a famous pass-coach, Morris of Jesus, to whom the derelicts of the college lecture-rooms went in gangs. . . . He was a really great amateur-actor; and he would try histrionic arts on his feeble-minded flock, passing before one of the feeblest, and exclaiming with a tragic gesture, 'Good God, that such a man should live!' The victim would be roused to greater efforts, and Morris's popularity grew. He also made humorous contracts with his clients: he agreed with them that they should pay double fees if they got through and nothing if they failed. . . . Also he made them sign a contract that they would work in his coaching-rooms eight hours a day, neither more or less. To such men this was really a contract of slavery; and he worked it like a master of galley-slaves. He was hospitably entertaining me once with a tête-à-tête supper in his rooms, to which he invited me that he 'might show me some of his methods'; after a cheerful meal of oysters and champagne, we were stretching our legs before the fire smoking our pipes, when about 11.30 p.m. I heard sighs and groans penetrating down from the room overhead, and I asked him what it meant. 'They are some of my eight-hours gang, doing their time', he replied. In a few minutes there was a knock on the door, and there entered a tall athletic youth with a haggard face and his hair wildly falling over his eyes, who gasped out piteously, 'Please, Sir, may I go home to bed?' Morris looked at his watch and sternly answered—'No. You have twenty minutes more to work before you go.'

Lewis R. Farnell, An Oxonian Looks Back, 1934.

TWO TRUTHS

🦁 *Two timeless truths about Oxford from Andrew Lang's* Oxford, *1882:*

This habit of carping, this trick of collecting notes, this inability to put a work through, this dawdling erudition, this horror of manuscripts, every Oxford man knows them, and feels those temptations which seem to be in the air. Oxford is a discouraging place. College drudgery absorbs the hours of students in proportion to their conscientiousness. They have only the waste odds-and-ends of time for their own labours. They live in an atmosphere of criticism. They collect notes, they wait, they dream; their youth goes by, and the night comes when no man can work.

That is the way of Oxford, a college is constantly rebuilding amid the protests of the rest of the University. There is no question more common, or less agreeable than this, 'What are you doing to your tower?' or 'What are you doing to your hall, library, or chapel?' No one ever knows; but we are always doing something, and working men for ever sit, and drink beer, on the venerable roofs.

ALL SOULS CANDIDATE, 1883

The term was only ten days old when the examination for the All Souls fellowship began. . . . The history papers suited me very well, as I was able to ramble round all manner of topics in ancient and modern times. My marked file shows me that I had a shot at the Greek conception of the State, the Roman legions under the Empire, the Anglo-Saxon conquest of Britain, the history of the Crusading States in the Levant, the social conditions of medieval Scotland, the Portuguese and Dutch colonies in the East Indies, the causes of the American War of Independence, and the claim of Napoleon to be the successor of Charlemagne. . . . There remained the paper of translation from five languages, where I found four of them easy enough. . . .

Charles Oman, *Memories of Victorian Oxford*, 1941.

PALSIED AND PUFFED

The difficult Rector of Lincoln College, Mark Pattison, who believed in research as the highest purpose of a university, was by no means satisfied with the effects of Oxford's mid-century reforms, as he demonstrated in his Memoirs, *1885:*

Since the reform of 1854 what superior talent there has been in Oxford has been pretty equally distributed among the colleges, as elections [to fellowships] have been made at haphazard, to no principle, and left to the chances of a vote. . . . In no common-room, so far as I know, is there now maintained a level of serious discussion, occupying itself with the great problems of speculation, or with the science or the literature of the day. Young M.A.'s of talent abound, but they are all taken up with the conduct of some wheel in the complex machinery of cram, which grinds down all specific tendencies and tastes into one uniform mediocrity. The men of middle age seem, after they reach thirty-five or forty, to be struck with an intellectual palsy, and betake themselves, no longer to port, but to the frippery work of attending boards and negotiating some phantom of legislation, with all the importance of a cabinet council. . . . Then they give each other dinners, where they assemble again with the comfortable assurance that they have earned their evening relaxation by the fatigues of the morning's committee. These are the leading men of our university, and who give the tone to it—a tone as of a lively municipal borough; all the objects of science and learning, for which a university exists, being put out of sight by the consideration of the material means of endowing them.

Our young men are not trained; they are only filled with propositions, of which they have never learned the inductive basis. From showy lectures, from manuals, from attractive periodicals, the youth is put in possession of ready-made opinions on every conceivable subject; a crude mass of matter, which he is taught to regard as real knowledge. Swollen with this puffy and unwholesome diet, he goes forth into the world regarding himself, like the infant in the nursery, as the centre of all things, the measure of the universe. He thinks he can evince his superiority by freely distributing sneers and scoffs upon all that does not agree with the set of opinions which he happens to have adopted from imitation, from fashion, or from chance.

THE OXFORD ETHOS

By the 1880s it was popular to talk of the Oxford ethos, the peculiar blend of traditions and influences that made the University, it was said, like no other. Here are three extracts to illustrate the conception:

I shall always remember the first essay that I ever took to Professor [Charles] Oman, and the devastating criticism that it evoked. He had told me to write an estimate of Cicero. He waved me into an armchair, and told me to read it to him. 'Marcus Tullius Cicero', I began, 'was born at Arpinum on January 3rd, B.C. 106.' 'No, never', cried my tutor, 'under any circumstances, begin an essay like that.' And he started me off on half a dozen different tracks. What did Cicero stand for? Was he a genuine politician? Was he a trimmer? Did he do good for the state or evil? 'Begin with an epigram, begin with a paradox, or begin with a demonstrably false premise and demolish it. But never, never, start off with such a dry and helpless statement as that "Marcus Tullius Cicero was born at Arpinum on January 3rd, B.C. 106." '

The whole of Oxford teaching was in that condemnation—ideas not facts, judgements not an index, life not death.

Arthur Waugh, *One Man's Road*, 1931.

The most permanent stamp of college reputation is the social stamp. This measure of worth often remains stationary under every variety of moral and intellectual change. To what stratum of society the undergraduates of any college belong, what degree of connection they have with the upper ten, what social status they maintain—this is the most patent fact about any college, and one which will never cease to be influential upon the choice of parents in this country, where the vulgar estimate of people by income and position is the universal and only standard of merit. As we judge of families in our neighbourhood, so when we are placing a boy at the University do we judge of colleges.

Mark Pattison, *Memoirs*, 1885.

Amidst all the coarseness and roughness of Oxford there runs a wholesome and manly dislike of everything that is sickly, mean and effeminate, and there is also a tendency to associate effeminacy with other failings. The suspicion is on the whole not unfounded, and young men who are fond of feathers, fans and crockery had perhaps better seek some other place than an Oxford college for the gratification of their peculiar tastes.

James Pycroft, *Oxford Memories*, 1886.

DIFFERENT APPROACHES

Lord C——, of Jesus College, who eventually took a second class, strangely and absurdly broke down in his divinity on the first occasion. He was asked as to a King of Israel. He replied rightly 'Saul'. 'Quite right, sir.' 'Afterwards called Paul.' 'Stay, we are speaking of the Old Testament and you say "Saul, afterwards called Paul." Am I to understand that this Saul, King of Israel, was the same person that was afterwards called Paul?' 'Yes, certainly.'

'Then' (shutting the book) 'it is quite needless to continue this examination.'

Lord C—— was followed out of the school by his friends . . . who said: 'How could you be such a fool? "Saul" was all right; why did you not leave off a winner? You must have seen there was something wrong.'

'Never mind', he said, 'if the examiner's ideas and mine differed as widely as all that, there was no use in humbugging any more about the matter.'

James Pycroft, Oxford Memories, 1886.

EXPECTANCIES

Many of my contemporaries have naturally passed away. It has been calculated that on average an Oxford man lives for 33 years after taking his B.A.

> Henry Robinson (B.A., St. Alban Hall, 1842, died 1887), in *London Society*, 1887.

THE PILGRIM

Of Oxford I feel small vocation to speak in detail. It must long remain for an American one of the supreme gratifications of travel. The impression it produces, the emotion it stirs, in an American mind, are too large and varied to be compassed by words. It seems to embody with undreamed completeness a kind of dim and sacred ideal of the Western intellect. . . . No other spot in Europe, I imagine, extorts from our barbarous hearts so passionate an admiration.

🦁 *Nevertheless Henry James (1843–1916), who was eventually to be given an honorary degree by the University, did record his Oxford sensations in some detail. Here are two examples, the first from* A Passionate Pilgrim, *1875, the second from* English Hours, *1905:*

We repaired in turn to a series of gardens and spent long hours sitting in their greenest places. They struck us as the fairest things in England and the ripest and sweetest fruit of the English system. Locked in their antique verdure, guarded, as in the case of New College, by gentle battlements of silver-grey, outshouldering the matted leafage of undisseverable plants, filled with nightingales and memories, a sort of chorus of tradition; with vaguely-generous youth sprawling bookishly on the turf as if to spare it the injury of their boot-heels, and with the great conservative college countenance appealing gravely from the restless outer world, they seem places to lie down on the grass in for ever, in the happy faith that life is all an endless summer afternoon. This charmed seclusion was especially grateful for my friend, and his sense of it reached its climax, I remember, on one of the last of such occasions and while we sat in fascinated *flanerie* over against the sturdy back of Saint John's. . . . 'Isn't it all a delightful lie?' he wanted to know. 'Mightn't one fancy this the very centre point of the world's heart, where all the echoes of the general life arrive but to falter and die? Doesn't one feel the air just thick with arrested voices? It's well there should be such places, shaped in the interest of factitious need, invented to minister to the book-begotten longing for a medium in which one may dream unwaked and believe unconfuted; to foster the sweet illusion that all's well in a world where so much is damnable, all right and rounded, smooth and fair in this sphere of the rough and ragged, the pitiful unachieved especially, and the dreadful uncommenced. The world's made—work's over. Now for leisure! England's safe—now for Theocritus and Horace, for lawn and sky! What a sense it all gives of the composite life of the country and of the essential furniture of its luckier minds!'

This delightful spot [All Souls] exists for the satisfaction of a small society of Fellows who, having no obligation save toward their own culture, no care save for fine learning as learning, and truth as truth, are presumably the happiest and most charming people in the world. The party, invited to lunch, assembled first in the library of the college—a cool, grey hall, of very great length and height, with vast wall-spaces of rich-looking book titles, and statues of noble scholars set in the midst. Had the charming Fellows ever anything more disagreeable to do than to finger these precious volumes and then to stroll about together in the grassy courts in learned comradeship, discussing their precious contents? Nothing, apparently, unless it were to give a lunch at Commemoration in the dining-hall of the college. When lunch was ready there was a very pretty procession to go to it. Learned gentlemen in crimson gowns, ladies in bright finery, paired slowly off and marched in a stately diagonal across the fine, smooth lawn of the quadrangle, in a corner of which they passed through a hospitable door. But here we cross the threshold of privacy; I remained on the further side of it during the rest of the day. . . .

WOMEN OF OXFORD

❧ *Victorian Oxford was overwhelmingly male, not to say misogynist, and women's attempts to enter the University were greeted by the majority of dons at best with condescension, at worst with graceless hostility. In the 1830s a woman competed for the Newdigate Poetry Prize under a pseudonym, and won it. In 1873 A. M. A. H. Rogers, who had been offered an exhibition by Worcester College, was rejected when the initials turned out to stand for Annie Mary Anne Henley. Gradually, though, the women edged their way in, until by the 1880s there were two women's colleges, and women could sit for most university examinations—though they only got a diploma when they passed, not a degree. This miscellany illustrates their progress down the century:*

A DEFINITION

A woman is a creature that cannot reason and pokes the fire from the top.

Richard Whately, Oriel College (1787–1863).

TWO KINDS

The Oxford female is only of two kinds—prim and brazen. The latter we will not describe; the former seem to live in perpetual fear of being winked at, and are indescribable.

Samuel Sidney, *Rides on Railways*, 1851.

BEYOND THEM

❧ *John Ruskin, in 1871, would not allow women to his lectures as Professor of Art:*

I cannot let the bonnets in, on any conditions this term. The three public lectures will chiefly be on angles, degrees of colour-prisms (without any prunes) and other such things, of no use to the female mind and they would occupy the seats in mere disappointed puzzlement.

❧ *Ten years later he went to tea at Somerville Hall, and wrote in a birthday book:*

So glad to be old enough to be let come and have tea at Somerville, and to watch the girlies play at ball.

INFERIOR TO US!

We have lived to witness the proposed revokal of the law of Nature, which is also the law of GOD concerning Woman—so far as Women's

Education is concerned. A new and hitherto unheard-of experiment it seems is to be tried in this place: nothing less than the education of young Women *like* young Men and *with* young Men. Has the University seriously considered the inevitable consequences of this wild project? . . . Will none of you have the generosity or the candour to tell [Woman] what a very disagreeable creature, in Man's account, she will inevitably become? If she is to compete successfully with men for 'honours', you must needs put the classic writers of antiquity unreservedly into her hands—in other words, must introduce her to the obscenities of Greek and Roman literature. Can you seriously intend it? Is it then a part of your programme to defile that lovely spirit with the filth of old-world civilization, and to acquaint maidens in their flower with a hundred abominable things which women of any age,—(and men too, if *that* were possible),—would rather a thousand times be without? . . .

I take leave of the subject with a short Allocution addressed to the other sex. . . . Inferior to us GOD made you: and our inferiors to the end of time you will remain.

J. W. Burgon, from a sermon preached at New College, 1884.

YOUNG MARRIEDS

For nine years, till the spring of 1881, we lived in Oxford, in a little house north of the Parks in what was then the newest quarter of the University town. . . . We had many friends, all pursuing the same kind of life as ourselves and interested in the same kind of things. Nobody under the rank of a Head of a College, except a very few privileged Professors, possessed as much as a thousand a year. The average income of the new race of married tutor was not much more than half that sum. Yet we all gave dinner-parties and furnished our houses with Morris papers, old chests and cabinets, and blue pots. The dinner-parties were simple and short. At our own early efforts of the kind there certainly was not enough to eat. But we all improved with time; and on the whole I think we were very fair housekeepers and competent mothers. Most of us were very anxious to be up-to-date and in the fashion, whether in aesthetics, in housekeeping or education. But our fashion was not that of Belgravia or Mayfair, which indeed we scorned! It was the fashion of the movement which sprang from Morris and Burne-Jones. Liberty stuffs, very plain in line, but elaborately 'smocked' were greatly in vogue, and evening dresses 'cut square', or with 'Watteau pleats' were generally worn, and often in conscious protest against the London 'low dress' which Oxford—young married Oxford—thought both 'ugly' and 'fast'. And when we had donned our Liberty gowns we went out to dinner, the husband walking, the wife in a bath chair, drawn by an ancient member of an ancient and close fraternity—the 'chairmen' of old Oxford.

M. A. Ward (Mrs. Humphry Ward), *A Writer's Recollections*, 1918.

ASTONISHED FRENCHMAN

A French visitor, Jacques Bardoux, went to a reception at Somerville in 1895 and reported himself 'astonished':

No barrack-like *lycées*, no grated convent, but villas covered with creepers and separated by flower-decked lawns. . . . The lady student who showed me over [the college] hardly troubled to conceal the disdain my astonishment caused her. Here is the gymnasium with its parallel bars, its horse, and its bicycles; the drawing-room, plain but comfortable, and provided with pianos and violoncellos. As we went up a delightful staircase she asked me about the most famous French geometrician. I muttered the name of M. Poincaré, but acknowledged my ignorance. My guide's disdain became more marked. After showing me one or two rooms, regular nests, covered with rugs and hangings, strewn with pictures and artistic objects, the student handed me over to the care of one of her friends, who hastened to speak to me about the chief manuscript of the 'Chanson de Roland'. I made my escape, sufficiently dumbfounded, but deeply interested by all I had seen.

J. Bardoux, tr. W. R. Barker, *Memories of Oxford*, 1899.

TOO MANY?

Mr. Gladstone had misgivings:

When Mrs. Gladstone was in Oxford a lady spoke of her visit as a 'pleasant surprise'; 'Not at all, not at all, ma'am,' said the old man in a tragic voice, 'there are far too many ladies in Oxford already.'

'T.R.' in the *Cornhill Magazine*, May 1908.

FACETIOUS EXCHANGE

Ye Somervillian students, Ye ladies of St. Hugh's,
Whose rashness and imprudence Provokes my warning Muse,
Receive not with impatience, But calmly, as you should,
These simple observations—I make them for your good.

Why seek for mere diplomas And commonplace degrees,
When now—unfettered roamers—You study what you please,—
While Man in like conditions Is forced to stick like gum
Unto the requisitions of a *curriculum*?

When Proctors fine and gate you, If walking thro'· the town
In *pupillari statu* Without a cap or gown:
When gauds that now delight you Away you have to throw,
And sadly go *vestitu In academico*:

When your untried impatience Is treated every day
By rules and regulations: When academic sway
Your study's sphere belittles, You'll find that life, I fear
Is not completely skittles, Nor altogether beer.

> A. D. Godley, *Virginibus*, in the *Oxford Magazine*, 1896.

You *horrid* A.G.! You unnatural man!
  I don't like your verses *a bit*;
Our JUST ASPIRATIONS you ruthlessly ban,
  And this, Sir, you fancy is wit!

I scorn your contempt, and disdain your advice;
  I don't see your logical *ergo*;
And though I could be most uncommonly nice,
  I am now most *indignantly*
                                          VIRGO.

> C. E. Brownrigg, *A Retort*, in the *Oxford Magazine*, 1896.

DAMN HER!

I spent all my time with a crammer,
And then only managed a gamma,
  But the girl over there,
  With the flaming red hair,
Got an alpha plus easily—damn her!

> Anon., *c.* 1900.

🐝 *But to balance the record, two suggestions of early liberation:*

I am the Dean, this Mrs. Liddell,
She plays the first, I second fiddle;
  She is the Broad,
  I am the High—
We are the University.

> C. A. Spring-Rice, *c.* 1880.

'Mrs. Jenkyns and Master Balliol'

  Footman's announcement of the Master of Balliol and his wife, *c.* 1850.

CONGRATS

🦁 *The Kaiser Wilhelm I, an admirer of Oxford, kept in touch with its affairs through the medium of Professor Max Müller, Professor of Comparative Philology, but was evidently unfamiliar with its structure. This is the telegram he sent the scholar, who detested all athletics, after an Oxford victory in the University Boat Race against Cambridge in the 1880s:*

MY FELICITATIONS TO YOU AND YOUR CREW ON YOUR GALLANT VICTORY

NATURAL SCIENCE VIVA, *c.* 1890

Examiner: What is Electricity?

Candidate: Oh, Sir, I'm sure I have learnt what it is—I'm sure I *did* know—but I've forgotten.

Examiner: How very unfortunate. Only two persons have ever known what Electricity is, the Author of Nature and yourself. Now one of the two has forgotten.

Falconer Madan, *Oxford Outside the Guide-Books*, 1923.

BIBLIOGRAPHICAL NOTE

🦁 *Some Victorian fiction set in Oxford:*

Faucit of Balliol, H. C. Merivale.
Blake of Oriel, A. Sergeant.
Hugh Heron, Ch. Ch., R. St. John Tyrwhitt.
Jack Harkaway at Oxford, S. B. Hemyng.
Peter Priggins, the College Scout, J. T. J. Hewlett.
The College Chums, C. Lister.
The Mysteries of Isis, H. J. W. Buxton.
Prince Maskiloff, A Romance of Modern Oxford, A. E. Evans.
Bertha Goodall, A Romance of the Cherwell, W. L. Courtney.

🦁 *(Compare page 376)*

A WALK WITH A DON

The story of the undergraduate's nerve-wracking walk with the taciturn don is an Oxford perennial (see for instance page 195). This is one of the few examples in which the undergraduate wins:

The undergraduate F. E. Smith [first Earl of Birkenhead, 1872–1930] was not insolent on principle, but refused to be patronized by anyone, even if they were older and more distinguished than himself. One of the older dons was well known in Oxford for a peculiar trait. It afforded him an oblique pleasure to take clever undergraduates out for long walks, and give them enough conversational rope on which to hang themselves. He would walk in dead silence, and sooner or later the young man would think that courtesy compelled him to say something. Usually it was a banal opening remark. The don would then quickly snub him. One youth, who had read a little of the *Divine Comedy* in translation, thought to please the don by expressing his interest in Dante, but was at once crushed by the reply 'Don't you think that Dante is rather provincial?'

Smith was forewarned of this habit, and when the invitation to the walk reached him, his plan of campaign was ready. The couple started walking down the High Street. They made for Shotover Hill in unbroken silence. They walked for an hour: not a word was exchanged on either side: the don could stand it no longer. 'They tell me', he began in a curious, rather high voice, 'they tell me you're clever, Smith. Are you?' 'Yes', replied Smith, and fell silent again. They turned round and made for Oxford. Not a word was exchanged on the homeward journey. They reached the college gate. F.E. paused and held out his hand. 'Good-bye, Sir,' he said, 'I've so much enjoyed our talk.'

The Second Earl of Birkenhead, F.E., 1959.

OXFORD BOOKS

. . . As we were leaving he hinted
    That a student could hardly do less
Than see how the volumes were printed
    At the time-honoured Clarendon Press.
So I went there with scholarly yearning,
    And I gathered from kind Mr. Gell,
Some books were to stimulate learning,
    And some were intended to sell. . . .

Anon., Oxford Magazine, 1892.

DON, *c.* 1894

Aged don: Would you kindly pump up my bicycle tyre?

Undergraduate: Certainly: I'd better pump up the other one, too, while I'm about it.

Don: Oh thank you: but *are they not connected*?

Told by E. F. Carritt, Fifty Years a Don, 1960.

OYEZ!

Oyez, oyez, oyez! All manner of persons who are suit and service to the Court Baron of the Warden and Scholars of St. Mary's College, Winchester, commonly called New College in Oxford, now to be holden, or have been summoned to appear at this time and place, draw near and give your attendance, every man answering his name.

Oyez, oyez, oyez! All manner of persons who have appeared this day at the Court Baron of the Warden and Scholars of St. Mary's College of Winchester in Oxford, commonly called New College, may now depart, keeping their day and hour on a new summons. God save the Queen and the Lords of this Manor!

Opening and closing Proclamations during the annual Progress of the
Warden and Fellows of New College around their estates, c. 1896.

TO CHRISTMINSTER

🦁 *In Thomas Hardy's* Jude the Obscure (*1896*) *Oxford figures as Christ-minster, the hero's magical but disillusioning city of dreams:*

Turning on the ladder Jude knelt on the third rung, where, resting against those above it, he prayed that the mist might rise.

He then seated himself again, and waited. In the course of ten or fifteen minutes the thinning mist dissolved altogether from the northern horizon, as it had already done elsewhere, and about a quarter of an hour before the time of sunset the westward clouds parted, the sun's position being partially uncovered, and the beams streaming out in visible lines between two bars of slaty cloud. The boy immediately looked back in the old direction.

Some way within the limits of the stretch of landscape, points of light like the topaz gleamed. The air increased in transparency with the lapse of minutes, till the topaz points showed themselves to be the vanes, windows, wet roof slates, and other shining spots upon the spires, domes, freestone-work, and varied outlines that were faintly revealed. It was Christminster, unquestionably; either directly seen, or miraged in the peculiar atmosphere.

The spectator gazed on and on till the windows and vanes lost their shine, going out almost suddenly like extinguished candles. The vague city became veiled in mist. Turning to the west, he saw that the sun had disappeared. The foreground of the scene had grown funereally dark, and near objects put on the hues and shapes of chimaeras. He anxiously descended the ladder, and started homewards at a run. . . .

The day came when it suddenly occurred to him that if he ascended to the point of view after dark, or possibly went a mile or two further, he would see the night lights of the city. It would be necessary to come back alone, but even that consideration did not deter him, for he could throw a little manliness into his mood, no doubt.

The project was duly executed. It was not late when he arrived at the place of outlook, only just after dusk; but a black north-east sky, accompanied by a wind from the same quarter, made the occasion dark enough. He was rewarded; but what he saw was not the lamps in rows, as he had half expected. No individual light was visible, only a halo or glow-fog over-arching the place against the black heavens behind it, making the light and the city seem distant but a mile or so. . . .

He had heard that breezes travelled at the rate of ten miles an hour, and the fact now came into his mind. He parted his lips as he faced the north-east, and drew in the wind as if it were a sweet liquor.

'You,' he said, addressing the breeze caressingly, 'were in Christminster city between one and two hours ago, floating along the streets, pulling round the weather-cocks . . . and now you are here, breathed by me—you, the very same.'

Suddenly there came along this wind something towards him—a message from the place—from some soul residing there, it seemed. Surely it was the sound of bells, the voice of the city, faint and musical, calling to him, 'We are happy here!'

'Ah, young man,' he observed, 'you'd have to get your head screwed on t'other way before you could read what they read there.'

'Why?' asked the boy.

'O, they never look at anything that folks like we can understand,' the carter continued, by way of passing the time. 'On'y foreign tongues used in the days of the Tower of Babel when no two families spoke alike. They read that sort of thing as fast as a night-hawk will whir. 'Tis all learning there—nothing but learning, except religion. And that's learning too, for I never could understand it. Yes 'tis a serious-minded place. Not but there's wenches in the streets o' nights . . . You know, I suppose, that they raise pa'sons there like radishes in a bed? And though it do take—how many years, Bob?—five years to turn a lirruping hobble-de-hoy chap into a solemn preaching man with no corrupt passions, they'll do it, if it can be done, and polish un off like the workmen they be, and turn un out wi' a long face, and a long black coat and waistcoat, and a religious collar and hat, same as they used to wear in the Scriptures, so that his own mother wouldn't know un sometimes. . . . There, 'tis their business, like anybody else's.'

'But how should you know—'

'Now don't you interrupt, my boy. Never interrupt your senyers. Move the fore hoss aside, Bobby; here's somat coming . . . You must mind that I be a-talking of the college life. 'Em lives on a lofty level; there's no gainsaying it, though I myself med not think much of 'em. As we be here in our bodies on this high ground, so be they in their minds—noble-minded men enough, no doubt—some on 'em—able to earn hundreds by thinking out loud. And some on 'em be strong young fellows that can earn a'most as much in silver cups. As for music, there's beautiful music everywhere in Christminster. You med be religious, or you med not, but you can't help striking in your homely note with the rest. And there's a street in the place—the main street—that ha'n't another like it in the world. . . .'

It was a windy, whispering, moonless night. To guide himself he opened under a lamp a map he had brought. The breeze ruffled and fluttered it, but he could see enough to decide on the direction he should take to reach the heart of the place.

After many turnings he came up to the first medieval pile that he had encountered. It was a college, as he could see by the gateway. He entered it, walked round, and penetrated to dark corners which no lamplight reached. Close to this college was another; and a little further on another; and then he began to be encircled as it were with the breath and sentiment of the venerable city. When he passed objects out of harmony with its general expression he allowed his eyes to slip over them as if he did not see them.

A bell began clanging and he listened till a hundred-and-one strokes had sounded. He must have made a mistake, he thought: it was meant for a hundred.

When the gates were shut, and he could no longer get into the quadrangles, he rambled under the walls and doorways, feeling with his fingers the contours of their mouldings and carving. The minutes passed, fewer and fewer people were visible, and still he serpentined among the shadows, for had he not imagined these scenes through ten bygone years, and what mattered a night's rest for once? High against the black sky the flash of a lamp would show crocketed pinnacles and indented battlements. Down obscure alleys, apparently never trodden now by the foot of man, and whose very existence seemed to be forgotten, there would jut into the path porticoes, oriels, doorways of enriched and florid middle-age design, their extinct air being accentuated by the rottenness of the stones. It seemed impossible that modern thought could house itself in such decrepit and superseded chambers.

GREATS

At the end of the nineteenth century the final Classical Honours School at Oxford—'Greats'—stood at the height of its prestige: to have passed through it was the hallmark of a superbly educated man, and its graduates went on to rule the nation and, in that heyday of British imperialism, half the world too. The examination comprised lengthy translations into and out of Latin and Greek, together with historical and philosophical questions like these, taken from the examination papers in Trinity term, 1899:

Sketch the history of the Syracusan democracy between the fall of Thrasybulus in 466 B.C. and the accession of Dionysius I in 406 B.C.

Is it a fact that thought begins not with the term but with the judgement?

Describe the circumstances which led to the Bank Charter Act of 1844.

What were the leading characteristics of fourth-century tyranny?

To what extent does history confirm Machiavelli's views on mercenary armies?

In what respects has Aristotle's advance in psychology enabled him to improve on the moral theories of Plato?

What account can be given of our perception of distance?

What is the ground of the obligation to veracity?

Trace the history of the principle of betterment in the English system of local taxation.

Describe the relations of Rome with Numidia at different periods of history.

🦂 *The general assumption was that a man who had mastered this range of thought and theory could master anything, even the Fuzzy-Wuzzies.*

A HEROIC PORTFOLIO

🦂 *The second half of the Victorian era saw an outpouring of heroic poems about Oxford. A portfolio of five:*

THE NOBILITY OF AGE

> I have known cities with the strong-armed Rhine
> Clasping their mouldered quays in lordly sweep;
> And lingered where the Maine's low waters shine
> Through Tyrian Frankfort; and been fain to weep
> 'Mid the green cliffs where pale Mozella laves
> That Roman sepulchre, imperial Treves.
> Ghent boasts her street, and Bruges her moonlight square;
> And holy Mechlin, Rome of Flanders, stands,
> Like a queen-mother, on her spacious lands;
> And Antwerp shoots her glowing spire in air.
> Yet have I seen no place, by inland brook,
> Hill-top, or plain, or trim arcaded bowers,
> That carries age so nobly in its look,
> As Oxford with the sun upon her towers.

F. W. Faber (1814–63), *Aged Cities*.

THE HALOED DREAM

> For ever ancient, and for ever new,
> Oxford, thy courts and cloisters are a bower
> Whence thought—earth-shaking, and earth-shaping—drew
> Promise and power;
> Here coming rulers, who will one day wield
> Old empire's rod, our England's finest flower,
> Practise their prentice-hands in mimic strife,
> And playing field,
> Learning the Master's touch and Maker's life;
> While Isis ripples with its storied stream,
> And every rill a hope and haloed dream.

Frederick William Ward (b. 1843), from *Oxford the Dreamer*.

A HEROIC PORTFOLIO

About the august and ancient Square,
Cries the wild wind; and through the air,
The blue night air, blows keen and chill;
Else, all the night sleeps, all is still.
Now, the lone Square is blind with gloom;
Now on that clustering chestnut bloom,
A cloudy moonlight plays and falls
In glory upon Bodley's walls;
Now wildlier yet, while moonlight pales,
Storm the tumultuary gales.
O rare divinity of Night!
Season of undisturbed delight;
Glad interspace of day and day!
Without, a world of winds at play,
Within I hear what dead friends say.

Lionel Johnson (1867–1902), from *Oxford Nights*.

THE GREAT THINGS

City of weathered cloister and worn court;
   Grey city of strong towers and clustering spires;
Where art's fresh loveliness would first resort;
   Where lingering art kindled her latest fires.

That is the Oxford, strong to charm us yet,
   Eternal in her beauty and her past.
What though her soul be vexed? She can forget
   Cares of an hour: only the great things last.

Only the gracious air, only the charm,
   And ancient might of true humanities;
These nor assault of man, nor time, can harm;
   Not these, nor Oxford with her memories.

Ill times may be; she hath no thought of time;
   She reigns beside the waters yet in pride.
Rude voices cry; but in her ears the chime
   Of full, sad bells brings back her old springtide.

Like to a queen in pride of place, she wears
   The splendour of a crown in Radcliffe's dome.
Well, fare she well! As perfect beauty fares;
   And those high places, that are beauty's home.

Lionel Johnson (1867–1902), from *Oxford*.

THE SECRET

Know you her secret none can utter?
 Her of the Book, the tripled Crown?
Still on the spire the pigeons flutter
 Still by the gateway flits the gown:
Still in the street, from corbel and gutter
 Faces of stone look down.

Once, my dear—but the world was young then—
 Magdalen elms and Trinity limes—
Lissom the oars and backs that swung then,
 Eight good men in the good old times—
Careless we and the chorus flung then
 Under St. Mary's chimes!

Still on her spire the pigeons hover;
 Still by her gateway haunts the gown;
Ah, but her secret? you, young lover,
 Drumming her old ones forth from town,
Know you the secret none discover?
Tell it when you go down. . . .
   A. T. Quiller-Couch (1863–1944), from *Alma Mater*.

A PROCTOR REMEMBERS

As in 1896 it was far easier than it has since become to distinguish in the streets between a 'member' and a 'non-member', and between a respectable female and 'a character', the problems of public discipline were simplified; and when the quick-sighted and alert senior 'bull-dog' whispered to his chief as they were patrolling in the dark: 'Member, Sir, with character,' he was not likely to be mistaken.

On some evening when the streets had been specially unquiet, I was patrolling with a strong 'posse' of Proctor's men; and about midnight I was yearning to get back to rest, as our students had all retired and the city was quiet; but Carfax was still noisy as I approached it, and I found there a low female, seemingly half-drunk, amusing a hilarious group of citizens by very indecent dancing and singing, while a city policeman looked on grinning. As there were no University men about, I thought it was more his duty than mine to stop this disorder, and I told him so. But he was sullen or afraid, and said that it was for me to arrest the woman if I liked. I did so promptly, and we bore her away swiftly to the Proctor's quarters at the end of the Broad followed by a crowd of citizens, booing, but otherwise respectful.

Lewis R. Farnell, *An Oxonian Looks Back*, 1934.

BELLOC

Boisterous, bellicose and gregarious, a formidable debater at the Oxford Union, Hilaire Belloc, who entered Balliol College in 1894, remained all his life an Oxford Man. His work is scattered through these pages, but here is some more concentrated stuff:

THE FRESHMAN'S VISION

> The Freshman ambles down the High,
>   In love with everything he sees,
> He notes the clear October sky,
>   He sniffs a vigorous western breeze.
>
> 'Can this be Oxford? This the place'
>   (He cries) 'of which my father said
> The tutoring was a damned disgrace,
>   The creed a mummery, stuffed and dead?
>
> 'Can it be here that Uncle Paul
>   Was driven by excessive gloom,
> To drink and debt, and, last of all,
>   To smoking opium in his room?
>
> 'Is it from here the people come,
>   Who talk so loud, and roll their eyes,
> And stammer? How extremely rum!
>   How curious? What a great surprise!
>
> 'Some influence of a nobler day
>   Than theirs (I mean than Uncle Paul's),
> Has roused the sleep of their decay,
>   And decked with light these ancient walls.
>
> 'O! dear undaunted boys of old,
>   Would that your names were carven here,
> For all the world in stamps of gold,
>   That I might read them and revere.

[299]

'Who wrought and handed down for me,
 This Oxford of the larger air,
Laughing, and full of faith, and free,
 With youth resplendent everywhere.'

From the 'Dedicatory Ode' in *Lambkin's Remains*, 1900.

O FOR OXFORD

O stands for Oxford. Hail! salubrious seat
Of learning! Academical Retreat!
Home of my Middle Age! Malarial Spot
which People call Medeeval (though it's not).
The marshes in the neighbourhood can vie
With Cambridge, but the town itself is dry,
And serves to make a kind of Fold or Pen
Wherein to herd a lot of Learned Men.
Were I to write but half of what they know,
It would exhaust the space reserved for 'O';
And, as my book must not be over big,
I turn at once to 'P', which stands for Pig.

Moral

Be taught by this to speak with moderation
Of places where, with decent application,
One gets a good, sound, middle-class education.

From *A Moral Alphabet*, 1899.

LINES TO A DON

Remote and ineffectual Don
That dared attack my Chesterton,
With that poor weapon, half-impelled,
Unlearnt, unsteady, hardly held,
Unworthy for a tilt with men—
Your quavering and corroded pen;
Don poor at Bed and worse at Table,
Don pinched, Don starved, Don miserable;
Don stuttering, Don with roving eyes,
Don nervous, Don of crudities;
Don clerical, Don ordinary,
Don self-absorbed and solitary;
Don here-and-there, Don epileptic;
Don middle-class, Don sycophantic,
Don dull, Don brutish, Don pedantic;

Don hypocritical, Don bad,
Don furtive, Don three-quarters mad;
Don (since a man must make an end),
Don that shall never be my friend.

Don different from those regal Dons!
With hearts of gold and lungs of bronze,
Who shout and bang and roar and bawl
The Absolute across the hall,
Or sail in amply billowing gown
Enormous through the Sacred Town,
Bearing from College to their homes
Deep cargoes of gigantic tomes;
Dons admirable! Dons of might!
Uprising on my inward sight
Compact of ancient tales, and port
And sleep—and learning of a sort.
Dons English, worthy of the land;
Dons rooted; Dons that understand.
Good dons perpetual that remain
A landmark, walling in the plain—
The horizon of my memories—
Like large and comfortable trees.

From *Verses*, 1910.

OXFORD AND THE G.O.M.

There is not a man who has passed through this great and famous
University that can say with more truth than I can say that I love her,
I love her, I love her from the bottom of my heart.

*Never was there a more devoted son of Oxford than W. E. Gladstone
(1809–98) who graduated from Christ Church in 1831, and nobody has
expressed his devotion more emotionally—or complacently:*

To call a man an Oxford man is to pay him the highest compliment
that can be paid to a human being.

*As an undergraduate Gladstone was not only the most powerful debater
the Oxford Union had ever known, but he aimed deliberately at that crown
of Victorian student achievement, a 'double first'—first class honours both
in classics and in mathematics. The classical examination came first, and
he recorded its challenge in his diary:*

November 13. I am cold, timid and worldly, and not in a healthy state of mind for the great trial of tomorrow: to which, I know, I am utterly & miserably unequal. . . . God grant that He who gave himself even for me may support me through it, if it be his will: but if I am covered with humiliation, O may I kiss the rod.

November 14. Went into the schools at ten, and from this time was little troubled with fear. Examined by Stocker in divinity: I did not answer as I could have wished: Hampden in science—a beautiful examination and with every circumstance in my favour . . . Then followed a very clever examination in history from Garbett—and an agreeable and short one in my poets from Cramer, who spoke very kindly to me at the close. . . . Everything was in my favour: the examiners kind beyond any thing: a good many persons there, and all friendly—at the end of the science of course my spirits were much raised, and I could not help at that moment giving thanks inwardly to Him without whom not even such moderate performances would have been in my power.

November 16. On going into the schools, found a moral Essay on a very fine but very difficult subject. Wrought hard for five hours.

November 17. Six hours in the Schools—with Latin prose, Greek verse, and Latin verse. I have been examined in Butler, Phaedo, Rhetoric, Ethics, Homer, Iliad & Odyssey, Virgil, Persius, Aristophanes, Herodotus, Thucydides.

At the head of one of his papers, 'Question in Moral Philosophy', Gladstone wrote this:

It will be my endeavour, in the consideration of this question, to adhere as closely as possible, in point of division and arrangement, to the order suggested by the form in which it has been proposed. The principal heads may be thus briefly stated—

I To discuss 1. the absolute ⎫ compatibility of virtue and self-denial.
            2. the relative ⎭

II To contrast the conclusion thus obtained with the Stoical doctrine of ἀπάθεια.

    Also with a particular reference to Aristotle.

III By considering the nature and office of conscience, to inquire how far the question whether it is factitious or innate is affected by anything hitherto stated.

❧ *His paper was fourteen pages long, and he concluded it very Glad-stonianly:*

As much has been done as time and hesitation would allow me. There is not even time to reperuse.

❧ *At his viva voce Gladstone was attended by crowds of admirers, and legend says that when the examiners, having questioned him on one philosophical subject, remarked that they would now leave that topic, Gladstone indig-nantly interrupted them—'No, if you please, sir, we will not leave it yet!' On 24 November he learnt that he had got his first, and 'wrote a long letter home'. The second examination followed two weeks later:*

December 9. In the Schools six hours with two papers of Algebra & Geometry in which I succeeded better than I had any right to expect.

December 10. Differential Calculus & Algebra papers: did the former but ill: the latter as well as any hitherto.

December 11. God be praised for this day of rest.

December 12. Had Differential Calculus & Hydrostatics. Did better than I expected in the latter, and worse in the former.

December 13. Had Mechanics & Optics. Did but ill in the former: wretchedly in the latter.

December 14. Principia in the morning: considerably disappointed in my performances. Astronomy! (which I had crammed for about half an hour,) during a short time in aftn.

❧ *That same evening he learnt that he had got his double first, and 'felt the joy of release':*

How much thankfulness was due, and how little paid!—It was an hour of thrilling happiness, between the past & the future, for the future was not I hope excluded: and feeling was well kept in check by the bustle of preparation for speedy departure. . . . Had tea with Bruce—& left Oxford on the Champion.

❧ *Oxford brought out the Tory in Gladstone:*

If I am to look back upon the education of Oxford as it was, it taught the love of truth, it provided men with those principles of honour which were nowhere perhaps so much required as amid the temptations of political controversy. It inculcated a reverence for what is ancient and free and great . . . Perhaps it was my own fault, but I must admit that I did not learn, when at Oxford, that which I have learned since, viz., to set a due value on the imperishable and the inestimable principles of human liberty.

🐾 *As G. W. Kitchin reported in* Ruskin in Oxford (1904):

Mr. Gladstone held that the distinctions of the outer world should have their echo in Oxford; that it was a lesson in the structure of society; that it protected poor men from temptations to high expenditure.

🐾 *Later, all the same, as Member of Parliament for the University, he was an influential supporter of University reform, and he was always fascinated by the academic progress of Oxford. When he spent some days there as an old man in 1890 he was treated like royalty, and a book was compiled to celebrate the occasion (C. R. L. Fletcher, Gladstone at Oxford). He enjoyed every moment of it, thoughtfully congratulated the Fellows of All Souls on their claret, when the Junior Fellow served port by mistake, responded happily to the salutes of the cabmen lined up in High Street, and addressed the Union on a topic of his own choice—whether or not Homer was acquainted with the Babylonian religion. Like Dr. Johnson, he wore his gown on all occasions, as Sir William Anson recalled in Fletcher's book:*

I am sure that he regarded the less frequent use of academical dress as a sign of decadence in university life. On one night of his visit he went with me to dine at the Club, a dining society of twelve persons. . . . The member who entertained the Club on that evening was Dr. Bellamy, who was then Vice-Chancellor. Mr. G. started with me in full academical dress. I remarked that we did not wear gowns at the Club dinner, and he replied that in the presence of the Vice-Chancellor he must wear his gown. I did not pursue the subject. . . . When we entered the drawing-room at St. John's, Dr. Bellamy said at once, after the first greetings, 'Mr. Gladstone, you must take off your gown.' 'But,' said Mr. G., 'in the presence of the Vice-Chancellor—' 'Oh, no,' said Dr. Bellamy, 'we make no account of Vice-Chancellors in the Club. You must take off your gown.' 'Well,' said Mr. Gladstone, sadly, 'in this lawless assembly I suppose I must conform to its rules.'

🐾 *Gladstone was also taken aback by changes in undergraduate behaviour:*

I was almost shocked with the spectacle of men in boating costume, indeed I may say in very scanty costume, in the High Street. Such a thing would have been impossible in my time. . . . I remember contemporaries—young men at Christ Church—who, when they were not hunting, made a point of promenading the High Street in the most careful attire. And some of them kept a supply of breeches which they only wore for that purpose, and *in which they never sat down lest any creases should appear*: I confess I think the undergraduates now seem to have passed to the other extreme.

� *But he loved it all to the last, and Oxford responded in kind. In the week of his death in 1898 the Union was to hold its cheerful Eights Week Debate, on the motion 'That the Better Half rules the world', but for it was substituted the motion 'That in view of Mr Gladstone's death, this House do adjourn'. F. E. Smith moved it, with the words:*

We came here with jests upon our lips, and they have been frozen before they could find expression.

✱ *The President, winding up the debate, quoted Gladstone's own last message to the University:*

There is no expression of Christian sympathy that I value more than that of the ancient university of Oxford, the God-fearing and God-sustaining university of Oxford. I served her, perhaps mistakenly, but to the best of my ability. My most earnest prayers are hers to the uttermost and the last.

THE COLOSSAL IDEA

    Cecil Rhodes, to Promising Rhodesian Child: I'll send you to Oxford, my boy.
    Promising Rhodesian Child, hastily retreating: Oh no you won't.

✱ *Cecil Rhodes the African imperialist, 'The Colossus', had been an undergraduate at Oriel, and cherished a passionate admiration for the Oxford ethos (though he had been an undistinguished and indeed intermittent student—it took him eight years to get his degree, and the Provost of Oriel grumbled that 'All the colleges send me their failures'). In 1902 his will, inspired by this fervour and by convictions of historical destiny, established a series of Oxford scholarships for colonial, American, and German students. Here are its relevant portions:*

Whereas I consider that the education of young Colonists at one of the Universities in the United Kingdom is of great advantage to them for giving breadth to their views for their instruction in life and manners and for instilling into their minds the advantage to the Colonies as well as to the United Kingdom for the retention of the unity of the Empire. . . . And whereas I also desire to encourage and foster an appreciation of the advantages which I implicitly believe will result from the union of the English-speaking peoples throughout the world. . . . [And whereas] a good understanding between England Germany and the United States of America will secure the peace of the world and educational relations form the strongest tie,

Now therefore I direct my Trustees . . . to establish for male students the Scholarships hereinafter directed to be established each of which shall be of the yearly value of £300 [£250 for German Scholars] and to be tenable at any College in the University of Oxford for three consecutive academical years. . . .

My desire being that the students who shall be elected to the Scholarships shall not be merely bookworms I direct that in the election of a student to a Scholarship regard shall be had to

(i) his literary and scholastic attainments
(ii) his fondness of and success in many outdoor sports such as cricket football and the like
(iii) his qualities of manhood truth courage devotion to duty sympathy for the protection of the weak kindliness unselfishness and fellowship and (iv) his exhibition during school days of moral force of character and of instincts to lead and take an interest in his schoolmates.

As mere suggestions for the guidance of those who will have the choice of students for the Scholarships I record that . . . my ideal qualified student would combine these four qualifications in the proportions of three-tenths for the first third two-tenths for the second three-tenths for the third and two-tenths for the fourth qualification. . . .

No student shall be qualified or disqualified for election to a Scholarship on account of his race or religious opinions.

In private conversation Rhodes defined his criteria for selection more frankly:

You know I am all against letting the scholarships merely to people who swot over books, who have spent their time over Latin and Greek. But you must allow for that element which I call 'smug', and which means scholarship. That is to stand for four-tenths. Then there is 'brutality', which stands for two-tenths. Then there is tact and leadership, again two-tenths, and then there is 'unctuous rectitude', two-tenths. That makes up the whole. You see how it works.

From The Review of Reviews, May 1902.

THE COLOSSAL IDEA

The plan got a mixed reception from the indigenes. The Oxford Union passed a motion regretting it, and Max Beerbohm expressed a popular view in his novel Zuleika Dobson (*1911*):

The President [of the Junta Club] showed much deference to his guest. . . . To all Rhodes Scholars, indeed, his courtesy was invariable. He went out of his way to cultivate them. And this he did more as a favour to Lord Milner than of his own caprice. He found these Scholars, good fellows though they were, rather oppressive. They had not—how could they have?—the undergraduate's virtue of taking Oxford as a matter of course. The Germans loved it far too little, the Colonials too much. The Americans were, to a sensitive observer, the most troublesome—as being the most troubled—of the whole lot. The Duke was not one of those Englishmen who fling, or care to hear flung, cheap sneers at America. Whenever any one in his presence said that America was not large in area, he would firmly maintain that it was. He held too, in his enlightened way, that Americans have a perfect right to exist. But he did often find himself wishing Mr. Rhodes had not enabled them to exercise that right in Oxford.

Mixed feelings were apparent in Compton Mackenzie's Sinister Street, *1914:*

'We don't altogether know what attitude to take up over the Rhodes Bequest,' said Maurice. Then boldly he demanded from the Warden what would be the effect of these imposed scholars from America and Australia and Africa.

'The speculation is not without interest,' declared the Warden. 'What does Fitzroy think?'

Fitzroy . . . said he thought the athletic qualifications were a mistake. 'After all, sir, we don't want the Tabs—I mean to say we don't want to beat Cambridge with the help of a lot of foreigners.'

'Foreigners, Fitzroy? Come, come, we can scarcely stigmatize Canadians as foreigners. What would become of the Imperial Idea?'

'I think the Imperial Idea will take a lot of living up to,' said Wedderburn, 'when we come face to face with its practical expression. Personally I loathe Colonials except at the Earls Court Exhibition.'

'Ah, Wedderburn,' said the Warden, 'you are luckily young enough to be able to be particular. I with increasing age begin to suffer from that terrible disease of age—toleration.'

On the other hand some seers rightly foresaw attractive sporting prospects:

If Mr. Rhodes's trust should be the means of our getting some gigantic Colonials—or even Boers, for he excludes no race—who can do

[307]

great things, say, at putting the weight, we may be able to wipe out Cambridge altogether! All Oxonians would agree that that would be a great achievement.

W. T. Stead, *The Last Will and Testament of Cecil John Rhodes*, 1902.

🦎 *The advent of the Rhodes Scholars created a new genre in Oxford fiction, later to reach a majestic climax in a comic masterpiece of the cinema, Laurel and Hardy's* Chump at Oxford. *The first novel about a Rhodes Scholar was* Downy V. Green *by G. L. Calderon (1902), whose hero was supposed to be the American grandson of the Victorian fictional character Verdant Green (page 246). Here Colonel and Mrs. Cheney, from the United States, are taking tea with an Oxford Professor:*

'I'm sure everybody in Oxford ought to be vurry grateful to Cecil J. Rhodes for what he's done for the place,' [Mrs. Cheney] said.

'Grateful? We all—recognize the compliment, I assure you. It's a great opportunity.'

'It is indeed, Professor. And now, sir, as a man of the world and a member of the Professorial Corpus, I should be glad if you would tell me exactly what good you think Cecil J. Rhodes expected his Will would do?'

'It is very plain, surely, my dear madam? Oxford is the representative of a—a very high form of culture; and Mr. Rhodes evidently wished the rest of the world to participate, so far as they were capable, *in* that culture.'

'Then you think you're goin' to civalize the world?'

'That is surely the only way of looking at it?'

'No, sir! You've gotten hold of the wrong end of the toastin' fork this time.'

'Sairey! Sairey! go stiddy!' interposed the Colonel.

'The object of the Cecil J. Rhodes's Will is to civalize Oxford by the infiltration of the American element.'

'Civilize Oxford? Ha, ha! Very good! With your permission, I shall repeat that at the High Table. How they will laugh!'

'You mustn't mind my wife, sir; she will have her joke.'

'Ring off, Colonel!'

'But, seriously, my dear madam, the idea of civilizing Oxford is what is vulgarly called "rather a large order", isn't it?'

'There I entirely agree with you, Professor . . . Civalization is Life! livin' better, doin' better, thinkin' better. And that's what I do not find in Oxford. No, sir; Oxford will need some tittavatin' before you can make it the hub of the Universe!'

🦎 *Though in fact Oxford was to have few more grateful sons than American*

War: Oxford at full moon, 1940

Peace: Oxford from Wytham Woods, 1945

Rhodes Scholars, still the theme of the American Oxonian scorned but eventually triumphant was so pervasive that our section on Mr. Rhodes's visionary scheme may properly end with a prime example, from James Childers's Laurel and Straw, *1927:*

Calthrope and his two friends sauntered into the quadrangle. Steele walked over to them.

'Pardon me, Calthrope, but may I speak with you a moment?'

The Englishman did not reply. He merely leaned back on his cane and dropped his left hand upon his hip. His yellow hair blew about his forehead and his jaw sagged open. He eyed Steele as one might consider an insect.

'Will you let me speak with you privately?' repeated Steele.

'Er-er-er-ah this will do. What jew desire? A few headache powders?'

Steele's fist doubled and his heart pounded. He bit his lip and swallowed. He had endured enough for one day, and entirely too much from this man; English or whatever he might be.

'Look here, Calthrope, I damn well know what you're sore about. Just let me tell you this—You're making a fool of yourself.' Steele's voice was low. But his face was splotched and there was an ugly twitching around the corner of his mouth. He moved closer. 'The person you're thinking about is too damn good for you even to look at. Furthermore, you're going to lay off this chatter about my friend in Paris. Get me?'

The Englishman's lip curled.

'Has your friend recovered yet?'

'All right, keep on. And I swear to God I'm going to crack you one.'

Calthrope stared coldly before him.

'How—how bally American.'

Almost before the last syllable was out the Englishman's mouth, Steele sunk his fist into it.

CHILDLIKE DONS

Besides founding his Oxford scholarships Cecil Rhodes the imperialist left £100,000 to his own college, Oriel, suggesting that his trustees should be consulted about its investment 'as the College authorities live secluded from the world and so are like children as to commercial matters'. W. T. Stead commented, in his Last Will and Testament of Cecil John Rhodes, *1902:*

Possibly Cecil Rhodes was thinking . . . of a story current in his day at Oriel—and current still—of John Keble, who was better at Christian poetry than at worldly calculation. One day Keble, who was Bursar, discovered to his horror that the College accounts came out nearly two thousand pounds on the wrong side. The learned and pious men of Oriel tried to find the weak spot, but it was not until expert opinion was called that they found that Keble, casting up a column, had added the date of the year to Oriel's debts!

IMPERIAL OXFORD

The climax of the British Empire did not profoundly affect Oxford, though half the Viceroys of India were Oxford men, and Cecil Rhodes called the University 'the energizing source of Empire'. Here, nevertheless, are a few snatches of imperialism:

INFLUENCE:

It is through the minds and the examples of those statesmen and administrators, who have imbibed their principles of life and action within her precincts, and have been trained in her schools and on her river or playing-fields, that the influence of the University is reflected on the outer world. Nor is it only the men like Lord Salisbury, Lord Rosebery and Mr. Gladstone, who guide the country at home, or like Lord Milner and Lord Curzon, who give their best work to Greater Britain, that are the true sons of the University; it is the plain, hard-working clergymen and civilians, also, who by their lives of honest and unselfish toil, hand on the torch of good conduct and high ideals which has been entrusted to them.

Cecil Headlam, *Oxford And Its Story*, 1912.

TONE:

Sons of Oxford have been among our greatest Empire builders and Empire rulers. . . . The tone of the place—its ideals—its merits and defects, are felt wherever the British flag flies.

George R. Parkin, *The Rhodes Scholarships*, 1913.

HOME THOUGHTS FROM THE EMPIRE:

Mind'st thou the days on the happy, happy Cherwell,
    Drowned a thousand fathom deep in crystal summer air?

IMPERIAL OXFORD

Floating on dream-river with the weed the paddles locking,
Slow invading ripples setting all the ripples rocking—
Mind'st thou that frog we put in Harry's stocking?
    Harry, as he went a'swimming
    Where the may-fly all were skimming,
Till he landed, mother-naked in the hospitable hay:
Does he mind it now, I wonder, District-judging, in Bombay?
'Emeritus', in the Oxford Magazine, 1906.

IMPERIAL THOUGHTS FROM HOME:

    What love in myriad hearts in every clime
      The vision of her beauty calls to pray'r:
    Where at his feet Himalaya sublime
      Holds up aslope the Arabian floods, or where
      Patriarchal Nile rears at his watery stair;
    In the broad islands of the Antipodes
    By Esperanza, or in the coral seas
      Where Buddha's vain pagodas throng the air;

    Or where the chivalry of Nipon smote
      The wily Muscovite, intent to creep
    Around the world with half his pride afloat,
      And send his battle to the soundless deep;
      Or with our pilgrim kin, and them that reap
    The prairie corn beyond cold Labrador
    To California and the Alaskan shore,
      Her exiled sons their pious memory keep. . . .
Robert Bridges, An Invitation to the Oxford Pageant, 1907.

THE EXILE:

    Not even this peculiar town
      Has ever fixed a friendship firmer,
    But—one is married, one's gone down,
      And one's a Don, and one's in Burmah.
Hilaire Belloc, from the 'Dedicatory Ode' in Lambkin's Remains, 1900.

RALPH'S CAREER

A funny parody of the Oxford Novel was Sandford of Merton, *allegedly
by Belinda Blinders, really by D. F. T. Coke, which appeared in 1903.
Some characteristic episodes:*

MAKING A FRIEND

His trained intuition told him at once what he should do. He did not approach the nearest group at random, but strolled into the porter's room and asked politely:—

'Please, sir, are any of the better men up here as yet?'

The porter walked to the door and shading his eyes against the sun, now almost sinking beneath the minarets of Merton,—

'That gentleman over there, sir,' he remarked, 'him with the frock coat and motor cap, he's Mr. Corser, captain of the Cricket team. He's a good sort, he is,' unconsciously punning on the ordinary and slang sense of the latter phrase.

Thanking him for his information, Ralph strolled across the grass to where the well-made young man stood chatting and laughing with a ring of friends. He was pleasant-faced, but slightly over-dressed, as is the habit of the Oxford youth, and his red tie was a shocking discord with his blue knitted gloves. Ralph approached him slowly, and holding out his hand to him,—

'My name is Ralph Sandford,' he said simply. 'I should like to be your friend.'

FORTUNE'S REBUFF

Each of the two looked forward to the French lectures, which were held in the class room at Brazenose, that beautiful name shortened by iconoclastic undergraduates into B.N.S.! Each was exceptionally early in arriving, and each tarried in departing. But though Ralph had often smiled at his divinity, had even proffered her his blotting-paper, he had not yet discovered her name. He had not, however, been introduced, and so naturally did not venture to address her. At last he hit upon a happy plan. As she walked down the class room, her many books beneath her lithe arm, Ralph neatly abstracted one and glanced at the title-page. On it, in charming writing, there was inscribed a double name followed by an address. So much he saw, enough to make his heart beat; but at that moment all-fickle Fortune turned her wheel. The other books slipped from under the fair charmer's arm, and fell heavily on Ralph Sandford's foot.

'Blow!' he cried.

It was his first oath!

TEMPTATION

It was only to be expected that a boy of such ingenuous countenance and such moral probity would at once become beloved by his fellows.

Unfortunately Ralph showed but little discrimination in his choice of friends. He had not been many days at Oxford before the following epistle reached him:—

'Dear Ralph Sandford,
    'Though we have never met, I like your face and think we shall be friends. Will you attend a party in my diggings next Monday at 9 p.m.?

'Your well-wisher,
'Ronald Dashgross.'

Ralph had foolishly accepted. It was a nicely-worded letter, save for the slang phrase for 'lodgings', and how should he know that 9 p.m. was too late for any righteous party?

Ralph was disillusioned as soon as he opened the door. Round a table sat some eight or nine young men, each inhaling the deadly fumes of nicotine, and in the centre of the table, surrounded by various fruits, there stood—a bottle! At once the full significance of the situation struck the lad. He had been enticed into an orgie! But he was of sterling merit, and steadfastly refused either to smoke or drink. Not so his comrades. As the bottle of strongest claret circulated, they became more and more riotous. Ralph sought an excuse to make his departure; but scarce had he determined that the most tactful method would be to say that he had a letter to write, when songs were called for. Ralph had ever loved music, and resolved to stay. Ah, Ralph! had you but known the nature of those songs, you would have fled the swifter! All the latest productions of the gilded gin-palaces were bellowed out—'Two Lovely Black Eyes', 'Ting-a-Ling', 'The Bogey Man', and 'Daisy'—each worse than the last, and all sung with relish. Ralph had never heard them before, nor had he even imagined such depravity.

RALPH'S TRIUMPH

    'Start!'
    The word sounded clear from the mouth of the Varsity captain of boats, and at once Ralph exerted the full force of his Herculean arms. His blade struck the water a full second before any other: the lad had started well. Nor did he flag as the race wore on: as the others tired, he seemed to grow more fresh, until at length, as the boats began to near the winning-post, his oar was dipping into the water nearly twice as often as any other.
    And now the climax of the race was reached, and Ralph put forth his full strength; his oar clashed against those of 'six' and 'eight',

the water foamed where his rowlock kept striking it, the boat shot forward, and slowly left St. Catharine's behind.

As the boat returned in triumph, she bounded down the stairs and threw herself upon Ralph's ample chest.

'Darling,' she cried so loud that all might hear, 'how strong you are! Alone you did it.'

Ralph flushed with pleasure. The remainder of the crew gallantly took up the cry.

'Ay, ay, it was Sandford who did it', they cried, and beat their oars against the house-boat as a token of applause.

D. F. T. Coke is also said to have been the author of the classic phrase, often attributed to Ouida, 'All rowed fast, but none so fast as stroke.'

BALLIOLISM

Long after the death of its great Master, Benjamin Jowett, Balliol College retained an especially potent reputation—in the late 1890s more than forty Balliol men sat in the House of Commons, and between 1878 and 1914 more than two hundred entered the Indian Civil Service. The college lore was full of pride and pungency, and was exuberantly celebrated.

BALLIOL MEN

Years ago when I was at Balliol,
  Balliol men—and I was one—
Swam together in winter rivers,
  Wrestled together under the sun.
And still in the heart of us, Balliol, Balliol,
  Loved already, but hardly known,
Welded us each of us into the others:
  Called a levy and chose her own.

Here is a House that armours a man
  With the eyes of a boy and the heart of a ranger
And a laughing way in the teeth of the world
  And a holy hunger and thirst for danger:
Balliol made me, Balliol fed me,
  Whatever I had she gave me again:
And the best of Balliol loved and led me.
  God be with you, Balliol men.

Hilaire Belloc, from *Verses*, 1910.

BALLIOL WIT

'The Masque of B-ll–l' was a broadsheet containing forty rhymes about contemporary members of the college, written by a group of undergraduates,

published in 1881, but soon withdrawn because of libel threats. This is a selection.

Benjamin Jowett, Master:

> First come I. My name is J-W-TT.
> There's no knowledge but I know it.
> I am Master of this College,
> What I don't know isn't knowledge.

S. L. Lee, later editor of the Dictionary of National Biography:

> I am featly-tripping L-E,
> Learned in modern history,
> My gown, the wonder of beholders,
> Hangs like a footnote from my shoulders.

Samuel Brearley, an American:

> No poor Britisher is nearly
> Half so fine a man as BR—RL-Y;
> But I cheerfully acknowledge
> Harvard's whipped by B-LL—L College.

George Nathaniel Curzon, later Viceroy of India and Chancellor of Oxford University:

> My name is G- -RGE N-TH-N—L C-RZ-N,
> I am a most superior person.
> My cheeks are pink, my hair is sleek,
> I dine at Blenheim once a week.

The lines destined most often to be quoted were the last: Curzon said they pursued him for the rest of his life.

THE ESSENTIAL BALLIOL

'What is the essential Balliol?' Michael demanded.

'Who could say so easily? Perhaps it's the same sort of spirit, slightly filtered down through modern conditions, as you found in Elizabethan England.'

Michael asked for a little more elaboration.

'Well, take a man connected with the legislative class, directly by birth and indirectly by opportunities, give him at least enough taste not to be ashamed of poetry, give him also enough energy not to be ashamed of football or cricket, and add a profound satisfaction with Oxford in general and Balliol in particular, and there you are.'

Compton Mackenzie, *Sinister Street*, 1914.

BALLIOL LIVING

I have sometimes thought that the life we led at Balliol half a century ago was a pattern, in miniature, of what a civilized western community ought to provide for us all. . . . We divided our days between sharpening our wits, exercising our bodies and talking to friends chosen by ourselves. We were under a gentle discipline, far less restrictive than the social and economic pressures of the world beyond the College walls; but nobody interfered with our freedom of thought and expression. We were mildly and rather self-consciously unconventional. . . . We had no slogans. We admired and envied originality. Our society was not classless, because birds of a feather, if uncaged, will always flock together. But 'class', with us, was a matter of affinity, and had nothing to do with who our fathers were or how much money they had. Surplus money, indeed, for the few that possessed it, was used for entertaining friends, but could never buy social success, while social failure, nine times out of ten, was your own fault. We lived under men we could, and did, look up to, and all our loyalties were spontaneous; we had no colonels, party chiefs, or 'bosses', towards whom our natural feelings had to be subdued by duty. As for power, we never even thought about it: a sure mark of Utopia.

L. E. Jones, *An Edwardian Youth*, 1956.

BALLIOL BALLADS

My friends gave this evening at Balliol a concert to their families. At the conclusion of their songs they went up to their rooms and still sang on. . . . The national airs of Oxford are better than the verses of Bruant. They were the 'Hundred Pipers', the song the Highlanders sang at Waterloo during the charge of the Cuirassiers; 'Mandalay', a favourite with the Indian troops, and a delightful ballad of 'Oh, my honey'. And the echo of the old walls sent back the refrain of the songs across the old, silent courts. I left them, as I leave them always, moved and charmed; a little jealous, however, of the care their forefathers took to surround their youth with things noble and beautiful.

J. Bardoux, tr. W. R. Barker, *Memories of Oxford*, 1899.

BALLIOL PHILOSOPHY

> The winter is dead, and the spring is a-dying
>> And summer is marching o'er mountain and plain,
> And tossing and tumbling and calling and crying
>> The Balliol rooks are above us again;

And watching them wheel on unwearied wings,
  I question them softly of vanished things:
Oh rooks, I pray you, come tell me true
  Was it better the old? Is it better the new?

    Caw, Caw, says every rook,
  To the dreamer his dreams, to the scholar his book;
  Caw, Caw, but the things for me
  Are the windy sky and the windy tree!

                    From *Balliol Songs*, an undated pamphlet.

BALLIOL BIGOTRY

*At Balliol was born the phrase 'effortless superiority', so often to be applied
to Oxford men, but the sense of privilege was less happily expressed in this
anonymous Balliol rhyme from the end of the nineteenth century:*

        The things that a fellow don't do,
        The things that a fellow don't do,
        They haven't been told to the Board School boy,
        They have not been revealed to the Jew.

*Julian Grenfell, the archetypal English gentleman, used to take delight
in cracking a stock-whip at the alien Philip Sassoon, when they were
members of the college together.*

WHERE THE GREAT MEN GO

Noon strikes on England, noon on Oxford town,
  —Beauty she was statue cold—there's blood upon her gown:
Noon of my dreams, O noon!
  Proud and godly kings had built her, long ago,
  With her towers and tombs and statues all arow,
With her fair and floral air and the love that lingers there,
  And the streets where the great men go.

              James Elroy Flecker (1884–1915), from *The Dying Patriot*.

UNIVERSITY ANIMALS

*Animals do not figure largely in the Oxford annals, but in 1900 a Mrs.
Wallace made amends in* Some Oxford Pets, *an academic bestiary. 'We
do not venture to presume', said the preface, 'that the domestic animals of a
University town are more highly gifted or developed than most of their*

kind', but five professors, nevertheless, contributed to the symposium, which was dedicated to The Soldiers of the Queen. Here is Provost Phelps of Oriel on the best-known of all Oxford dogs, 'Oriel Bill':

The more he lived among us, the more he caught the humour of the place. Never would he venture into college, save on the day in Summer term when the college was photographed. Then he strode in, settled himself on a table in the midst of the group, and faced the camera with placid courage. He knew every member of the college, and would go with an Oriel man anywhere, but to all others he turned a deaf ear. His attendance at college matches, whether cricket or football, was unfailing, and thereby hangs a tale. He had been with the eleven to the Keble ground, the day was hot, and his energy exhausted by encouraging applause. Seeing a hansom just starting for the town, he jumped in and was driven home.

🦁 *This is C. R. L. Fletcher the historian on 'Buttery Dick', a Magdalen cat:*

Age and changes came to Dick as to all of us: but even Royal Commissions were powerless against our friend. He took no advantage of the statute enabling Fellows to marry or live out of the College— indeed he never showed the slightest inclination to change his good old bachelor ways. In the late eighties—his own very late teens—he made few new acquaintances among the undergraduates: he slept longer hours on the top of the Hall mantel-piece, seldom even awaking to listen to lectures delivered in the Hall, and finally passed away in his twentieth year.

🦁 *And this is a cat of Corpus Christi:*

Suddenly, all but full-grown, in the fine flower of beauty and buoyancy, Tom appeared in Common Room . . . He became, though from our human point of view he never attained his majority, almost at a bound, 'The Senior Fellow'; through many years, even to extreme old age, he was known by that title; whenever any other name but Tom has to be applied to him, he is known by that title to us still. . . . In Common Room he was equal to every occasion, he has been known to taste and even to condemn the port.

🦁 *Truest to the Oxford spirit, though, is the poem addressed to 'An Oxford Poodle', which ends with the lines:*

> Ah Puffles, dear detested friend,
> Vice, and virtues, mingled blend.
> I hate thee—yet I love thee more,
> A four-legged angel—and a bore.

DON, *c.* 1900

Of the successful man who is a Don by accident I confess an ignorance
that borders on dislike. . . . [In] every possible way he keeps a firm
connection with the great outer world. He knows the female cousins
of all the undergraduates of his college, and many of them have been
mildly in love with him in a punt. He is often in London, where he
is very academic, and would wish to appear merely well-informed.
When he meets London friends in Oxford, he is anxious to prove that
he at least is not a mere Don; yet his friends can only wonder that
there is now no such thing as an Oxford point of view, but only an
Oxford drawl. His sitting-room is magnificent, and like style, con-
ceals the man. . . . His books are noble up to the year 1800—abundant
and select, often old, always fine; but after the year 1800 a certain
timidity of taste may be observed. Of course his friends' books are
there, with the books which you are expected to know in country
houses. For the rest, he has overcome the difficulty of selection by
not selecting. . . . He is a brilliant host, suave, considerate—with
comprehensive views—and ready to make allowances for those who
are not Dons. Perhaps he is in the main a summer bird. Then he
shows that he is a gallant as well as a scholar and a man of the world.
He is the figure-head of his college barge during The Eights, and
with an eye-glass, that is a kind of sixth sense, he surveys woman-
kind, and sees that it is good.

Edward Thomas, *Oxford*, 1903.

UNDERGRADUATE, *c.* 1900

There was Lonsdale. Lonsdale really possessed the serene perfection
of a great work of art. Michael [Fane] thought to himself that almost
he could bear to attend for ever Ardle's dusty lectures on Cicero in order
that for ever he might hear Lonsdale admit with earnest politeness
that he had not found time to glance at the text the day before, that
he was indeed sorry to cause Mr. Ardle such a mortification, but that
unfortunately he had left his Plato in a saddler's shop, where he had
found it necessary to complain of a saddle newly made for him.

'But I am lecturing on Cicero, Mr. Lonsdale. The Pro Milone was
not delivered by Plato, Mr. Lonsdale.'

'What's he talking about?' Lonsdale whispered to Michael.

'Nor was it delivered by Mr. Fane,' added the Senior Tutor dryly.
Lonsdale looked at first very much alarmed by this suggestion,

then seeing by the lecturer's face that something was still wrong, he assumed a puzzled expression, and finally in an attempt to relieve the situation he laughed very heartily and said:

'Oh, well, after all, it's very much the same.' Then, as everybody else laughed very loudly, Lonsdale sat down and leaned back, pulling up his trousers in gentle self-congratulation.

'Rum old buffer,' he whispered presently to Michael. 'His eye gets very glassy when he looks at me. Do you think I ought to ask him to lunch?'

Compton Mackenzie, *Sinister Street*, 1914.

WELL PLAYED

The aims of the O.U.D.S. seems to be to get as many blues as possible into the cast of a Shakespearian production, with the idea, perhaps, of giving Oxford its full money's worth. I remember well the sensation made by the most famous of all university athletes,—a 'quadruple blue', who played on four university teams, was captain of three of them, and held one world's record. The play was 'The Merchant of Venice', and the athlete in question was the swarthy Prince of Morocco. Upon opening the golden casket his powers of elocution rose to unexpected heights. Fellows went again and again to hear him cry, 'O hell! What have we here?'

John Corbin, *An American at Oxford*, 1902.

The athlete-actor was C. B. Fry of Wadham.

FAME: AN AMERICAN ILLUSION

'It's a wonderful place is Oxford. You English gentlemen arrivin' here from variously located parts of the country must feel fair cowed when you think of all the famous men who have lived in this little town before you.'

No one offered to interrupt his monologue.

'You must feel a thrill in your bones when you say to yourselves, "I'm walkin' the streets which John Ruskin has walked; I am livin' on the very same ward as has once contained Arnold, Froude and Newman . . ." 'I would give a thousand dollars,' said Downy, 'to have been up here with Newman!'

None of the undergraduates looked as if he would have given a

Greek grammar for it. Only one of them showed signs of life; he cleared his throat and moved uneasily in his chair for a few moments.

'Which Newman do you mean?' he asked.

'Why, Newman, sir; the Newman.'

'Do you mean W. G. Newman who fielded point, or T. P. Newman who broke the roof of the pavilion in the M.C.C. match?'

All the freshmen glared at Downy.

'Neither, sir, neither; Cardinal Newman, the eminent divine!'

'Never heard of him!' said the bold freshman, and went on with his egg.

George Leslie Calderon, *Downy V. Green*, 1902.

GLORY: AN AMERICAN RESPONSE

The thousand years of English glory stretch across the English sky from 900 to 1900 in a luminous tract where the stars are sown in multitudes outnumbering those of all the other heavens; and in Oxford above all other places one needs a telescope to distinguish them. . . . What strikes one with the sharpest surprise is not the memories of distant times, however mighty, but those of yesterday, of this forenoon, in which the tradition of their glory is continued. The aged statesman whose funeral eulogy hardly ceased to echo in the newspapers, the young hero who fell in the battle of the latest conquest, died equally for the honour of England, and both are mourned in bronze which has not yet lost its golden lustre beside the inscriptions, forgotten themselves in the time-worn lettering of the tablets on the walls, or the brasses in the floors. Thick as the leaves in Vallombrosa, they strew the solemn place. . . .

Of course we have only to live on a few centuries more and our universities can eclipse this splendour. . . .

W. D. Howells, in the *North American Review*, 1906.

COLLEGE SPIRITS, 1907

The colleges were still all-important, in early twentieth-century Oxford, and their reputations greatly varied. Here 'A Graduate', writing in the National Review *in 1907, advises prospective parents of Oxford students where to send their sons:*

Christ Church is the chief resort of the aristocracy. This is, in many ways, good for Christ Church, and not bad for the aristocracy, but a less pleasant feature is that the presence of this same aristocracy attracts to the House an entirely undesirable element of the Greeks and Hebrews.

All the world knows that there are Medes and Ethiopians at *Balliol*, but they are readily assimilated, and now form a recognized feature of the College. What is commonly called culture is far more generally diffused [there] than anywhere else.

University: If the majority of its members can hardly be described as intellectual, they are, at all events, good fellows.

Trinity is the proud possessor of a certain element of muscular Christianity. It contains a large number of undergraduates from the best public schools.

At *Magdalen* the average man is very pleasant, but he is sadly uniform. There appears to exist a certain repression of the individuality.

The average *New College* man is not intolerant of learning, but he is possibly more prone to admire it in others than to seek it for himself. The ornamental futility [of the college] is infinitely attractive, but seldom leads to great results.

🦁 '*A Graduate's*' conclusion was that if your son was clever, send him to Balliol: if he was not clever, send him to Balliol anyway.

THE CONCERT BEGAN

The moving spirit in those Balliol concerts fifty years ago was Dr. Ernest Walker. Thin, bent, etiolated, with a despondent black beard and a falsetto voice, Dr. Walker looked anything but the 'live wire' that he, in fact, was. I was one of the Master's dinner-party on a Sunday evening when he escorted his guests, as was his custom, to the middle of the front row. His principal guest was the irrepressible old Irish Lord Chancellor, Lord Ashbourne, who had been plainly reluctant to abandon the Master's port for an hour of high-brow music. Ernest Walker was to open the programme with a Beethoven Sonata and sat at the piano a few feet from Lord Ashbourne, waiting for silence. The old Irishman went on talking. Dr. Walker fixed him with a hostile and impatient eye, hands raised, wrists drooping above the keyboard, like a dog begging. In the silence of the expectant Hall, Lord Ashbourne's voice rang out:

THE CONCERT BEGAN

'That young man looks as if a glass of woine would do him good.'
Ernest Walker continued to eye the speaker, unmoved.

'Oi fancy he would loike us to stop conversing,' added Lord Ashbourne, as loudly as before, but turning to the Master. The Master
made a sign of assent; Lord Ashbourne settled himself in his chair;
the thin, pale hands stopped drooping, pounced, and the concert began.

L. E. Jones, *An Edwardian Youth*, 1956.

A CHANGE OF POSITION

We are told . . . that one day Lord Randolph Churchill was sent for
by the Warden of Merton to be rebuked for some delinquency. It was
winter, and the interview began with the Warden standing before the
fireplace and the undergraduate in the middle of the room. By the
time the next culprit arrived Lord Randolph was explaining his
conduct with his back to the fire and the Warden was a somewhat
embarrassed listener in a chilly corner. Such are the tales.

W. S. Churchill, *Lord Randolph Churchill*, 1906.

INSTINCT

Wherever philosophical insight is combined with literary genius and
personal charm, one says instinctively, 'That man is, or ought to be,
an Oxford man.'

G. W. E. Russell, *Seeing and Hearing*, 1907.

ACADEMIC PUNTERS

*From the betting-book of All Souls' College, 1873 to 1919, edited by
Sir Charles Oman in 1938:*

1876: Mowbray bets Anson and Buchanan 1/- each that Absalom did
not suggest that there were any number short of 10 righteous men in
Sodom.

He didn't.

1882: Hardinge bets Reichel half a crown that the word used by
Achilles' ghost in *Odyssey* xi is καταφθιμένοισιν as against ἀποφθι-
μένοισιν.

Hardinge was right.

[323]

1883: Reichel bets Fletcher 1/- that the name of the second 'Bretwalda' in the Saxon Chronicle is not Ceolred but Ceawlin.

🐉 *The bet was undecided.*

1884: Doyle bets Wakeman 2/6 that the name of the old Cortina guide, as spelt by Baedeker, is Sandro, not Santo, Siorpaes.

🐉 *It was Santo.*

1886: Wakeman bets Oman an even five shillings that the railway line Turin–Vercelli–Milan–Venice is a double line and not a single one.

🐉 *It was.*

1886: Hardinge bets Oman 6d that there is no such thing as an allusion to 'the variegated beauty of the gall-bladder' to be found in the 'Prometheus Vinctus'.

🐉 *But there is.*

1887: Fletcher bets Raleigh 1/- that certain persons known as the 'Bonnymuir Rioters' were beheaded alive in Scotland, not earlier than 1819.

🐉 *Raleigh won: they were beheaded dead.*

1896: Hardinge bets Doyle 1/- that Admiral Benbow was a black man.

🐉 *He was white.*

1898: Doyle bets Ker one shilling that 'a rose red city half as old as time' is Palmyra.

🐉 *It is Petra.*

1898: Edgeworth bets Robertson that not more than 50,000 persons will be killed in any war in which the European Powers are engaged on opposite sides during the first twenty-five years of the XX century.

🐉 *Five million persons were to die in the First World War, 1914–18.*

1902: Oman bets Wilbraham one shilling that no one in the smoking-room tonight knows who is president of the Helvetic Federation.

🐉 *Nobody did.*

1904: Asquith bets Malcolm 1/- that twice round his (Asquith's) stomach is less than the combined perimeter of Malcolm's stomach and head.

❧ *Twice round Asquith's stomach was further.*

1907: Steel-Maitland bets Edgeworth 1/- that with three exceptions no monument exists within a radius of five miles from the centre of Rome, built between 100 BC and 300 AD, that possesses an arch which has a lateral thrust.

❧ *The British Ambassador in Rome was appointed referee, and decided for Edgeworth.*

A DIVINITY VIVA, *c.* 1910

The first candidate appeared to me rather older than most undergraduates; grave and composed, he seemed to radiate dignity and wisdom. Almost, I felt, the examiners should apologise for subjecting such a one to this elementary test.

'Where, Mr. X,' asked the Chairman, 'was St. Paul born?'

A long pause followed; evidently the candidate was considering the question in all its bearings; finally he replied confidently.

'Ephesus, Sir.'

The Chairman shook his head, and posed another question.

'Can you tell me, Mr. X, to what place St. Paul was travelling at the time of his conversion?'

This time the pause was even more prolonged. I had the feeling that the whole roomful of people was trying to force the correct answer into the candidate's mouth. . . . At length he looked up and replied, though with a little less assurance.

'To Ephesus, I believe.'

A second time the Chairman wearily shook his head, and asked yet another question.

'Where, Mr. X, was St. Paul shipwrecked? . . .'

This time . . . the pause and the silence became almost unbearable. . . . At long last the candidate made his decision.

'I think, Ephesus,' he said.

'Thank you, Mr. X, you may go,' said the Chairman, and long before I could recover my equanimity I was seated at the table. Somehow the first question appeared to come to me from a great distance: 'Mr. Gresham,' said one of the other examiners—for I suppose that the Chairman had exhausted his repertoire—'Mr. Gresham, can you tell me where the goddess Diana was worshipped?'

I felt, I remember, that some horribly subtle and ingenious trap had been laid for me; I must on no account fall into it. . . . I could feel the tension in the room, and I could imagine the eager faces of my companions sitting behind me. Finally in a voice so low that I fancy the examiners can hardly have heard me I murmured timidly:

'Was it not Ephesus, Sir?'

'Thank you, Mr. Gresham; it was, and you may go.'

J. C. Masterman, *To Teach the Senators Wisdom*, 1952.

ZULEIKA'S OXFORD

The emblematic novel of Edwardian Oxford was Zuleika Dobson (*1911*), *by Max Beerbohm of Merton College. Miss Dobson, visiting her uncle, the Warden of Judas, conquers young Oxford so utterly that the entire student body drowns itself in the Isis for love of her:*

That old bell, presage of a train, had just sounded through Oxford station; and the undergraduates who were waiting there, gay figures in tweed or flannel, moved to the margin of the platform and gazed idly up the line. Young and careless, in the glow of the afternoon sunshine, they struck a sharp note of incongruity with the worn boards they stood on, with the fading signals and grey eternal walls of that antique station, which, familiar to them and insignificant, does yet whisper to the tourist the last enchantments of the Middle Age.

At the door of the first-class waiting-room, aloof and venerable, stood the Warden of Judas. An ebon pillar of tradition seemed he, in his garb of old-fashioned cleric. Aloft, between the wide brim of his silk hat and the white extent of his shirt-front, appeared those eyes which hawks, that nose which eagles, had often envied. He supported his years on an ebon stick. He alone was worthy of the background.

Came a whistle from the distance. The breast of an engine was described, and a long train curving after it, under a flight of smoke. . . . Into the station it came blustering, with cloud and clangour. Ere it had yet stopped, the door of one carriage flew open, and from it, in a white travelling-dress, in a toque a-twinkle with fine diamonds, a little and radiant creature slipped nimbly down to the platform.

A cynosure indeed! A hundred eyes were fixed on her, and half as many hearts lost to her. The Warden of Judas himself had mounted on his nose a pair of black-rimmed glasses. Him espying, the nymph darted in his direction. The throng made way for her. She was at his side. . . .

'My dear Zuleika', he said, 'welcome to Oxford. . . .'

Through those slums which connect Oxford with the world, the landau rolled on towards Judas. Not many youths occurred, for nearly all—it was the Monday of Eights Week—were down by the river, cheering the crews. There did, however, come spurring by, on a polo pony, a very splendid youth. His straw hat was encircled with a riband of blue and white, and he raised it to the Warden.

'That', said the Warden, 'is the Duke of Dorset, a member of my College. He dines at my table tonight'. . . .

As the landau rolled into 'the Corn', another youth—a pedestrian, and very different—saluted the Warden. He wore a black jacket, rusty and amorphous. His trousers were too short, and he himself was too short: almost a dwarf. His face was as plain as his gait was undistinguished. He squinted behind spectacles.

'And who is that?' asked Zuleika.

A deep flush overspread the cheek of the Warden. 'That,' he said, 'is also a member of Judas. His name, I believe, is Noaks.'

'Is he dining with us tonight?' asked Zuleika.

'Certainly not,' said the Warden. 'Most decidedly not.'

'Listen, you fools,' cried the Duke. But through the open window came the vibrant stroke of some clock. He wheeled round, plucked out his watch—nine!—the concert!—his promise not to be late!—Zuleika!

All other thoughts vanished. In an instant he dodged beneath the sash of the window. From the flower-box, he sprang to the road beneath. (The façade of the house is called, to this day, Dorset's Leap.) Alighting with the legerity of a cat, he swerved leftward in the recoil, and was off, like a streak of mulberry-coloured lightning, down the High.

The other men had rushed to the window, fearing the worst. 'No', cried Oover. 'That's all right. Saves time!' and he raised himself on to the window-box. It splintered under his weight. He leapt heavily but well, followed by some uprooted geraniums. Squaring his shoulders, he threw back his head, and doubled down the slope.

There was a violent jostle between the remaining men. The Mac-Quern cannily got out of it, and rushed downstairs. He emerged at the front door just after Marraby touched ground. The Baronet's left ankle had twisted under him. His face was drawn with pain as he hopped down the High on his right foot, fingering his ticket for the concert. Next leapt Lord Sayes. And last of all leapt Mr. Trent-Garby, who, catching his foot in the ruined flower-box, fell headlong, and was, I regret to say, killed. The MacQuern overtook Mr. Oover at St. Mary's and outstripped him in Radcliffe Square. The Duke came in an easy first.

Youth, youth!

'Zuleika!' he cried in a loud voice. Then he took a deep breath, and, burying his face in his mantle, plunged. . . .

There was a confusion of shouts from the raft, of screams from the roof. Many youths—all the youths there—cried 'Zuleika!' and leapt emulously headlong into the water. 'Brave fellows!' shouted the elder men, supposing rescue work. The rain pelted, the thunder pealed. . . . From the towing-path—no more din there now, but great single cries of 'Zuleika!'—leapt single figures innumerable through rain to river. The arrested boats of the other crews drifted zigzag hither and thither. The dropped oars rocked and clashed, sank and rebounded, as the men plunged across them into the swirling stream. . . .

All along the soaked towing-path lay strewn the horns, the rattles, the motor-hooters, that the youths had flung aside before they leapt. Here and there among these relics stood dazed elder men, staring through the storm.

'Dead?' gasped the Warden. 'Dead? It is disgraceful that I was not told. What did they die of?'

'Of me.'

'Of you?'

'Yes. I am an epidemic, grand-papa, a scourge, such as the world has not known. Those young men drowned themselves for love of me.'

He came towards her. 'Do you realize, girl, what this means to me? I am an old man. For more than half a century I have known this College. To it, when my wife died, I gave all that there was of heart left in me. For thirty years I have been Warden; and in that charge has been all my pride . . .' He raised his head. 'The disgrace to myself is nothing. . . . It is because you have wrought the downfall of Judas that I am about to lay my undying curse on you.'

'You mustn't do that!' she cried. 'It would be a sort of sacrilege. I am going to be a nun. Besides, why should you? I can quite well understand your feeling for Judas. But how is Judas more disgraced than any other College? If it were only the Judas undergraduates who had—'

'There were others?' cried the Warden. 'How many?'

'All. All the boys from all the Colleges.'

The Warden heaved a deep sigh. 'Of course,' he said, 'this changes the aspect of the whole matter. I wish you had made it clear at once.

You gave me a very great shock,' he said, sinking into his arm-chair, 'and I have not yet recovered. . . .'

Max Beerbohm, Zuleika Dobson, 1911.

VARSITY TALK

Edwardian Oxford spoke in slang, particularly in the idiom which altered the ends of words to -er. Here is a compendium sentence culled freely from the diaries of a Worcester College freshman, Willie Elmhirst, in 1911 (and first published, as A Freshman's Diary, *in 1969):*

After Toggers brekker went to divvers leccer, then to eat at the Ugger; saw the Britter watching rugger, then to tea at Jaggers, passing the Magger in the street; found my sitter full of freshers, so retired to my bedder early.

The young man means that after attending a breakfast for the Torpids racing crew, he went to a divinity lecture, followed by lunch at the Oxford Union. At the rugby match he observed a well-known Oxford character (said to be the world's first authority on Basque verbs) known as the British Workman, and after passing the Master of his own college in the street on his way to tea at Jesus, he returned home to find his sitting-room so full of other freshmen that he retired to his bedroom—he had to get up early, perhaps, for Commuggers (Holy Communion).

Desmond Coke, in his comic novel Sandford of Merton (1903), *invented lots more, including:*

A cugger of cocoa.
A sigger-sogger round the piano.
Playing a pragger jogger on someone.
The Vigger-Chagger of the Varsity.

A JONES AT OXFORD

It was to see the great cedar that a party of us first visited those delectable gardens [of Wadham.] They were well guarded, for to visit them it was necessary to pass through the Warden's own Lodgings, and to write your name and College in a copy-book, to which a pencil was attached by a string, that lay upon a table in the passage. There were several of us, and I came last, and to save time scribbled 'Jones' in the book, without initials. On our return we were confronted by the Warden, lame little Dr. Wells, looking displeased.

'Gentlemen,' he said, 'I did not think that men from my own College, when taking advantage of a privilege, would have sunk so low as to write "Jones" in my book.'

I made a stiff little bow. 'Unfortunately for me, sir, it happens to be my name.'

The Warden made a very low bow indeed. 'Sir,' he said, 'I humbly beg your pardon.'

L. E. Jones, *An Edwardian Youth*, 1956.

THE OXFORD MAN, 1913

One can tell an Oxford man . . . almost at a glance. He is full of opinions and of the ability to defend them. He sets his mind at yours with a conscious briskness which seems to foreknow victory. He uses his culture like a weapon always drawn—with a flourish wonderfully easy and graceful. . . . His trim dress and carefully poised voice, his brisk movements and the precise elaboration of his speech, all seem deliberately and yet easily accomplished, all part of his armoury, weapons fit for his confident attack upon the universe. Not even the world-old traditions, the crumbling towers and quiet quadrangles of his own city, have succeeded in luring him into the past. Not even the study of Literae Humaniores has seduced him into abstraction; not even the rich seclusion of the Thames Valley has availed to enervate him: he has pressed all into the service of his own ambitions.

Charles Tennyson, *Cambridge from Within*, 1913.

BIBLIOGRAPHICAL NOTE

Monfries, J. D. C., *Ye Oxford booke of drivel*. Oxf., 1914.

THE PURPOSE OF EDUCATION

Gentlemen, you are now about to embark upon a course of studies which will occupy you for two years. Together, they form a noble adventure. But I would like to remind you of an important point. Some of you, when you go down from the University, will go into the Church, or to the Bar, or to the House of Commons, or to the

Home Civil Service, or the Indian or Colonial Services, or into various professions. Some may go into the Army, some into industry and commerce; some may become country gentlemen. A few—I hope a very few—will become teachers or dons. Let me make this clear to you. Except for those in the last category, nothing that you will learn in the course of your studies will be of the slightest possible use to you in after life—save only this—that if you work hard and intelligently you should be able to detect *when a man is talking rot*, and that, in my view, is the main, if not the sole, purpose of education.

> J. A. Smith, Professor of Moral Philosophy, opening a lecture course in 1914 (quoted by Harold Macmillan in *The Times*, 1965).

PREMONITIONS

As, in the first decades of the twentieth century, Europe postured and bickered towards disaster, some sense of premonition, perhaps of elegy, seemed to enter Oxford writing. Here for instance is Alfred Noyes (1880–1953) sadly revisiting his old college:

Timid and strange, like a ghost, I pass the familiar portals,
　　Echoing now like a tomb, they accept me no more as of old;
Yet I go wistfully onward, a shade thro' a kingdom of mortals
　　Wanting a face to greet me, a hand to grasp and to hold.

Hardly I know as I go if the beautiful City is only
　　Mocking me under the moon, with its streams and its willows agleam,
Whether the City of friends or I that am friendless and lonely,
　　Whether the boys that go by or the time-worn towers be the dream. . . .

Whether all these or the world with its wars be the wandering shadows!
　　Ah! sweet over green-gloomed waters the May hangs, crimson and white;
And quiet canoes creep down by the warm gold dusk of the meadows,
　　Lapping with little splashes and ripples of silvery light.

Others like me have returned; I shall see the old faces tomorrow,
　　Down by the gay-coloured barges, alert for the throb of the oars,
Wanting to row once again, or tenderly jesting with sorrow
　　Up the old stairways and noting the strange new names on the doors. . . .

City of dreams that we lost, accept now the gift we inherit—
  Love, such a love as we knew not of old in the blaze of our noon,
We that have found thee at last, half City, half heavenly Spirit,
  While over a mist of spires the sunset mellows the moon.

From *Forty Singing Seamen*, 1907.

Even Max Beerbohm, in his froth of a novel Zuleika Dobson (*1911*),
tinged his fun with a certain wistfulness:

Oxford, that lotus-land, saps the will-power, the power of action. But,
in doing so, it clarifies the mind, makes larger the vision, gives, above
all, that playful and caressing suavity of manner which comes of a
conviction that nothing matters, except ideas, and that not even ideas
are worth dying for, inasmuch as the ghosts of them slain seem worthy
of yet more piously elaborate homage than can be given to them in
their heyday. If the colleges could be transferred to the dry and bracing
top of some hill, doubtless they would be more evidently useful to the
nation. But let us be glad there is no engineer or enchanter to compass
that task. *Egomet*, I would liefer have the rest of England subside
into the sea than have Oxford set on a salubrious level. For there is
nothing in England to be matched with what lurks in the vapours of
these meadows, and in the shadows of these spires—that mysterious,
inenubilable spirit, spirit of Oxford. Oxford! The very sight of the
word printed, or sound of it spoken, is fraught for me with most
actual magic.

And in 1913 the Newdigate Prize Poem, Oxford, *by M. Roy Ridley,*
ended with the following words:

> So now, our Queen and Mother, take of us,
> Thy sons, our thanks and homage; for we pass
> On a far journey: as the gates swing wide,
> For the last time we turn, and give to thee
> High salutation and a long farewell.

In the following year hostilities with Germany began, and 14,500 members
of the University of Oxford went to war.

Coming to Terms
1914 – 1945

*The Great War shattered Oxford, and obliged the
University to come to terms with the world. The first
Government subsidy was accepted in 1920, and there-
after, though the memoirs do not often show it, Oxford
gradually shed the consequence of its Victorian apogee—
moving ever nearer, in fact, as the autonomy of the
colleges weakened, the social order shifted, and the
insular arrogance of the old place lost its meaning, to the
studium generale of poor diligent scholars which it had
been in the beginning.*

STAND IN THE TRENCH, ACHILLES!

Was it so hard, Achilles,
   So very hard to die?
Thou knowest and I know not—
   So much the happier I.

I will go back this morning,
   From Imbros over the sea;
Stand in the trench, Achilles,
   Flame-capped, and shout for me!

Patrick Shaw-Stewart, Balliol College, killed in action, 1917.

Some 2,700 Oxford men were killed in the Great War, at a time when the student population was about 3,000, and Oxford was haunted by ever-absent friends:

Sweet as the lawn beneath his sandalled tread,
Or the scarce rippled stream beneath his oar,
So gently buffeted it laughed the more,
His life was, and the few blithe words he said.
One or two poets read he, and re-read;
One or two friends with boyish ardour wore
Close to his heart, incurious of the lore
Dodonian words might murmur overhead.
Ah, demons of the whirlwind, have a care,
What, triumphing your triumphs, ye undo!
The earth once won, begins your long despair
That never, never is his bliss for you.
He breathed betimes this clement island air
And in unwitting lordship saw the blue.

George Santayana, *The Undergraduate Killed in Battle*, 1915.

Leaving our rooms, we often checked our way
   As on a sudden wonderingly we saw
The summer grass sweet-smelling, where it lay
   Through the stone archway of our corridor:
Like men that, gazing from some hidden place,
   Might catch a glimpse of Beauty unaware,
And hear her very tones, and see her face,
   And so in adoration linger there.

> They are terrible now, those corridors. The hue
> Of midmost summer blazes on the Hill;
> The Garden laughs to heaven; at the west side
> The Tree puts on its glory of strange blue:
> But the corridors are grey and very still
> Like monasteries where all the monks have died.

Victor Gollancz, My Dear Timothy, 1952.

🦋 *In every college the tragic memorial slabs went up, and to many Oxonians, even to some of a later generation, the real Oxford, like the real England, died with those young men in the holocaust:*

Even now it is their standards that this city, half regretfully, half mockingly, still fitfully aspires to: their fair frank forms we instinctively look for, and fail to find, on the cross-benches of the Union; their particular culture which is only now disappearing from the Oxford scene, like the smile of the Cheshire Cat—the substance mostly gone, the shadow now dissolving, until soon there will only be memories and legends, or 30 feet of names on a college memorial, to remind us what it was.

James Morris, Oxford, 1965.

🦋 *Harold Macmillan, a future Prime Minister, had started his career at Balliol before the war, and after four years in the army could not bring himself to return:*

I did not go back to Oxford after the war. It was not just that I was still a cripple. There were plenty of cripples. But I could not face it. To me it was a city of ghosts. Of our eight scholars and exhibitioners who came up in 1912, Humphrey Sumner and I alone were alive. It was too much.

His Oxford, the spacious, innocent Oxford of the Edwardian age, was gone for ever:

I know well that I look back through somewhat rose-tinted spectacles. Yet the picture is vivid and real to me. Or is it, perhaps, after all, a dream?

From The Times, 1975.

🦋 *George Santayana on Oxford survivors:*

These young men are no rustics, they are no fools; and yet they have passed through the most terrible ordeal, they have seen the mad heart of this world riven and unmasked, they have had long vigils before battle, long nights tossing with pain, in which to meditate on the spectacle; and yet they have learned nothing. The young barbarians want to be again at play. . . . They are going to shut out from view everything except their topmost instincts and easy habits, and to trust to luck. Yet the poor fellows think they are safe! They think that the war—perhaps the last of all wars—is over!

Only the dead are safe; only the dead have seen the end of war.

Soliloquies in England, 1922.

Evelyn Waugh on a Trench-shocked Don:

He was, I now recognize, a wreck of the war in which he had served gallantly. No doubt a modern doctor would have named, even if he could not cure, his various neuroses. It was as though he had never cleaned himself of the muck of the trenches. His conspectus of history was narrowed to the few miles of the Low Countries where he had fought, and the ultimate, unattainable frontier towards which he had gazed through his periscope over the barbed wire. He was obsessed by the Rhine and it was the first, sharp difference between us that I was ignorant of its course.

He had a kind of rough geniality which found expression in coarse soldiers' language and quickly gave place to a frustrated pugnacity. He had been a fellow of All Souls before 1914 and must then have been a young man of more polished manners, for he was of perfectly respectable origins; but all were blown and gassed away in two years' fighting. As Dean of the college he seemed often to fancy himself in command of a recalcitrant platoon. He had binges like a subaltern on leave, got grossly drunk when he dined out and was sometimes to be seen as St. Mary's struck midnight, feeling his way blindly round the railings of the Radcliffe Camera believing them to be those of the college. When crapulous, as he normally was when conducting college business, he fell into violent rages. He was a misogynist to such an extreme degree that he refused to have women at his lectures. The college porter had instructions to repel them. If one slipped in, he drove her out, crimson faced, by his obscenities. He had one associate, whom he referred to as 'that little hack', a Platonist and fellow infantryman.

A Little Learning, 1964.

W. B. Yeats on Ghosts:

> Midnight has come, and the great Christ Church Bell
> And many a lesser bell sound through the room;
> And it is All Souls' Night,
> And two long glasses brimmed with muscatel
> Bubble upon the table. A ghost may come;
> For it is a ghost's right,

His element is so fine
Being sharpened by his death,
To drink from the wine-breath
While our gross palates drink from the whole wine. . . .

. . . Names are nothing. What matters who it be,
So that his elements have grown so fine
The fume of muscatel
Can give his sharpened palate ecstasy
No living man can drink from the whole wine.
I have mummy truths to tell
Whereat the living mock,
Though not for sober ear,
For maybe all that hear
Should laugh and weep an hour upon the clock.

From All Souls' Night, 1920.

STANDARDS

🦁 *For all the traumatic shock of the Great War, much of the old Oxford survived. George Nathaniel Curzon, first Marquess Curzon of Kedleston (page 315), was Chancellor of the University throughout the war, and did not let standards slip. When in 1921 Queen Mary was awarded an honorary degree, Curzon was asked to approve in advance the menu of the luncheon to which she was to be entertained at Balliol. He returned it to the Bursar with a single comment written in the corner:*

Gentlemen do not take soup at luncheon.

C. S. LEWIS

🦁 *C. S. Lewis, poet, critic, and theologian, resumed his Oxford career after being wounded in the Great War. These extracts from letters trace his progress from undergraduate at University College to newly-admitted Fellow of Magdalen—where he remained until his translation to a Professorship at Cambridge thirty years later:*

People talk about the Oxford manner and the Oxford life and the Oxford God knows what else; as if the undergraduates had anything to do with it. . . . The real Oxford is a close corporation of jolly, untidy, lazy, good-for-nothing, humorous old men, who have been electing their own successors ever since the world began and who intend to go on with it. They'll squeeze under the Revolution or leap over it when the time comes, don't you worry.

We are old, disillusioned creatures now, and look back on the days of 'buns and coffee' through a long perspective, and only seldom come out of our holes; the young men up from school in immaculate clothes think we have come to clean the windows when they see us. It happens to everyone here. In your first year you drink sherry and see people; after that your set narrows, you haunt the country lanes more than the High, and cease to play at being the undergraduate of fiction—there will be no revelry by night.

I had almost forgotten, if I had ever known, that 'prizemen' have to read portions of their compositions at our ceremony of Encaenia. . . . I have had a good lesson in modesty from seeing my fellow-prizemen. I was hardly prepared for such a collection of scrubby, beetle-like, bespectacled oddities. . . . It brings home to one how little I know of Oxford; I am apt to regard my own set, which consists mainly of literary gents, with a smattering of political, musical and philosophical—as being central, normal, representative. But step out of it, into the athletes on one side or the pale pot-hunters on the other, and it is a strange planet. . . .

My formal 'admission' at Magdalen . . . was a formidable ceremony, and not entirely to my taste. Without any warning of what was in store for me, the Vice-President ushered me into a room where I found the whole household . . . [Herbert] Warren (the President) was standing, and when the V.P. laid a red cushion at his feet I realized with some displeasure that this was going to be a kneeling affair. Warren then addressed me for some five minutes in Latin. I was able to follow some three-quarters of what he said; but no one had told me what response I was to make, and it was with some hesitation that I hazarded *do fidem* as a reply. This appeared to fill the bill. I was then told in English to kneel. When I had done so, Warren took me by the hand and raised me with the word, 'I wish you joy.' It sounds well enough on paper, but it was hardly impressive in fact; and I tripped over my gown in rising. I now thought my ordeal at an end; but I was never more mistaken in my life. I was sent all round the table and every single member in turn shook my hand and repeated the words: 'I wish you joy.' You can hardly imagine how odd it sounded by the twenty-fifth repetition. English people have not the talent for graceful ceremonial. They go through it lumpishly and with a certain mixture of defiance and embarrassment, as if everyone felt he was being rather silly, and was at the same time ready to shoot down anyone who said so.

Letters of C. S. Lewis, ed. W. H. Lewis, 1966.

WAUGH

🏃 *Evelyn Waugh, whose fiction was to epitomize the jazzier side of Oxford life in the 1920s, entered Hertford College in 1922. His first Oxford writing appeared in the student magazine* Isis, *for which he reported the debates of the Oxford Union:*

Mr. M. A. Thomson (Exeter) was almost wholly fatuous and successful.

Mr. J. L. Parker (New College) was much in favour of blood, iron, Bismarck, France and all those sorts of thing.

Mr. H. Lloyd-Jones (Jesus) gave the impression of having been suddenly stirred from a deep slumber by the previous speaker's mention of Wales.

Mr. R. H. Bernays (Worcester) was, as always, vehement, long-winded, biblical, homely, and not ineffective. He quoted French with an accent for which he thought he need not apologize.

Mr. Alfonso de Zulueta (New College) told an enchanting story about a school where the matron was wanton.

Mr. S. F. Villiers-Smith (New College) made the sort of speech which one associates with aged colonels.

Mr. Nobbs (Wadham) actually used the expression 'made the Empire what it is'.

Mr. H. J. V. Wedderburn (Balliol) addressed some of Shakespeare's more unrestrained love poems to the President in a most shameless manner.

Mr. D. J. Dawson (Christ Church) was brief almost to the point of insignificance.

Mr. I. B. Lloyd (of Exeter) said something in a foreign tongue of which I happened to know the meaning, but could not see the interest.

I detest all that Mr. A. Gordon Bagnall (St. John's) says always.

🏃 *Waugh's principal Oxford novel was* Brideshead Revisited, *in which many of his friends and enemies could be recognized, at least by themselves. Here are three particularly Oxford extracts:*

My cousin Jasper . . . was in his fourth year and, the term before, had come within appreciable distance of getting his rowing blue; he was secretary of the Canning and president of the J.C.R.; a considerable person in college. He called on me formally during my first week and stayed to tea; he ate a very heavy meal of honey-buns, anchovy toast, and Fuller's walnut cake, then he lit his pipe and, lying back in the basket-chair, laid down the rules of conduct which I should follow. . . . 'You're reading History? A perfectly respectable school. The very worst is English literature and the next worst is Modern Greats. You want either a first or a fourth. There is no value in anything between. Time spent on a good second is time thrown away. . . . Clothes. Dress as you do in a country house. Never wear a tweed coat and flannel trousers—always a suit. And go to a London tailor; you get better cut and longer credit. . . . Clubs. Join the Carlton now and the Grid at the beginning of your second year. If you want to run for the Union—and it's not a bad thing to do—make your reputation *outside* first, at the Canning or the Chatham, and begin by speaking on the paper. . . . Keep clear of Boar's Hill. . . .' The sky over the opposing gables glowed and then darkened; I put more coal on the fire and turned on the light, revealing in their respectability his London-made plus-fours and his Leander tie. . . . 'Don't treat dons like schoolmasters; treat them as you would the vicar at home. . . . Beware of the Anglo-Catholics—they're all sodomites with unpleasant accents. . . .'

AN OXFORD SUNDAY

It was the last Sunday of term; the last of the year. . . . I walked down the empty Broad to breakfast, as I often did on Sundays, at a tea-shop opposite Balliol. The air was full of bells from the surrounding spires and the sun, casting long shadows across the open spaces, dispelled the fears of night. The tea-shop was hushed as a library; a few solitary men in bedroom slippers from Balliol and Trinity looked up as I entered, then turned back to their Sunday newspapers. I ate my scrambled eggs and bitter marmalade with the zest which in youth follows a restless night. I lit a cigarette and sat on, while one by one the Balliol and Trinity men paid their bills and shuffled away, slip-slop, across the street to their colleges. It was nearly eleven when I left, and during my walk I heard the change-ringing cease and, all over the town, give place to the single chime which warned the city that service was about to start.

None but church-goers seemed abroad that morning; undergraduates and graduates and wives and tradespeople, walking with that unmistakable English church-going pace which eschewed equally both haste and idle sauntering; holding, bound in black lamb-skin and white celluloid, the liturgies of half a dozen conflicting sects; on their way to St. Barnabas, St. Columba, St. Aloysius, St. Mary's, Pusey House, Blackfriars, and heaven knows where besides; to restored Norman and revived Gothic, to travesties of Venice and Athens; all in the summer sunshine going to the temples of their race. Four proud infidels alone proclaimed their dissent; four Indians from the gates of Balliol, in freshly-laundered white flannels and neatly pressed blazers, with snow-white turbans on their heads, and in their plump, brown hands bright cushions, a picnic basket and the *Plays Unpleasant* of Bernard Shaw, making for the river.

In the Cornmarket a party of tourists stood on the steps of the Clarendon Hotel discussing a road map with their chauffeur, while opposite, through the venerable arch of the Golden Cross, I greeted a group of undergraduates from my college who had breakfasted there and now lingered with their pipes in the creeper-hung courtyard. A troop of boy scouts, church-bound, too, bright with coloured ribbons and badges, loped past in unmilitary array, and at Carfax I met the Mayor and Corporation, in scarlet gowns and gold chains, preceded by wand-bearers and followed by no curious glances, in procession to the preaching at the City Church. In St. Aldates I passed a crocodile of choir boys, in starched collars and peculiar caps, on their way to Tom Gate and the Cathedral. So through a world of piety I made my way to Sebastian.

REPROACH FROM A FOURTH-YEAR MAN, 1924

Jasper would not sit down; this was to be no cosy chat; he stood with his back to the fireplace and, in his own phrase, talked to me 'like an uncle'.

'I expected you to make mistakes your first year. We all do. I got in with some thoroughly objectionable O.S.C.U. men who ran a mission to hop-pickers during the long vac. But you, my dear Charles, whether you realize it or not, have gone straight, hook line and sinker, into the *very worst set in the University*. . . . None of these people you go about with pull any weight in their own colleges, and that's the real test. They think that because they've got a lot of money to throw about, they can do anything.

'And that's another thing. I don't know what allowance my uncle makes you, but I don't mind betting you're spending double. All *this*,' he said, including in a wide sweep of his hand the evidence of profligacy about him. It was true; my room had cast its austere winter garments, and, by not very slow stages, assumed a richer wardrobe. 'Is *that* paid for?' (the box of a hundred cabinet Partagas on the side-board) 'or those?' (a Lalique decanter and glasses) 'or *that* peculiarly noisome object?' (a human skull lately purchased from the School of Medicine, which, resting in a bowl of roses, formed, at the moment, the chief decoration of my table. It bore the motto '*Et in Arcadia ego*' inscribed on its forehead).

'Yes,' I said, glad to be clear of one charge, 'I had to pay for the skull.'

'You can't be doing any work. Not that that matters, particularly if you're making something of your career elsewhere—but are you? Have you spoken at the Union or at any of the clubs? Are you connected with any of the magazines? Are you even making a position in the O.U.D.S.? And *your clothes!*' continued my cousin. 'When you came up I remember advising you to dress as you would in a country house. Your present get-up seems an unhappy compromise between the correct wear for a theatrical party at Maidenhead and a glee-singing competition in a garden suburb.

'And drink—no one minds a man getting tight once or twice a term. In fact, he ought to, on certain occasions. But I hear you're constantly seen drunk in the middle of the afternoon.'

He paused, his duty discharged. . . .

'I'm sorry, Jasper,' I said.

In 1964 Waugh looked back on his Oxford years in an autobiography, A Little Learning, *from which these portraits of some contemporaries are taken:*

HAROLD ACTON, HISTORIAN AND CONNOISSEUR

Slim and slightly oriental in appearance, talking with a lilt and resonance and in a peculiar vocabulary that derived equally from Naples, Chicago and Eton, he set out to demolish the traditional aesthetes who still survived here and there in the twilight of the 90's and also the simple-living, nature-loving, folk-singing, hiking, drab successors of the 'Georgian' poets. Harold brought with him the air of connoisseurs of Florence and the innovators of Paris, of Berenson and of Gertrude Stein, Magnasco and T. S. Eliot; above all of the three Sitwells who were the objects of his admiration and personal affection. He was vividly alive to every literary and artistic fashion, exuberantly appreciative, punctilious, light and funny and energetic. He loved to shock and then to conciliate with exaggerated politeness. He was himself shocked and censorious at any breach of his elaborate and idiosyncratic code of propriety. The one quality he despised, traditionally characteristic of aesthetic Oxford, was languor.

🜨 *For Acton's own view of his Oxford self, see page 345.*

For Acton's own view of his Oxford self, see page 345.

ROBERT BYRON, TRAVEL WRITER AND HISTORIAN

He was short, fleshy and ugly in a painfully ignominious way. His complexion was yellow. He had a marked resemblance, which he often exploited at fancy-dress parties, to Queen Victoria at the time of her jubilee. He dealt with his ill looks, as others have done, by making them grotesque. He affected loud tweeds, a deer-stalker hat, yellow gloves, horn-rimmed pince-nez, a cockney accent. He leered and scowled, screamed and snarled, fell into rages that were sometimes real and sometimes a charade—it was not easy to distinguish. Wherever he went he created a disturbance, falling down in the street in simulated epilepsy, yelling to passers-by from the back of a motor car that he was being kidnapped. Robert was poor and determined not, Heaven knows, in any obsequious fashion, to force his way into the worlds of power and fashion; and he succeeded. Works of art were quite strange to Robert and when he encountered them he was excited to irrational outbursts of adoration or reprobation; either: 'Why does no one know about this?' (when everyone who cared, did) or 'Trash. Muck. Rubbish.' (of many established masterpieces).

ALFRED DUGGAN, HISTORICAL NOVELIST

Alfred was a full-blooded rake of the Restoration. He was very rich then with the immediate disposal of a fortune greater than any of our contemporaries. He was, moreover, the stepson of the Chancellor of the University, Lord Curzon. This connection irked the authorities, who otherwise would have summarily sent him down. We were often drunk, Alfred almost always. He came up with a string of hunters; he kept an account at the Macpherson's garage for day- and night-chauffeurs. Whether in the saddle in the late mornings or at 'the 43' (Mrs. Meyrick's night club in Gerrard Street) in the early mornings, Alfred was always tight; never violent, always carefully and correctly dressed, always polite, he lived in an alcoholic daze. The vultures of Balliol won what were for them great sums of money from him at cards. He paid punctiliously. When the card party broke up, Alfred would climb out of his window to his waiting car and be driven to London.

BRIAN HOWARD

Brian, when he came up, determined to eschew the arts and to pose as a sportsman. . . . More than this, in the intensely snobbish era which immediately succeeded my own, he contrived to make himself more than the entertainer, the animator, almost the arbiter, of the easy-going aristocrats whom he set himself to reform in his romantic model, like the youthful D'Israeli inspiring 'Young England'. 'Put your trust in the Lords' was the motto on the banner in his rooms on his birthday and there are many placid peers today who may ascribe most of their youthful fun to Brian. Sometimes he embarrassed them as, when Trinity hearties broke up a party he was at and impelled the guests to the gate, he threatened: 'We shall tell our fathers to raise your rents and evict you'. . . . He was an incorrigible homosexual, subject to a succession of delusions, and died by suicide at the time when he at last became rich. . . . At the age of nineteen he had dash and insolence, a gift of invective and repartee, a kind of ferocity of elegance that belonged to the romantic era of a century before our own. Mad, bad and dangerous to know.

THE AESTHETE

The most spectacular Oxford undergraduate of the early 1920s was Harold Acton, a half-American Etonian whose home was in Florence. These are some Oxford glimpses from his Memoirs of an Aesthete, *1948:*

Most freshmen at 'The House' coveted rooms in Tom Quad, Peckwater or Canterbury, but a room with a balcony overlooking Christ Church meadow appealed more to me. Externally, Meadow Buildings are grimly Victorian Gothic and internally sombre; but I painted my rooms lemon yellow and filled them with Victorian bric-à-brac—artificial flowers and fruit and lumps of glass, a collection of paperweights imprisoning bubbles that never broke and flowers that never faded.

Back to mahogany was my battle-cry. . . . I filled my room with Early Victorian objects, I bought a grey bowler, wore a stock and let my side-whiskers flourish. Instead of the wasp-waisted suits with pagoda shoulders and tight trousers affected by the dandies, I wore jackets with broad lapels and broad pleated trousers. The latter got broader and broader. Eventually they were imitated elsewhere and were generally referred to as 'Oxford bags'.

Aquarium, my first volume of poems, was published during my second term, and its red, black and yellow striped cover met me everywhere like a challenge. For a book of poems it had a prompt success. Since

I was free from false modesty, as from everything false, and possessed a resonant voice, I never faltered when I was asked to read them, but shouted them lustily down a megaphone. Nor would I tolerate interruptions. The megaphone could also be brandished as a weapon. . . . My poems made many friends. I read them from my balcony to groups in Christ Church meadow.

The invasion of Jazz had begun, under the aegis of David Greene, and an influx of London West-enders who converted any party into a night club: David's sister Olivia, with minute pursed lips and great goo-goo eyes; Elizabeth Ponsonby, always ready for a lark; 'Cara', a trim *garçonne* of indeterminate sex; Gracie Ansell, an Edwardian dame who had remained a sugar-baby; and the momentary matinée idol in the role of eternal adolescent, Tom Douglas, bringing albums of the latest American victrola records and distributing, among the favoured few, photographs of himself in *Fata Morgana*. . . .

Cocktails were substituted for Amontillado, and parties moved from college to college as from night club to night club. Conversation was stifled by the gramophone, and the talkative devised a special basic English in which to shoot wise-cracks at each other in the style of Noël Coward, while couples clung to ether forlornly, swaying to some raucous Blues. . . . *Rhapsody in Blue* seeped through the Gothic twilight of Oxford and gave us all the fidgets.

SPOONER

From 1903 to 1924 the Warden of New College was the sensible, beloved, but comical William Spooner, 'The Spoo'. Altogether he spent sixty-two years at the college. He was an albino, and suffered from a slight speech impediment which, affectionately exaggerated by generations of undergraduates, gave his name to the English language in the word 'Spoonerism'—a usage which entered the Oxford English Dictionary during his own lifetime.

SPOONER'S DECISION

Towards the end of my last year at New College [1904] I was honoured with an invitation to wait upon the Warden in his lodgings at 10 a.m. . . .

'Ah, Mr. Woolley,' began the Warden, 'Quite so. I think that when you came up to Oxford you had every intention of taking Holy Orders?'

I murmured something unintelligible and waited.

'And I am afraid that you have quite abandoned the idea?'

'Oh rather', I said, hurriedly, 'yes, quite, Mr. Warden, quite given it up.'

'And what do you propose to do?'

'Well,' I answered, 'I want to be a schoolmaster, I've done a little at odd times and like it awfully, so I think of going in for it permanently.'

'Oh, yes; a schoolmaster, really; well, Mr. Woolley, *I* have decided that you shall be an archaeologist.'

There was no more to be said.

Leonard Woolley, Dead Towns and Living Men, 1920.

Sir Leonard Woolley became 'Woolley of Ur', excavator of the great Chaldean city in Iraq.

SPOONER'S WARNING

One by one we were taken up to him for a little conversation. When my turn came he put to me a few conventional questions—what School was I reading? Had I pleasant rooms? What was the general nature of my interests?—and when these were duly answered he went on to say: 'I hope you will have a happy time at Oxford. I am sure you will. But, if you will take the advice of an old man'—I am a little doubtful about that introductory sentence, not at all about the words that followed—'*beware of the lure of men and women*'. This surprising injunction, quickly and quietly uttered, impressed itself indelibly upon my mind, and though I have never been sure what exactly was the warning it was intended to convey, I have always done my best to follow the general line it seemed to indicate.

John Sparrow, quoted in William Hayter, Spooner, 1977.

SPOONER'S TUTORIAL

His oddities were numerous. I was to dine with him alone one night, and arriving rather early, was asked to go straight to his bedroom and talk to him while he finished dressing. I found him struggling with his tie. We chatted a little. The door to an adjoining room was slightly ajar, and I heard a low drone coming from it. Presently the tie was satisfactorily adjusted, and we were just about to leave when 'Wait a minute,' he said, 'I'd forgotten.' He went to the door I have spoken of, opened it wider, put his head in, and said, almost peevishly, 'Very bad, very bad indeed. Write for next week on the Epistle of St. Paul to the Ephesians.' Then we went down and dined.

Victor Gollancz, My Dear Timothy, 1952.

SPOONER'S HOUSE

College tradition avers that Mrs. Spooner . . . thought that an Edwardian head of house should live on an episcopal scale, and certainly the reconstruction that followed [Spooner's election to the Wardenship] created an almost princely mansion. There were sixteen bedrooms, and servants' quarters that included a larder, a game-larder, a scullery, a pantry, a servants' hall, a housekeeper's room, besides the huge fourteenth-century kitchen, with its stone-flagged floor and great coal-burning range, and various other annexes. A row of extra bedrooms was built on the roof to accommodate the eleven indoor servants, and there was another row on the second floor for the Spooner children. . . . The reception rooms were on a grand scale, and there were stabling for four horses, a harness room and a big coach-house in the fourteenth-century Warden's Barn, which also contained quarters for the gardener and the groom.

William Hayter, *Spooner*, 1977.

Sir William Hayter, a Warden of New College too, lived in the house himself from 1958 to 1976: but by then it was smaller again.

SPOONER'S TABLE

Dinner table was filled with things. Elaborate flower pieces, two enormous silver mugs, tankards, I suppose, little clear glass pitchers of water, old silver, large salt cellars and pepper mills, silver dishes with the dessert around the flowers, candles and silver mugs. Two perfect maids in dark dresses with white vest fronts and collars and black ties, white caps. Thick brown soup, fish, partridges, sort of omelet, gelatine dessert with fruit in it, melon, apples, bananas, grapes. . . .

Mrs. W. W. Campbell, 1903, quoted in William Hayter, *Spooner*, 1977.

SPOONER'S MEMORIAL

Thanks largely to Spooner's intervention, a plaque was erected in New College Chapel recording the names of New College men who had died fighting on the enemy side in the First World War. Its wording may thus stand as his memorial, as well as theirs:

IN MEMORY OF THE MEN OF THIS COLLEGE WHO COMING FROM A FOREIGN LAND ENTERED INTO THE INHERITANCE OF THIS PLACE AND RETURNING FOUGHT AND DIED FOR THEIR COUNTRY IN THE WAR 1914–1919

Prinz Wolrad-Friedrich zu Waldeck-Pyrmont
Freiherr Wilhelm von Sell
Erwin Beit vom Speyer

In the sermon I have just preached, whenever I said Aristotle, I meant St. Paul.

'Mr. Coupland, you read the lesson very badly.'
'But, Sir, I didn't read the lesson.'
'Ah, I thought you didn't.'

'Do come to dinner tonight to meet our new Fellow, Casson.'
'But Warden, I *am* Casson.'
'Never mind, come all the same.'

Spooner: Ah, let me see, what is your initial?
Undergraduate: V.
Spooner: And V. stands for what?
Undergraduate: Victor.
Spooner (after a short pause): Victor what?

Kinquering Kongs their Tikles Tate.

You have tasted a whole worm. You have hissed my mystery lectures. You will leave by the town drain.

Which of us has not felt in his heart a half-warmed fish?

Undergraduates recur.

BIBLIOGRAPHICAL NOTE

Titles of some Oxford student periodicals, with dates of first issue:

The Aunt, 1919.
The Barge, 1900.
The Best Man, 1906.
The Birch, 1795.
The Boost, 1920.
The Bulldog, 1896.
The Buller, 1913.
The Bump, 1898.
The Bust, 1910.
The Censor, 1813.
The Chaperon, 1910.
The Clown, 1891.
The Comet, 1886.
The Cornstalker, 1898.
The Crier, 1925.

The Ephemeral, 1893.
The Farrago, 1816.
The Fritillary, 1894.
The Goat, 1919.
The Harlequin, 1866.
The Infant, 1919.
The Inspector, 1804.
The Jester, 1902.
The Jokelet, 1886.
The Loiterer, 1789.
The Magnum, 1908.
The May Bee, 1900.
The Meteor, 1911.
The New Rattle, 1890.
The Octopus, 1895.
The Pipe, 1900.
The Proctor, 1896.
The Right Thing, 1916.
The Scarlet Runner, 1899.
The Spout, 1919.
The Squeaker, 1893.
The Squib, 1908.
The Umbrella, 1905.
The X, 1898.

Many magazines failed after the first issue, among them Ye Tea-potte, Oxenforde (*1898*), *and* Hush (*1920*), *which consisted in its entirety of eight blank pages. Among those whose early writings appeared in Oxford student magazines, at one time or another, were W. H. Auden, Max Beerbohm, Hilaire Belloc, John Betjeman, Laurence Binyon, Lewis Carroll, Grahame Greene, Richard Hughes, Compton Mackenzie, Louis MacNeice, William Morris, Alfred Noyes, A. T. Quiller-Couch, Stephen Spender, Algernon Swinburne, Evelyn Waugh and Oscar Wilde.*

TO BE FREUD

The first savant from a German University to address an Oxford audience since the War, Dr. Emil Busch, of the University of Frankfurt, discussed before a crowded audience in the Grand Jury Room, Town Hall, Oxford, on Saturday night, his views of Freudian psychology. . . . A little man, in early middle life, with a large head, keen pale blue eyes, blonde wavy moustache, and a quick impatient manner, and staccato sentences, [Dr. Busch] would pass for French more easily than German. He admires Freud, and was almost too kindly in giving him a very fair exposition.

At the same time, his own view is that repressed sexual desire and submerged memory do not themselves supply the whole explanation of exceptional cases of human psychology. He believes rather in the existence of two or more personalities against a mental background yet to be defined, a screen on which the figures of our mental cinema move. Somewhere in the background will be discovered, he believes, a link which welds the personalities together in unity, and this is the ego.

We were a wall, and neither Hume nor Freud had allowed enough for the cement which held the bricks together. There was somewhere a mental background, against which, so to speak, all these things must be ranged, a screen on which the figures of our mental cinema moved, the strongest thing that was in us. It was a sort of microcosm, an inter-consciousness. . . .

In reply to questions, Dr. Busch said it was possible that the discoveries of Prof. Einstein would influence the research, but it could hardly affect yet what was so far only the indefinite speculation of a few savants. In any case, a mathematical discovery would only affect certain applications of the principle.

He did not agree that man was the slave of circumstance. Man could, he believed, influence his career largely by auto-suggestion. Asked whether the background was something additional to the sum total of the combined personalities, he thought not, as it was apparently more in the nature, so to speak, of a fluid.

From the *Oxford Chronicle*, 17 March 1922.

In this, one of the best of Oxford hoaxes, Dr. Busch was impersonated by George Edinger of Balliol, later a distinguished journalist, and several heads of colleges were among the audience, invited there by the 'Home Counties Psychological Association'. Psychiatry was all the rage then— as another Oxford humorist put it:

> Joy was it in that dawn to be Freud,
> But to be Jung was very Heaven.

SUNSHINE SKETCHES

My private station being that of a university professor, I was naturally deeply interested in the system of education in England. I was therefore led to make a special visit to Oxford and to submit the place to a searching scrutiny. Arriving one afternoon at four o'clock, I stayed at the Mitre Hotel and did not leave until eleven o'clock next morning. The whole of this time, except for one hour spent in addressing the undergraduates, was devoted to a close and eager study of the great university.

On the strength of this basis of experience I am prepared to make the following positive and emphatic statements. Oxford is a notable university. It has a great past. It is at present the greatest university in the world; and it is quite possible that it has a great future. Oxford trains scholars of the real type better than any other place in the world. Its methods are antiquated. It despises science. Its lectures are rotten. It has professors who never teach and students who never learn. It has no order, no arrangement, no system. Its curriculum is unintelligible. It has no president. It has no state legislature to tell it how to teach, and yet—it gets there. Whether we like it or not, Oxford gives something to its students, a life and a mode of thought which in America as yet we can emulate but not equal.

If anyone doubts this let him go and take a room at the Mitre Hotel (ten and six for a wainscoted bedroom, period of Charles I) and study the place for himself.

I understand that the key to this mystery is found in the operations of the person called the tutor. It is from him, or rather with him, that the students learn all that they know: one and all are agreed on that. Yet it is a little odd to know just how he does it. 'We go over to his rooms,' said one student, 'and he just lights a pipe and talks to us.' 'We sit round with him,' said another, 'and he simply smokes and goes over our exercises with us.' From this and other evidence I gather that what an Oxford tutor does is to get a little group of students together and smoke at them. Men who have been systematically smoked at for four years turn into ripe scholars.

Now, the principal reason why I am led to admire Oxford is that the place is little touched as yet by the measuring of 'results' and by [the American] passion for visible and provable 'efficiency'. The whole system at Oxford is such as to put a premium on genius and to let mediocrity and dullness go their own way. . . . Of all the various reforms that are talked of at Oxford, and of all the imitations of American methods that are suggested, the only one worthwhile, to my thinking, is to capture a few millionaires, give them honorary degrees at a million pounds sterling apiece, and tell them to imagine that they are Henry VIII. I give Oxford warning that if this is not done the place will not last another two centuries.

Stephen Leacock, *My Discovery of England*, 1922.

IN THE SCHOOLS, 1922

*Among the undergraduates who came up to New College soon after the
First World War was a rich financier's son, Peter Ralli. According to
C. M. Bowra (Memories, 1966), he did no work at all, but his response
to one question in his History finals, 1922, became legendary. Written in
a large flamboyant hand, it said in its entirety:*

Her subjects wanted Queen Elizabeth to abolish tunnage and pound-
age, but the splendid creature stood firm.

OUR MAN IN BALLIOL

Now I look back, there seems something a little bizarre about my
Oxford days. . . . A small affair of what might have become espionage
began innocently enough in early 1924. I had read a book of short
stories by Geoffrey Moss called *Defeat* about the occupied zones of
Germany. Moss described the attempt of the French authorities in
their zone to establish a separatist Palatine Republic between the
Moselle and the Rhine. German criminals had been brought in from
Marseilles and other ports—pimps, brothel-keepers, thieves from
French prisons—to support the collaborators. Even one of the
ministers had served a prison sentence. French troops held the crowds
back while unarmed German police were beaten unconscious. . . .

I was easily aroused to indignation by cruelties not my own, and
the idea of experiencing a little danger made me write to the German
Embassy in Carlton Gardens and offer my services as a propagandist.
The *Oxford Outlook* was at my disposal, for I was the editor, and to
the *Oxford Chronicle*, a city paper, I was a regular contributor, if only
of the five-shilling love poems.

I had not expected the promptitude of the German response.
Coming back one early evening to my rooms in Balliol I found my
armchair occupied, my only bottle of brandy almost finished and a fat
blond stranger who rose and introduced himself, 'Count von Bernstorff'. He was the first secretary of the German Embassy, a man
who loved luxury and boys. . . .

My days after that seemed to be filled by Germans—there was a
very pretty Countess von Bernstorff, the diplomat's cousin, who left
a scented glove behind in my room to be added to my adolescent
harem of inanimate objects, a young man with a long complicated
title, who claimed a nobler and longer descent than the Hohenzollerns,

and a mysterious wizened narrow figure with a scarred face, Captain P., whose full name I have now forgotten. . . .

It was an odd schizophrenic life I lived during the autumn term of 1924. I attended tutorials, drank coffee at the Cadena, wrote an essay on Thomas More, studied the revolution of 1688 'from original sources', read papers on poets to the Ordinary and the Mermaid, attended debates at the Union, got drunk with friends; then 'Cross a step or two of dubious twilight, come out on the other side, the novel'. There another life began, where I exchanged last letters with the woman I loved, who was engaged to another man, wrote a first novel never to be published, the unhappy history of a black child born to white parents, and prepared plans with Bernstorff for espionage. . . .

Graham Greene, *A Sort of Life*, 1971.

EVENINGS AT THE INN

All old Oxford men can remember pleasant occasions, incidents not to be forgotten, in taverns either in the town or in the country round. There is one which I remember always with affection; its name shall never be revealed by me either to senior or junior members of the University, nor will any of the company which frequented it ever reveal its name, for they all went to the wars and none came back. It was Proctor-proof then (and I hope it still is), and, as further security, it had three separate exits by which a man might flee suddenly and obscurely. It was kept by 'Old Mother'—who allowed the use of her own ancient and private parlour, in which, indeed, she presided, filling a large horse-hair chair, her fat, bare red arms folded on her lap, and always on the table before her was a glass sometimes of stout, sometimes of gin. It cannot be said that she led the conversation and wit, for her conversation lumbered along and was full of 'Lor' now's!' and 'Fancy that's!' and of anecdotes, without beginning or end, of Jim and Jack, personages unknown to any of us. But she had a wonderful good humour and a great variety of enormous laughter, so that she was a kind of sounding-board which enlarged and reverberated the merriment of the company. Peace to her memory!

L. Rice-Oxley, *Oxford Renowned*, 1925.

SLIGGER AND THE COLONEL

No memoir of Oxford in the 1920s was complete without a mention of two dons, George Alfred Kolkhorst of Exeter, called 'the Colonel', and F. F. ('Sligger') Urquhart of Balliol. Though they do not seem as interesting in retrospect as they must have been in person, in deference to their celebrity here are John Betjeman's portraits of them both, from My Oxford, *1977.*

We nicknamed him 'Colonel' Kolkhorst, as he was so little like a colonel. . . . He wore a lump of sugar hung from his neck on a piece of cotton 'to sweeten his conversation', and at some of his parties would be dressed in a suit made entirely of white flannel, waistcoat and all. . . . He carried a little ear trumpet 'for catching clever remarks', but would swiftly put it away and yawn if they were not clever.

The don who dominated Balliol was 'Sligger' Urquhart, who held court in summer on a lawn of the garden quad. . . . He liked people to be well-born, and if possible Roman Catholic, and he gave reading parties in Switzerland.

Kolkhorst was once caught by the proctors spitting on passers-by from the top of Magdalen Tower. It was to Urquhart that Evelyn Waugh was alluding when he was heard one night chanting loudly in a quadrangle: 'The Dean of Balliol lies with men.'

PHELPS OF ORIEL

Lancelot Phelps (1853–1936), for sixty-four years a member of Oriel College, was once listening to a sermon when the preacher, quoting an episode from the Bible, interrupted himself with a rhetorical question:

Preacher: And what application, we may ask ourselves, does this Biblical incident have to our own times?

Phelps, instantly and loudly from the congregation: None whatever, sir.

A lifelong bachelor, Phelps lived in college as a Fellow, and took a cold bath in his bedroom every morning. An undergraduate living in neighbouring rooms reported that on frosty mornings he could be heard muttering to himself:

Be a man, Lancelot, be a man.

For fifteen years he was Provost of the college, and J. I. M. Stewart tells us (in My Oxford, 1977) *that he devised an intricate technique for cutting short undergraduates' tea-parties at the Provost's Lodging:*

He would lead the conversation towards some athletic topic, from this to the college games field, and from this again to the subject of badgers—which he would aver, quite baselessly, to have established a set endangering the cricket pitch. He would then recall Sir Thomas Browne's holding in debate whether or not badgers have longer legs on one side than the other, this the more readily to scamper round hills. Next, he would suddenly recall that a portrait of a badger hung somewhere in the Lodging, from which the truth of this matter might conceivably be verified. The picture would be located after a walk through the ramifying house; the badger would be seen to be equipped as other quadrupeds are; and then one would discover that the picture hung beside the Provost's front door, which stood open before one.

AN INQUIRY

Say, usher, is this a purely literary establishment, or can I get a snack here?

American visitor to Christ Church college porter, c. 1925.

THE RAPE OF OXFORD

🦡 *In 1922 William Morris started a motor-car factory at the suburb of Cowley, and Oxford became an industrial city at last. Elsewhere housing developments extended the city ever deeper into Matthew Arnold's country-side, and contemporary connoisseurs thought the University city spoilt:*

Oxford today, far from being the peerless city of our imagination, has little claim to be regarded as better than the rest of our semi-manufacturing, semi-commercial slums. The awful villa residences, the ramshackle modern street rows, petrol stations, and shops with their large commercial signs, have all but obliterated the peculiar atmosphere of one of the unique towns of the world, not to mention the nineteenth century buildings of which Keble is perhaps the crown. To sweep the mining camp away from the University is at present impossible, though much may slowly be done; but it would be ridiculous for those who have eyes to pretend that Oxford as a town today is greatly superior to Croydon or Burslem.

From the Architectural Review, 1929.

THE RAPE OF OXFORD

And now we come down Cumnor Hill. What an approach to the city of learning! What learned architecture! Here the half-timbered villa holds its own boldly beside the bogus-modern, here the bay windows and stained glass front door survey the niggling rock garden and arid crazy paving. . . . The Scholar Gypsy must wash his bronzed face in birdbaths and sleep under the shade of stone toadstools if he is still to roam the slopes of Cumnor Hill.

John Betjeman, *An Oxford University Chest*, 1938.

But not everyone was despondent:

The Past had them by the throat. Dark towers, and old half-lit stone-work, winding, built in, glimpsy passages; the sudden spacious half-lighted gloom of a chanced-on quadrangle; chiming of clocks, and the feeling of a dark and old empty town that was yet brimming with hidden modern life and light, kept them almost speechless. . . .

'This must be the heart,' said Clare. 'Oxford certainly has its points. Whatever they do to the outside, I don't see how they can spoil all this.'

John Galsworthy (1867–1933), *The End of a Chapter*.

WOMEN IN OXFORD

In the twentieth century women were accepted by Oxford at last, and took their places as fully equal members of the University. It was a very long haul. Their most influential, and most surprising, champion was Lord Curzon, who as Chancellor of the University urged that women should be eligible for degrees. Even he, though, was anxious that a degree should not lead to a vote:

To give a woman a degree is to enable her to obtain the reward of her industry or her learning. As such, it is an extension of private liberty. To give her a vote is to give her the right to govern others, and is the imposition of a public duty. Even if an academic degree were undesirable, it would do no harm but to the woman herself. But if women proved to be unfit to exercise the Parliamentary franchise, the injury would be done not to the individual female voter, but to the nation at large, since, once given, the privilege could never be withdrawn. . . . There is all the difference in the world between giving women an opportunity of increasing and improving their natural powers, and granting them a share in political sovereignty.

Principles and Methods of University Reform, 1909.

❧ *Oxford degrees for women became a fact in 1920, but this did not mean that women undergraduates were socially emancipated:*

A woman undergraduate may not enter men's rooms, either in college or in lodgings, without obtaining leave from the Principle of her Society or her representative. She must have a chaperone approved by the Principal or her representative.

A woman undergraduate must obtain leave before accepting invitations for the evening, or for mixed parties. She may not be out in the evening without permission, and must report her return.

Mixed theatre parties may not be arranged except in reserved seats. There must be at least two women in the party, and permission must be obtained beforehand.

Mixed parties may not be held in cafés before 2 p.m. or after 5.30 p.m. Between these hours they are permitted, provided that permission has been obtained beforehand, and that there are at least two women in the party.

From *Intercollegiate Rules for Women*, 1924.

❧ *Dilys Powell, later to become a celebrated cinema critic, made fun of these absurd rules in* Isis *in 1924:*

'Please,' I said, 'what is the Chaperone System for?'

'Why,' she replied with a quick smile, 'I *am* glad you have asked me that. I feel I *ought* to talk to my students about these things.'

'I should be very grateful,' I said, 'if you would talk to me.'

'We, you know—but are you *quite* comfortable there? Sure? Do you feel the draught from this window? No? Well, I think, don't you, that the reputation of our college is a great thing.'

'Reputation?' I murmured.

'Well, you know, we don't *want* to be mixed up in anything *unpleasant*, do we?'

'Unpleasant?' I repeated.

'You see, we can't have our students running wild. We want steady, responsible people at the University.'

'Please,' I said, 'is a café steadier at 2 p.m. than at 1.55 p.m.?'

'Ah, but *lunch* is so expensive, and we don't want our young men to run into debt up here, do we?'

'Then,' I said, 'is that why "there must be at least two women in the party", and why "mixed theatre parties may not be arranged except in reserved seats"?'

'Now, I'm so glad you've asked me that. Won't you have another cushion? I like my students to feel *quite* at home when they come and talk to me. Well, don't you think it's much *nicer* for girls to go out alone? So much—so much—'

'Please,' I interrupted, 'what does Nice mean?'

'Oh, but it's just—just *Nice*, isn't it? . . .'

Dilys Powell, in Isis, *1924.*

Dr. Lewis Farnell, Rector of Exeter (see page 298) was Vice-Chancellor then, and he did not sympathize with the advance of Woman:

During the war our overworked officers and men had been in the habit of taking coffee or chocolate or other café stuff, when it was possible, about eleven in the morning. . . . This trench-habit was carried home with them when the peace came, and eagerly caught up by our lazy and self-indulgent boys and girls. . . .

I put it to my Proctors—who were two of my very best—that we should put the cafés out of bounds between ten and one, as we did the bars. I pointed out that, though gay conversation with an 'under-graduette' might be more improving than conversation with a bar-maid, cups of coffee and chocolate were not morally so superior to a glass of beer as to justify this general waste of valuable morning time. . . . Then while we were discussing preliminary difficulties, the officials of the women's colleges got wind of our intentions . . . and very soon we received an earnest and anxious petition, signed, I think, by all the lady Heads; begging us not to do anything so severe against their poor girls, who could not stand the strain of going from nine to one without sustenance. We realized that a bi-sexual university has its own special difficulties. Two of us were of the opinion that it might be even good for the girls to practise such austerity. I also put it to the Proctors, in the wise words of Antony's Enobarbus, 'Under a compelling occasion, let women die.'

But the odds were against Dr. Farnell, and he dropped the proposal—'to my perpetual regret'. Another cynic about women at Oxford was Christopher Hobhouse, who attacked them in a notorious passage in his book Oxford, *1939:*

Though their numbers are so small, a casual visitor to Oxford might well gain the impression that the women form an actual majority. They are perpetually awheel. They bicycle in droves from lecture to lecture, capped and gowned, handle-bars laden with note-books, and note-books crammed with notes. Relatively few men go to lectures, the usefulness of which was superseded some while ago by the invention of the printing press. The women, docile and literal, continue to flock to every lecture with medieval zeal, and record in an hour of longhand scribbling what could have been assimilated in ten minutes in an arm-chair. Earnestly they debate the merits of their teachers—the magnetism of X, the eloquence of Y, the spirituality of Z—as though these insignificant pedants were so many Abelards. . . .

Very few of the women take the least pains to be attractive or even mature. Fifty years have not mellowed them; they still care nothing for appearance or comfort. They run no tailors' bills in the High Street, but deck themselves in hairy woollens and shapeless tweeds . . . their hair is braided into stringy buns. Their domestic background is equally repellent. Instead of a quiet pair of rooms . . . each girl has a minute green-and-yellow bed-sitter opening off an echoing shiny corridor. Instead of deep sofas and coal fires, they have convertible divans and gas stoves. Instead of claret and port, they drink cocoa and Kia-Ora. Instead of the lordly breakfasts and lunches which a man can command in his own rooms, they are fed on warm cutlets and gravy off cold plates at a long table decked with daffodils.

Hobhouse described women dons as presenting 'a terrifying caricature of the medieval tutor', but there were women, too, who disliked the style of the University's feminization:

Somerville smelt frousty to me. I disliked the ugliness of most of the public rooms, and I disliked the glass and the crockery and the way in which the tables were set. I disliked the food, and, more still, the way in which it was served. . . . And I disliked the dowdiness of the dons, and more still that of the other girls. . . . I could not bear the cloister-ishness of the place, and felt irritated by the cautious way in which we were shut off from contact with men, the air of forced brightness and virtue that hung about the cocoa-cum-missionary-party-hymn-singing girls, and still more the self-conscious would-be naughtiness of those who reacted from this into smoking cigarettes and feeling wicked. And I disliked the slightly deprecating and dowdy, and again very self-conscious, atmosphere of ladylike culture that hung about the dons at play. . . . For most of the place, quite unaware that I was watching the awkward adolescence of something infinitely worth-while, I had nothing but intolerant contempt.

The Viscountess Rhondda, *This Was My World*, 1933.

But it was infinitely worthwhile, and the style was presently to change. The presiding genius of the women's movement was Annie Rogers, who had fired some of the most telling shots of the original campaign nearly fifty years before, and who had lived to see absolute victory. She was a formidable figure. As a writer in the magazine The Ship *recalled in 1937,*

. . . it was to Miss Rogers that we fled to borrow sugar in the war days when visitors came unexpectedly, sugar lent with the brief comment: 'Men, I suppose; parsons? I thought so.'

🦋 *But when success came she was magnanimous, as she showed in her memoir* Degrees by Degrees *(1938):*

Women now held an assured position in the University. . . . We were in the wilderness—not with very great murmuring and discontent— for more than 40 years, but when we entered the promised land we came in peacefully, not with shouts and blowing of rams' horns, and Jericho opened its gates and welcomed us with smiles.

🦋 *Still, an earlier generation of Oxford men remembered with a barely-concealed nostalgia the days when there were women* of *Oxford, so to speak, but not* in *it:*

There were no women. Ours was an entirely masculine, almost monastic, society. We knew of course that there were women's colleges with women students. But we were not conscious of either. Their colleges were situated on the suburban periphery. Their students never came into our college rooms. . . . They were not members of the Union. They joined no political societies. If they came to lectures they were escorted by a chaperone or duenna. For practical purposes they did not exist.

Harold Macmillan, in The Times, 1975.

Woman-free were our lives . . . we had no truck with girls in our courts and quadrangles and would certainly have regarded their daily and casual invasion of these sanctuaries as an interruption and a bore. Let them come, less as fellow-creatures than as a distant species, lightly touched by mystery, to Eights Week or to Commemoration Balls, but never to disturb our brave masculine preoccupations!

L. E. Jones, An Edwardian Youth, 1956.

BETJEMANESQUE

🦋 *John Betjeman, one of the best-known undergraduates of the 1920s, was sent down from Oxford when he failed to pass the then compulsory divinity examination ('Divvers'). He told the story of his Oxford career in his verse autobiography* Summoned by Bells *(1960), from which these extracts are taken:*

Balkan Sobranies in a wooden box,
The college arms upon the lid; Tokay
And sherry in the cupboard; on the shelves
The University Statutes bound in blue,
Crome Yellow, *Prancing Nigger*, Blunden, Keats.
My walls were painted Bursar's apple-green;
My wide-sashed windows looked across the grass
To tower and hall and lines of pinnacles.
The wind among the elms, the echoing stairs,
The quarters, chimed across the quiet quad
From Magdalen tower and neighbouring turret-clocks,
Gave eighteenth-century splendour to my state.

'Harry Strathspey is coming if he can
After he's dined at Blenheim. Hamish says
That Ben has got twelve dozen Bollinger.'
'And Sandy's going as a matelot.'
'I will not have that Mr. Mackworth Price;
Graham will be so furious if he's asked—
We do *not* want another ghastly brawl'—
'Well, don't ask Graham, then.' 'I simply must.'
'The hearties say they're going to break it up.'
'Oh no, they're not. I've settled *them* all right,
I've bribed the Boat Club with a cask of beer.'
Moon after parties; moon on Magdalen Tower,
And shadow on the place for climbing in . . .
Noise, then the great, deep silences again.

On tapestries from Brussels looms
    The low late-'20s sunlight falls
In those black-ceilinged Oxford rooms
    And on their silver-panelled walls;
ARS LONGA VITA BREVIS EST
Was painted round them—not in jest.

And who in those days thought it odd
    To liven breakfast with champagne
And watch, in Canterbury Quad,
    Pale undergraduates in the rain?
For, while we ate Virginia hams,
Contemporaries passed exams.

Failed in Divinity! Oh count the hours
Spent on my knees in Cowley, Pusey House,
St. Barnabas', St. Mary Mag's, St. Paul's,
Revering chasubles and copes and albs!
Consider what I knew of 'High' and 'Low' . . .
Failed in Divinity! O, towers and spires!
Could no one help? Was nothing to be done?
No. No one. Nothing. Mercilessly calm,
The Cherwell carried under Magdalen Bridge
Its leisured puntfuls of the fortunate
Who next term and the next could still come back. . . .
    Outside, the sunny Broad,
The mouldering busts round the Sheldonian,
The hard Victorian front of Exeter,
The little colleges that front the Turl,
The lean acacia trees in Trinity,
Stood strong and confident, outlasting me.

In 1938 Betjeman published An Oxford University Chest, *an Oxford pot-pourri, which included this skit on conversation in a senior common room of the day:*

'Read the latest Dornford Yates, Professor?'

'I have not yet given myself that pleasure, H.J., doubtless you, as a philosopher, have to keep in touch with all the latest authorities.'

'Come, come, Professor. That is hardly fair. I am, *nescio-quid*, able for a moment to allow myself the leisure of reading lay literature.'

'Lay literature. Is it something to do with poultry? Enlighten me, Mr. Domestic Bursar.'

'I think H.J. is referring to a humorous writer. Mr. Yates rejoices in that reputation.'

'Thank you. Thank you. No doubt you are referring to a philosopher of Humour. . . .'

'Not at all, Professor. I am referring to the humour of philosophy.'

'Surely humour, H.J., is an affection of the will. As far as I recollect —pardon a blundering Numismatist attempting to correct the University Lecturer in Experimental Philosophy—as far as I recollect Schopenhauer defines it as such.'

'No, Professor, that is his definition of joy. He says joy and sorrow are not ideas of the mind.'

'Too subtle, too subtle, H.J. I must accuse myself of what Theophrastus calls ἀναισθησία καὶ βραδυτὴς ψυχῆς.'

And so on till well towards midnight.

COMING TO TERMS · 1914–1945

🦎 *As Matthew Arnold celebrated the Oxford countryside, so Betjeman in later life captured better than anyone else the essence of North Oxford, the ample Victorian suburb of dons, wistaria, and preparatory schools:*

> Belbroughton Road is bonny, and pinkly burst the spray
> Of prunus and forsythia across the public way,
> For a full spring-tide of blossom seethed and departed hence,
> Leaving land-locked pools of jonquils by a sunny garden fence.
>
> And a constant sound of flushing runneth from windows where
> The toothbrush too is airing in this North Oxford air.
> From Summerfields to Lynam's, the thirsty tarmac dries,
> And a Cherwell mist dissolveth on elm-discovering skies. . . .
>
> And open-necked and freckled, where once there grazed the cows,
> Emancipated children swing on old apple boughs,
> And pastel-shaded book rooms bring New Ideas to birth
> As the whitening hawthorn only hears the heart beat of the earth.
>
> From *New Bats in Old Belfries*, 1945.

🦎 *And best of all, uniquely perhaps, Betjeman caught the hint of poignancy that lingers always in academic Oxford, whose constant stream of young life, passing exuberantly year by year, leaves many pools behind:*

Sudden Illness at the Bus Stop

> At the time of evening when cars run sweetly,
>   Syringas blossom by Oxford gates.
> In her evening velvet with a rose pinned neatly
>   By the distant bus-stop a don's wife waits.
>
> From that wide bedroom with its two branched lighting
>   Over her looking-glass, up or down,
> When sugar was short and the world was fighting
>   She first appeared in that velvet gown.
>
> What forks since then have been slammed in places?
>   What peas turned out from how many a tin?
> From plate-glass windows how many faces
>   Have watched professors come hobbling in?
>
> Too much, too many! so fetch the doctor,
>   This dress has grown such a heavier load
> Since Jack was only a Junior Proctor,
>   And rents were lower in Rawlinson Road.
>
> From *Old Lights for New Chancels*, 1940.

THE OXFORD VOICE

When you hear it languishing
and hooing, cooing, sidling through the front teeth,
   the oxford voice
     or worse still
   the would-be oxford voice
you don't even laugh any more, you can't.

For every blooming bird is an oxford cuckoo nowadays,
you can't sit on a bus nor in the tube
but it breathes gently and languishingly in the back of your neck.

And oh, so seductively superior, so seductively
   self-effacingly
   deprecatingly
     superior.—
We wouldn't insist on it for a moment
   but we are
     we are
   you admit we are
     superior.—

D. H. Lawrence, Pansies, 1929.

1930s: AN INDIAN VIEW

[A] universal feature of social life is 'eleven o'clock coffee'. A Fleet
Street journalist, if he spent a week-end in Oxford, would probably
speak of small tables neatly arranged—a radiogram or an orchestra
striking jazz—the *naive* freshman looking intriguedly into the blue
eyes of the little Scandinavian, who has come to Oxford only 'to
learn English'—the third-year man so fed up and terribly blasé—the
straight-haired undergradu*ette* with her cap and gown and twenty-page
essay on 'domesticity and its advantages', blushing at her own thoughts,
as she admires from the distance the physique of some 'hearty' with
his sweater and scarf and college tie, or the frail beauty of some *nice*
boy in the corner—Buchmanites hot-gospelling—Union speakers
coining epigrams—the intellectual Tory assuring the intellectual
Socialist in low whispers that 'the purpose of industry is not to make
profits, but only to make industry pay'—and all over a cup of coffee!

Crossing the quad there is already the distant murmur of voices.
Going up the staircase, chattering and occasional laughter is heard.

There is no need to knock. In the room itself, which is enveloped in smoke, are some twenty or thirty people, dressed perhaps a little more carefully than usual—the women seldom so smartly turned out as the men. . . . On a table are bottles of James Brown's delightful sherry—both dark and pale—with a varied assortment of odd glasses. 'Help yourself' is the only idea that the general environment seems to suggest. Now and then someone winds the gramophone. The music varies with the individual who happens to be nearest the records—sometimes the long-drawn strains of a sentimental waltz, sometimes the shrieks of some coloured band, 'producing the type of music that burns holes in the carpet', sometimes even the vulgar notes of a licentious rumba. Someone from the *Isis* staff is probably lurking about, ready to jot down incautious utterances for publication in their next issue, as 'heard in the High'. . . .

In the Raleigh, the Durham, and the Chatham they are discussing politics. In the Williams and the Harwicke they are mooting over *Leslie* v. *Shield* or whether the snail had anything to do with Lord Atkin's judgement in *Donoghue* v. *Stevenson*. In the Lotus and the Morley someone is always reading a paper on something. In their individual rooms small groups of friends are still discussing whether sex is worthy of discussion, whether the nationalization of banks is in the interest of the working classes, or even whether the National Government had anything to do with *Cavalcade*. Somewhere near Holywell the voice of the Bach choir is heard, somewhere in the Giler, if it is Saturday, the voice of the proctor: 'Name and college, please'. What an immense range of activity! How impossible it is to generalize. Yet all these are the various elements that comprise the great university, and form some part of the social life of some section of the undergraduate world.

D. F. Karaka, *The Pulse of Oxford*, 1933.

1930s: A CAMBRIDGE VIEW

🐝 *In 1933 a muscular young original from Cambridge, Tom Harrisson, appeared in Oxford and, settling down at a riverside inn near the city, proceeded to write a blistering* Letter to Oxford, *which he published privately. He did not like 'Oxen':*

Oxen form naturally into small herds (flocks). . . . Within these carefully organized units of intellectualism, these mass-masturbation meetings, they can dogmatize the angels out of paradise. But they are dimly conscious that outside their unit, in the wider world of UnOxen, they couldn't get away with it, and they are afraid, deadly afraid, of committing themselves to anything—that is obvious from everything said and written in Ox. They are picadors, bad ones; never a man with a sword who brings the bull with perfect posture to the centre of the ring.

What is the effect of Oxparadisity? It is to exaggerate certain tendencies out of all sense, reduce others beyond existence. The primary emphasis is on talk, and on ideas. Talk, talk, talk; a seething whirlpool of words and ideas and Thingsthatmatter. . . . Unreality, the cult of unreason, the decay of positive feeling or positive faith, the what-is-there-left-to-believe-in-cry? You silly bastards. You Oxen, you Wet Oxen. You have nothing to do with life, you are unreal, obscene and exceedingly funny.

Oxen (and Oxwomen) are the worst and most inefficient sexual units I have ever known. Their intellectual fervour has extended to sex. Sex has become supra-mental rather than physico-mental. Masturbation is dominant. Perversion flourishes. And everywhere Oxen try to deny sex; they hate the inevitability of sex in themselves; along with their bladders and themselves. . . . The common denominator of all sex activities is ignorance and incompetence and timidity. There never were worse sex-hounds than Oxen.

ESSENTIALS FOR SUCCESS

Among the essentials for 'success' in Oxford, the following are profoundly important: Possessing (at least) one pair of plus fours; a repertoire of pornographic stories; some skill, legendary or otherwise, at golf; a Morris car; a sneer on your face; and an exhaustless capacity for suppurating self-conceit.

J. G. Sinclair, *Portrait of Oxford*, 1931.

PITMAN GOES HOME

A Yorkshire miner, A. A. Eaglestone, having taken an Oxford degree on a Ruskin College scholarship, describes pseudonymously his return home:

Goodbye, university! We run out; the meadows, Godstowe, and the low wall of the nunnery wheel to one's left shoulder. Now all Oxford falls away, a flicker of the conscious, a dithyrambic dancing to the wheels . . . sunshot shadows with the cattle and the clover—C. E. M. Joad, in a blue shirt, speaking—Big Tom Tower bell groaning out its plaint—Mr. Harold Acton faintly mincing—gang of roaring 'hearties' on the tavern door-step—handkerchief of blackberries on the slope of Cumnor—ukelele prinking through the cleft of Holywell—pitter-patter footsteps in the hollow Camera—old port, biscuits, waiters, butlers—white stone, grey stone, greenstone, cloisters—black and white of Proctors—sober gait of 'bull dogs'. . . .

Now headstocks, coal ranged in many wagons, flicker of pulleys, ochre of fumes, the clatter of clogs. A porter still stands before the long grille of the cinema. At the four lane-ends are men who sat with me on the same benches at school. The bus fills up at the bottom of the hill. Here of all people is the horse-keeper. He grins broadly; 'How do?'

'How do you do?'

'Home again, lad?'

'Yes.'

'For good?'

'AYE. . . .'

Roger Dataller, *A Pitman Looks at Oxford*, 1933.

EINSTEIN IN OXFORD

On two successive summers (1931 and 1932) we had Einstein living with us in Christ Church and he dined on most evenings. He was a charming person, and we entered into relations of easy intimacy with him. He divided his time between his mathematics and playing the violin; as one crossed the quad, one was privileged to hear the strains coming from his rooms. In our Governing Body I sat next to him; we had a green baize table-cloth; under cover of this he held a wad of paper on his knees, and I observed that through all our meetings his pencil was in incessant progress, covering sheet after sheet with equations. His general conversation was not stimulating. . . . I am afraid I did not have the sense that, so far as human affairs were concerned, I was in the presence of a wise man or a deep thinker. Rather I had the idea that he was a very good man, a simple soul and rather naive about worldly matters. He had his little fund of amusing stories on an unsophisticated level. Victor Cazalet was dining with us on one occasion during the first summer, and happened to return during the second. He had a proper respect for great people, and on the second occasion he drew from his pocket a little diary. 'Oh, Dr. Einstein,' he said, 'will you be very kind and tell us that story again that you told us last year about the bank director and the cow?'

R. F. Harrod, *The Prof*, 1959.

NOT A SPECIALITY

Scholarly Visitor: Are you interested in incest, Professor?
Gilbert Murray (*1866–1957*), *Regius Professor of Greek*: In a general way.

SOMETHING ABOUT THE PLACE

'There's something about this place,' said Peter presently, 'that alters all one's values.' He paused, and added a little abruptly: 'I have said a good deal to you one way and another, lately; but you may have noticed that since we came to Oxford I have not asked you to marry me.'

'Yes,' said Harriet, her eyes fixed upon the severe and delicate silhouette of the Bodleian roof, just emerging between the Sheldonian and the Clarendon Building. 'I had noticed it.'

'. . . But I will ask you now, and if you say No, I promise you that this time I will accept your answer. Harriet; you know that I love you; will you marry me?'

The traffic lights winked at the Holywell Corner: Yes; No; Wait. Catte Street was crossed and the shadows of New College walls had swallowed them up before she spoke;

'Tell me one thing, Peter. Will it make you desperately unhappy if I say No?'

'Desperately? . . . My dear, I will not insult either you or myself with a word like that. I can only tell you that if you will marry me it will give me very great happiness.'

They passed beneath the arch of the bridge and out into the pale light once more.

'Peter!'

She stood still; and he stopped perforce and turned towards her. She laid both hands upon the fronts of his gown, looking into his face while she searched for the word that should carry her over the last difficult breach.

It was he who found it for her. With a gesture of submission he bared his head and stood gravely, the square cap dangling in his hand.

'*Placetne, magistra?*'

'*Placet.*'

The Proctor, stumping grimly past with averted eyes, reflected that Oxford was losing all sense of dignity. But what could he do? If

Senior Members of the University chose to stand—in their gowns, too!—closely and passionately embracing in New College Lane right under the Warden's windows, he was powerless to prevent it.

Dorothy L. Sayers, *Gaudy Night*, 1935.

INJUNCTION, 1930s

NO CHARS-A-BANC ALLOWED HERE

Notice, Queen's Lane.

MAURICE BOWRA

🦋 *The most considerable 'character' of Oxford in the 1930s was Maurice Bowra, who returned from the First World War to be a Fellow of New College and later Warden of Wadham. He was not a very inspiring writer, but with his squat aggressive form, his bullet-head, his brilliantly catty tongue and his often bawdy humour, he seems to have had a liberating effect upon his younger contemporaries, and in return, by innumerable memoirs and by word of mouth, they made him the best-known Oxford don of his day. Here are some characteristic recollections:*

HATES AND PLEASURES

Immensely generous, Bowra entertained a great deal at Wadham. . . . The dinner-parties were of six or eight, good college food, lots to drink, almost invariably champagne, much laughter and gossip, always a slight sense of danger. This faint awareness of apprehension was by no means imaginary, because the host could easily take offence (usually without a visible sign, except to an expert) at an indiscreet word striking a wrong, anyway personally unpleasing note, in dialogues which were, nevertheless, deliberately aimed at indiscretion. Bowra's reaction was likely to be announced a day or two later.

'What so-and-so said the other night has just come back as Bad Blood'. . . .

The impact on myself, as an undergraduate, of Bowra's personality and wit is not easy to define, so various were its workings. If the repeated minor shock from this volcano took many forms, their earliest, most essential, was a sense of release. Here was a don— someone by his very calling, anyway to some extent, suspect as representative (in those days) of authority and discipline, an official promoter of didacticism—who so far from directly or indirectly attempting to expound tedious moral values of an old-fashioned kind, openly praised the worship of Pleasure. . . .

Anthony Powell, in Maurice Bowra: A Celebration, 1974.

THE ATTACK

Perhaps it would be Mr. B——. M——. The attack, to begin with, would be manœuvre. Mr. Bowra would collect himself. Builded for war, his compactly constructed form would no longer offer a perch to any propitiatory dove. One watched the guns swinging out over his sides. One visualised the stokers working inside him, preparing for even more knots. . . . Suddenly Mr. Bowra's guns would deliver the preliminary broadside. 'Of all men the most *boastful*, the most inexplicably *vain*, and *wicked* . . . etc.' At the end, as you can guess, Mr. B——. M——. was done for indeed.

Brian Howard, in Cherwell, 1926.

GRAND OPERA

Every time one entered that small, ground-floor room, in the Warden's Lodgings, where he was accustomed to receive visitors, one's spirits sank afresh. No trace of personality did it reveal; the furniture was as severely arranged as in a provincial French salon or a Harley Street waiting-room; no cigarette ends sullied the ash trays, and the blank expanse of the table top was unrelieved by any scrap of paper or empty glass or casually laid aside book. The curtains and chair covers were provocatively undemonstrative, and the few pictures on the walls were tributes to his friendship with the artists responsible rather than the outcome of any process of aesthetic selection. Everything was protectively neutral. And then the door would be flung open and the whole atmosphere was at once changed and re-charged.

So great was the warmth and excitement which the Warden's presence immediately generated that the bleakness of the setting seemed not just irrelevant but justified. Any more elaborate décor, hinting at personal taste or revelatory of character . . . could only, one now realized, have proved distracting. Like Verdi's *Falstaff*, Maurice was a grand opera that needed no overture.

Osbert Lancaster, in Maurice Bowra: A Celebration, 1974.

BOWRA TALK

After meeting an exotically dressed woman lecturer in French Literature: All the colours of the Rimbaud.

About his affection for a rather plain young lady: Buggers can't be choosers.

The sort of man who'd give you a stab in the front.

Awful shit: never met him.

The Master of Balliol has been ill but unfortunately is getting better. Otherwise deaths have been poor for the time of the year.

I'm a man more dined against than dining.

THREE POETS

Three dominant poets of the 1930s, poets who set the taste of half a generation, were at Oxford. Here are examples of their varied responses:

Louis MacNeice, Merton College, on the Oxford education:

> . . . certainly it was fun while it lasted
>   And I got my honours degree
> And was stamped as a person of intelligence and culture
>   For ever wherever two or three
> Persons of intelligence and culture
>   Are gathered together in talk
> Writing definitions on invisible blackboards
>   In non-existent chalk.
> But such sacramental occasions
>   Are nowadays comparatively rare;
> There is always a wife or a boss or a dun or a client
>   Disturbing the air.
> Barbarians always, life in the particular always,
>   Dozens of men in the street,
> And the perennial if unimportant problem
>   Of getting enough to eat.
>
>
> But in case you should think my education was wasted
>   I hasten to explain
> That having once been to the University of Oxford
>   You can never really again
> Believe anything that anyone says and that of course is an asset
>   In a world like ours;
> Why bother to water a garden
>   That is planted with paper flowers?
>
> From *Autumn Journal*, 1938.

THREE POETS

Stephen Spender, University College, on undergraduate life:

Oxford was not as I had imagined it would be: I soon discovered that I was a new boy among public-school boys, who thought that not to come from a public school was as ridiculous as to be a foreigner.

I took revenge on them for disappointing me, by becoming self-consciously their opposite. I became affected, wore a red tie, cultivated friends outside the college, was unpatriotic, declared myself a pacifist and a Socialist, a genius. I hung reproductions of paintings by Gauguin, Van Gogh and Paul Klee on my walls. On fine days I used to take a cushion into the quadrangle, and sitting down on it read poetry.

Affectation is an aping of hidden, outrageous qualities which are our real potentialities. I aped my own exhibitionism, effeminacy, rootlessness and lack of discipline.

One day the other freshmen decided that the time had come when they should break up my rooms. They decided this not out of enthusiasm but on principle, because it was the correct thing to do. I was sitting in a chair reading Blake when about a dozen of them trooped in, equipped with buckets and other clanking instruments of room-breakers and throwers-into-rivers. I could not decide on the most suitable way of receiving them, so I went on reading, very conscious of course that I was reading poetry. They were as embarrassed as I. They stood about in an awkward semi-circle. One of them said: 'What's the big idea, Spender?' For reply I read aloud a few lines of Blake. I achieved the result: they simply changed their minds and left the room, shrugging their shoulders as though to indicate that I was too crazy for their treatment. . . .

The hearties despised the aesthetes, and regarded anyone who showed any tendency to interest himself in the arts as an aesthete. At Univ. the two or three college aesthetes were certainly sickly young men. They called one another 'dear' and burned incense in their rooms.

World Within World, 1951.

W. H. Auden, Christ Church, on Oxford itself:

> Nature invades: old rooks in each college garden
> Still talk, like agile babies, the language of feeling,
> By towers a river still runs coastward and will run,
>   Stones in those towers are utterly
>   Satisfied still with their weight.
>
> Mineral and creature, so deeply in love with themselves
> Their sin of accidie excludes all others,
> Challenge our high-strung students with a careless beauty,
>   Setting a single error
>   Against their countless faults.

Outside, some factories, then a whole green county
Where a cigarette comforts the evil, a hymn the weak,
Where thousands fidget and poke and spend their money:
    Eros Paidagogos
    Weeps on his virginal bed.

And over this talkative city like any other
Weep the non-attached angels. Here too the knowledge of death
Is a consuming love, and the natural heart refuses
    A low unflattering voice
    That sleeps not till it find a hearing.

🦎 *And Auden again on Oxford donnishness:*

The Oxford Don: I don't feel quite happy about pleasure.

From The Orators, 1932.

A FAMOUS DEBATE

OXFORD UNION SOCIETY

Question for Debate

'That this House will in no circumstances fight for its King and Country'

Moved by Mr. K. H. DIGBY, St. John's

Opposed by Mr. K. R. F. STEEL-MAITLAND, Balliol

Mr. D. M. GRAHAM, Balliol, Librarian, will speak third

THE HON. QUINTIN HOGG, Christ Church and All Souls, Ex-President, will speak fourth

Mr. C. E. M. JOAD, Balliol, will speak fifth.

Tellers

For the Ayes	For the Noes
Mr. M. Beloff, C.C.C.	Mr. R. G. Thomas, Brasenose

Agenda for the Oxford Union debate of 9 February 1933.

🦎 *This was the most celebrated debate in the history of the Oxford Union, for the passing of the motion by 275 votes to 153 was widely taken to demonstrate the degeneracy of Oxford, and so of Young England. The national press gave the debate enormous publicity, but varied in its responses. Here are three editorial views:*

Daily Express: There is no question but that the woozy-minded Communists, the practical jokers, and the sexual indeterminates of

Oxford have scored a great success in the publicity that has followed this victory. . . . Even the plea of immaturity, or the irresistible passion of the undergraduate for posing, cannot excuse such a contemptible and indecent action as the passing of that resolution.

Manchester Guardian: The obvious meaning of this resolution [is] youth's deep disgust with the way in which past wars 'for King and Country' have been made, and in which, they suspect, future wars may be made; disgust at the national hypocrisy which can fling over the timidities and follies of politicians, over base greeds and communal jealousies and jobbery, the cloak of an emotional symbol they did not deserve.

The Times: Children's Hour.

While this is how one private citizen reacted, in a letter to the Isis:

Dear Sir

I don't know much about the Oxford Union, but I judge from the report that the majority of its members have declared that they will not endanger their precious skins in fighting for their country, that the Union consists chiefly of aliens and perverts. It is a pity that the sweet creatures' names are not published. The police would find them useful.

Joseph Banister

Within the University, where the debate had been considered nothing special, the furore caused surprise and resentment:

No one could nowadays accuse those who refuse to fight of wanting to save their skins. The experts tell us that in the event of war the entire city population of England will in a space of a few days be literally annihilated. . . . If a few lunatics make war, the remainder will suffer. And most of them will not even know what it is all about.

If anyone calls us cowards, we will gladly black his eye.

Editorial in the *Isis*, February 1933.

At a later Union debate, Randolph Churchill, Winston Churchill's son, having induced many life members to come to Oxford to vote, proposed a motion to expunge the 'King and Country' decision from the society's records, but this only angered student opinion more, and he was defeated by 750 votes to 138. It was all hypothetical anyway. When the time came, not so many years on, this very generation of Oxonians fought as willingly as ever for King and Country, and often died.

BIBLIOGRAPHICAL NOTE

🐵 *Some twentieth-century fiction set in Oxford:*

Death at the President's Lodging, Michael Innes.
Murder M.A., A. Kennington.
Obsequies at Oxford, R. B. Montgomery.
Landscape with Dead Dons, R. Robinson.
Murder at Pentecost, Dorothy Sayers.
Coffin in Oxford, G. Butler.
Death Lives Next Door, G. Butler.
Who Dies?, S. P. B. Mais.
The Mummy Case Mystery, D. Morrah.
Little Victims, R. Rumbold.
The Day they Burned Miss Termag, D. Balsdon.
We Have Been Warned, N. Michison.

🐵 (*Compare page* 290.)

'THE PROF'

🐵 *'The Prof' was the universally-known nickname of F. A. Lindemann (1886–1933), later Lord Cherwell, Professor of Experimental Philosophy and the father of modern physics in Oxford. A rich man, born in Germany, he lived in Christ Church and stamped his personality memorably upon that not easily dented college:*

He had been assigned a spacious set [of rooms] in Meadow Buildings, overlooking the Christ Church meadow, with a distant view of the trees that bordered the river, whose name he later took for his title. He had done them up lavishly, with all regard to comfort, and installed a bathroom. White paint was used throughout, and the undergraduates used to refer to them as 'The White City'. The contents were hideous. Among all the objects there was not one that paid the remotest tribute to the human desire for something pleasing to look at. . . . I do not think he realized how ugly everything was; he just had no visual feeling at all. . . . In his dining-room he had two large oil paintings of Edwardian or late Victorian date. One represented two nude figures reposing in lush grass, the other two large-sized kittens playing round a box. I sometimes chaffed him about them.

'Prof,' I said, 'you are an important person in this ancient university, and you just cannot have these daubs hanging in your dining-room; they bring discredit on the whole of Christ Church.' To which he invariably replied somewhat acidly, 'I suppose that you want Picasso'. . . . He thought that . . . supposed differences in artistic quality were just a lot of mumbo-jumbo, devised by the alleged experts, to make themselves important and give themselves something to do.

He had a favourite conversational trope for teasing the humanists, claiming that scientists had normally a very good general knowledge of humane matters, but pointing out some appalling item of ignorance on the part of the humanist on a most elementary scientific point. 'You do not realize,' he said, 'what abysmal ignorance that shows; it would be as though a scientist did not know. . . .' At this point his discourse took one of two lines. To a general audience he might refer to some well-known fact like the date of William the Conqueror, about which it would be absurd to suppose a scientist ignorant. In more sophisticated company he would say, 'it would be as though a scientist did not know the significance of the battle of Tours'. No doubt he had it in mind that the humanist himself might not be so sure about that battle.

R. F. Harrod, *The Prof*, 1959.

Apart from his failure in interior decoration, Lindemann lived in great comfort. His personal needs were attended to by his servant Harvey and an assistant, and his cars driven by a chauffeur. . . . Devoted to foreign travel, he would set out on leisurely voyages of exploration in a Rolls-Royce or a Mercedes, driven by his chauffeur and often accompanied by his friend and protégé Bolton King. He travelled in patrician comfort, his progress resembling that of some English milord in the eighteenth century. He complained in the Common Room that there was no second-class on the Golden Arrow. An astonished chorus of voices protested:

'Oh, but Prof, you surely never travel second-class?'

'No, but I mean that one had to have one's servant with one.'

Earl of Birkenhead, *The Prof in Two Worlds*, 1961.

'The Prof' was to become Churchill's scientific adviser in the Second World War, and was thought by some to be a visionary saviour of his country, and by others to be a bloodthirsty villain.

THE CARRUTHERS SYNDROME

As the social structure of England changed, so Oxford increasingly debated how best to reform its own entry, still predominantly upper-crust. Everybody but a lunatic core wished to see the University opened to people from all social backgrounds: a more intractable problem concerned someone we may call, borrowing the name from Ronald Knox, Carruthers—the agreeable, stalwart, dependable, salt-of-the-earth young English gentleman who was not, as it happened, very clever. Was there a place for him in a truly competitive University? In these miscellaneous extracts we observe Carruthers both in the theory and in the flesh:

CARRUTHERS AND THE PUNDITS

A workman ought to have a vulgar prejudice against Oxford . . . the business of Oxford is to make a few scholars and a great many gentlemen.

> Bernard Shaw, refusing an invitation to lecture at Ruskin College, 1899.

It is as desirable that Oxford should educate the future country squire, or nobleman, or banker, or member of Parliament, or even the guardsman, as it is that it should sharpen the wits of the schoolmaster or the cultivated artisan.

> George Nathaniel Curzon, *Principles and Methods of University Reform*, 1909.

CARRUTHERS BORN TO RULE

By the time of Elizabeth Oxford appears to have learned the lesson of the Norman Conquest, that Englishmen need governors. It endeavoured to supply them on national lines by becoming the nursery of the rich. It has succeeded ever since to a certain extent in subjecting the sons of the rich to a more or less voluntary discipline as a preparation for those places on the quarter-deck of the vessel of state which it inculcates, by means chiefly of an appeal to experience, that they have every right to expect. . . . This is the Oxford system to dispossess or disinherit which will require the most strenuous and unremitting labour on the part of the ages.

> Thomas Seccombe, *In Praise of Oxford*, 1910.

CARRUTHERS THE SCHOLAR

I have got quite recently *one* pupil. . . . I questioned him about his classical reading, and our dialogue went something like this:

Self: Well S., what Greek authors have you been reading?

S. (cheerfully): I can never remember. Try a few names and I'll see if I get on to any.

Self (*a little damped*): Have you read any Euripides?

S.: No.

Self: Any Sophocles?

S.: Oh yes.

Self: What plays of his have you read?

S. (*after a pause*): Well—the *Alcestis*.

Self (*apologetically*): But isn't that by Euripides?

S. (*with the genial surprise of a man who finds £1 where he thought there was only a 10/- note*): Really. Is it now? Then by Jove I *have* read some Euripides.

. . . However, he is one of the cheeriest, healthiest, and most perfectly contented creatures I have ever met with.

C. S. Lewis, Letters, 1923.

CARRUTHERS THE SPORTSMAN

Lionel Hedges had come up from Tonbridge [to Trinity College] with a tremendous reputation as a schoolboy cricketer, having already played for Kent. A seedy looking middle-aged gentleman called on him one morning of a match. Imagining him to be a reporter, Lionel said to him brusquely, 'I have nothing to say to you.' The man tried to expostulate but Lionel repeated, 'I have nothing to say to you.' It only afterwards transpired that the seedy man was not a reporter but his tutor, with whom he was not otherwise acquainted.

Christopher Hollis, Oxford in the Twenties, 1976.

ORNAMENTAL CARRUTHERS

Mordaunt: That blasted gramophone again! I wish somebody would tell me what you ought to do with a man like Carruthers. . . . He's quite charming, adequately athletic, and has none of the squalid vices. But I'm supposed to teach him, you see; and he's got the sort of mind you just can't establish any contact with. . . . Ought we to take on those people at all, when they're just nature's thirds?

Beith: That's the examination system; thank God, my men are all doing research. In my department, the whole thing works itself out quite simply; if a man's a dud, you spot it in less than a term, and you just tell him to go and read something else. . . . Education's nonsense, I've always said, unless there's some kind of bond between tutor and pupil.

Massingham: I haven't any bond at all with my pupils. If they seem hopeless, I'm very, very rude to them, and tell them they will plough. Which makes them so angry that they quite often pass, simply to spite me. I think kindness is wasted on the young. . . . No, but seriously, Beith, if Carrutherses are prepared to pay for the privilege of being uneducated here, why shouldn't we let them? All these ornamental young men form a kind of puddle, in which you and your friends can make a culture of pure students. They don't do much harm, and when they've finished playing here they can go out like good little boys and govern the Empire.

Ronald Knox, *Let Dons Delight*, 1939.

❦ *Carruthers survived at Oxford until the Second World War, which he won.*

SPARED

At home, as in no other city, here
summer holds her breath in a dark street
the trees nocturnally scented, lovers like moths
go by silently on the footpaths
and spirits of the young wait
cannot be expelled, multiply each year.

In the meadows, walks, over the walls
the sunlight, far-travelled, tired and content,
warms the recollections of old men, touching
the hand of the scholar on his book, marching
through quadrangles and arches, at last spent
it leans through the stained windows and falls.

This then is the city of young men, of beginning,
ideas, trials, pardonable follies,
the lightness, seriousness and sorrow of youth.
And the city of the old, looking for truth,
browsing for years, the mind's seven bellies
filled, become legendary figures, seeming

stones of the city, her venerable towers;
dignified, clothed by erudition and time.
For them it is not a city but an existence;
outside which everything is a pretence:
within, the leisurely immortals dream,
venerated and spared by the ominous hours.

Keith Douglas, Merton College, killed in action, 1944.

❦ *Douglas's poem, 'Oxford', was written in 1941, and only the spirits of the young were there. Oxford had gone to war again, this time in no spirit of heroic exaltation, like Patrick Shaw-Stewart (see page 335), nor with any mystic sense of elegy, but in the certain knowledge that the world,*

England, and the University of Oxford would never be the same again. Once more the young men went off to fight—of the Trinity boat crew which won the Eights Week races in 1939, all but two died—and once more the college masons prepared their chapel walls for the memorial slabs. This time, though, the issues were starker, and it seemed to some that this might really be the end of Oxford's long, tangled, sometimes silly and sometimes splendid story. Harold Nicolson, the writer and politician, who loved Oxford so much that just the name of the place on a marmalade pot gave him a frisson of loyalty, came to Oxford in June 1942, the very middle of the war, and recorded his thoughts in his diary that night:

I dine at All Souls, where there are only three Fellows present. We have coffee afterwards in the quad, and the sun sinks gently over St. Mary's and the Bodleian. I gaze with love at those dear buildings, wondering whether they will be assailed by one of the Baedeker raids, and whether I shall ever see them again. . . . I return in the lovely warm night to the Randolph. Dear Oxford.

But Oxford survived, and the spire of St. Mary's church, where Universitas Oxoniensis *had come into being eight centuries before, was spared by the ominous hours to welcome the fighting students home from victory, and to beckon their numberless successors into the old mysteries.*

DOMINUS ILLUMINATIO MEA

An Oxford Glossary

Obsolete usages are marked with an asterisk (*).

Act, The*: old name for Encaenia, the annual ceremony at which honorary degrees are bestowed.

Austins*: category of disputation, originally conducted at the convent of the Augustinian friars.

Battels: college fees and expenses, origin unknown but endlessly debated.

Blue: award of colours for representing the University at a major sport—half-blue for lesser recreations like cycling or archery.

B.N.C.: Brasenose College.

Bocardo*: prison formerly over Oxford's North Gate, named after a particularly baffling kind of syllogism (or possibly vice versa).

Broad, The: Broad Street.

Bulldogs: University (plain clothes) policemen.

Bumping Race: kind of rowing race devised for Oxford's narrow river, in which each boat tries to bump the one ahead.

Chancellor: honorific head of University, elected for life by Convocation.

Chest, University: University treasury, named after medieval charity chests.

Christ Church [sic]—never Christ Church College, though often The House: often written Ch: Ch:

Class: category of honours degree—first, second, third, or formerly fourth.

Come Up: take up residence at the University, hence 'Is he up yet?', or 'Are you staying up?'

Commoner: member of a college who is not a Scholar.

Congregation: legislative body of the University, composed mainly of academic and administrative staff.

GLOSSARY

Convocation: theoretical assembly of all Oxford M.A.s, whose only functions are to elect the Chancellor and the Professor of Poetry. Members must vote in person, on the spot.

Corn, The: Cornmarket Street.

Don: University teacher, from the Latin *dominus*.

Eights Week: Inter-college rowing regatta, every June.

Encaenia: annual ceremony at which honorary degrees are bestowed.

Frideswide, Saint: otherwise misty patron saint of Oxford.

Gated: confined to college for a misdemeanour.

Gentleman-Commoner*: category of undergraduate confined to those of supposedly aristocratic birth.

Giler*: St. Giles' street.

Go Down: leave the University, either temporarily or for good, hence 'sent down' = expelled.

Greats: second and final part of classical honours course.

Hebdomadal Council: administrative council of the University, meeting weekly in term time.

High, The: High Street.

House, The: Christ Church.

Isis: poetical name for Thames at Oxford.

Littlego*: slang for Moderations, one of the examinations for a bachelor's degree.

Manciple: college functionary responsible for food.

Matriculate: have one's name entered on books of a college, from Latin *matricula*, rolls.

Meadows, The: properly Christ Church meadow, a fenny expanse of land between that college and the river.

Moderations: 'Mods', in some subjects the first of the two examinations for a bachelor's degree.

New College: [sic]—never abbreviated to 'New'.

Newdigate Prize: principal University prize for poetry, awarded annually since 1806 for a poem on a set theme.

Oak, Sporting the: closing the outer door of a college room, to express a desire for privacy.

Other Place, The*: Cambridge.

GLOSSARY

O.U.D.S.: Oxford University Dramatic Society, pronounced 'Owds'.

Parks, The: [sic]—never in the singular: the University park, named from the gun parks established there during the Civil War.

*Ploughed**: failed in an examination.

*Plucked**: failed in another examination.

Preliminaries: 'Prelims'; in some subjects the first of the two examinations for a bachelor's degree.

Proctors: University officers, with disciplinary powers, elected each year in rotation from the colleges, and first heard of in 1248.

Randolph, The: principal Oxford hotel, on the corner of Beaumont Street and St. Giles'.

Responsions: University entrance examination.

Rusticated: temporarily expelled from the University.

St. Edmund Hall: *sic*—not St. Edmund's, though sometimes Teddy Hall.

Sconcing: jovial student rite entailing compulsory over-drinking for the commission of social solecisms.

Scout: college servant.

Screwing (*in this sense): fastening the outer doors of college rooms from the outside, as a practical joke.

*Servitor**: category of undergraduate who worked their way through college by serving richer fellow-students.

Student: of Christ Church, a Fellow; of any other college, a student.

*Terrae-Filius**: licensed jester of University ceremonies.

*Theatre, The**: the Sheldonian, not a theatre really, though symbolically based upon a classical amphitheatre, but a ceremonial assembly hall.

Torpids: college rowing races, in February or March, for less advanced oarsmen.

*Tufts**: gentleman-commoners, from the gold tassels they wore on their caps, hence 'tuft-hunters'.

Union, The: University debating society and attendant club, not to be confused with a more recent institution, the Student Union.

Vice-Chancellor: chief administrative officer of the University, appointed by Congregation for four-year term.

Visitor: formal, non-resident, head of college.

Viva voce: oral examination, 'viva' for short.

Acknowledgements

The editor and publishers gratefully acknowledge permission to reproduce copyright material in this book.

Sir Harold Acton: from *Memoirs of an Aesthete* (Methuen & Co. Ltd.). Reprinted by permission of David Higham Associates Ltd., on behalf of the author.

The *Architectural Review*: from an editorial in Volume 65, 1929. Reprinted by permission of the Architectural Press Ltd.

W. H. Auden: from 'The Orators', *The English Auden*, edited by Edward Mendelson. Reprinted by permission of Faber & Faber Ltd., and Random House, Inc.

Sir Max Beerbohm: from *Zuleika Dobson* (Copyright 1911 by Dodd, Mead & Company, copyright renewed 1938 by Max Beerbohm). Reprinted by permission of William Heinemann Ltd., and Dodd, Mead & Company, Inc.

Hilaire Belloc: 'The Dodo' from *Cautionary Verses*. Reprinted by permission of Gerald Duckworth & Co. Ltd. 'The Freshman's Vision' and 'The Exile' are extracts from the 'Dedicatory Ode' in *Lambkins Remains*; 'O for Oxford' from *A Moral Alphabet*; 'Lines to a Don' from *Verses* (1910); and 'Balliol Men' from *To the Balliol Men Still in Africa*. All published by Gerald Duckworth & Co. Ltd., and reprinted by permission of A. D. Peters & Co. Ltd.

Frank Benson: from *My Memoirs*. Reprinted by permission of Ernest Benn Ltd.

Sir John Betjeman: extract from *Summoned by Bells*, and poems from *The Collected Poems*, reprinted by permission of John Murray (Publishers) Ltd., and Houghton Mifflin Company. Extract from Sir John Betjeman's contribution to *My Oxford*, edited by Ann Thwaite, by permission of Robson Books Ltd., Publishers. Extracts from *An Oxford University Chest* (John Miles, Publisher, Ltd.) reprinted by permission of Curtis Brown Ltd.

Second Earl of Birkenhead: extract from *F.E.* (Eyre & Spottiswoode Ltd.) reprinted by permission of Hughes Massie Ltd. Extract from *The Professor in Two Worlds*, reprinted by permission of Collins Publishers.

C. M. Bowra: from *Memories*. Reprinted by permission of George Weidenfeld & Nicholson Ltd.

ACKNOWLEDGEMENTS

Robert Bridges: from 'An Invitation to the Oxford Pageant, July 1907', *The Poetical Works of Robert Bridges*. Reprinted by permission of Oxford University Press.

E. F. Carritt: from *Fifty Years a Don* (unpublished). Reprinted from a copy donated to the Central Library, Oxford, dated 1960, by permission of Michael J. Carritt.

James Childers: from *Laurel and Straw* (D. Appleton & Co.).

W. S. Churchill: from *Lord Randolph Churchill* (Odhams Press). Reprinted by permission of The Hamlyn Publishing Group Ltd.

Earl Curzon of Kedleston: from *Principles and Methods of University Reform*. Reprinted by permission of Oxford University Press.

Roger Dataller: from *A Pitman Looks at Oxford* (J. M. Dent & Sons, Ltd.). Reprinted by permission of Curtis Brown Ltd.

Keith Douglas: 'Oxford' from *The Complete Poems of Keith Douglas* (Oxford University Press). Reprinted by permission of J. C. Hall and Marie J. Douglas.

The *Daily Express*: from an editorial on 13 February 1933. Reprinted by kind permission of the *Daily Express*.

L. R. Farnell: from *An Oxonian Looks Back*, published by Martin Hopkinson. Reprinted by permission of The Bodley Head.

John Galsworthy: from *The End of a Chapter*. Reprinted by permission of William Heinemann Ltd., and Charles Scribner's Sons.

Victor Gollancz: from *My Dear Timothy* (Copyright © 1952 by Victor Gollancz Ltd.). Reprinted by permission of Victor Gollancz Ltd., and Simon & Schuster, a Division of Gulf and Western Corporation.

Graham Greene: from *A Sort of Life* (Copyright © 1971 by Graham Greene)(The Bodley Head Ltd.). Reprinted by permission of Laurence Pollinger Ltd., on behalf of the author, and Simon and Schuster, a Division of Gulf and Western Corporation.

The *Guardian*: from an editorial in the *Manchester Guardian* of 13 February 1933. Reprinted by permission of Guardian Newspapers Ltd.

Thomas Hardy: from *Jude the Obscure*. Reprinted by permission of the Trustees of the Hardy Estate and Macmillan London and Basingstoke.

Tom Harrisson: from *A Letter to Oxford* (The Hate Press). Reprinted by permission of the Executors of Tom Harrisson.

R. F. Harrod: from *The Professor*. Reprinted by permission of Macmillan London and Basingstoke.

William Hayter: from *Spooner*. Reprinted by permission of W. H. Allen & Co. Ltd., as publishers.

Cecil Headlam: from *Oxford and its Story* from the Medieval Towns Series. Reprinted by permission of J. M. Dent & Sons Ltd., publishers.

ACKNOWLEDGEMENTS

Vyvyan Holland: from *Son of Oscar Wilde* (Rupert Hart-Davis Ltd.). Reprinted by permission of Curtis Brown Ltd., on behalf of the Estate of Vyvyan Holland.

Christopher Hollis: from *Oxford in the Twenties*. Reprinted by permission of William Heinemann Ltd., as publishers.

Aldous Huxley: from *Oxford Poetry 1914–16* (Blackwells). Reprinted by permission of Chatto & Windus Ltd., on behalf of Mrs. Laura Huxley.

Isis: from an editorial dated 15 February 1933. Letter dated 23 February 1933 from Joseph Banister. Reprinted by permission of *Isis, The Oxford University Magazine*.

L. E. Jones: from *An Edwardian Youth*. Reprinted by permission of Macmillan London and Basingstoke.

D. F. Karaka: from *The Pulse of Oxford* (J. M. Dent & Sons Ltd.).

G. W. Kitchin: from *John Ruskin at Oxford*. Reprinted by permission of John Murray (Publishers) Ltd.

Ronald Knox: from *Let Dons Delight* (Sheed & Ward Ltd.). Reprinted by permission of A. P. Watt & Son, on behalf of the Estate of the late Ronald Knox.

Osbert Lancaster: from *Maurice Bowra: A Celebration*, edited by Hugh Lloyd-Jones. Reprinted by permission of Gerald Duckworth & Co. Ltd.

D. H. Lawrence: from *The Complete Poems of D. H. Lawrence*, edited by V. de Sola Pinto and F. Warren Roberts (Copyright © 1964, 1971 by Angelo Ravagli and C. M. Weekly). (William Heinemann Ltd.). Reprinted by permission of Laurence Pollinger Ltd., the Estate of the late Mrs. Frieda Lawrence Ravagli, and of The Viking Press, Inc.

Stephen Leacock: from *My Discovery of England* (Copyright 1922 by Dodd, Mead & Company, Inc. Copyright renewed 1949 by George Leacock). Reprinted by permission of Dodd, Mead & Company, Inc., and the Canadian Publishers, McClelland & Stewart Ltd., Toronto. Reprinted also from *The Bodley Head Leacock* by permission of The Bodley Head.

C. S. Lewis: from *Letters of C. S. Lewis* (Copyright © 1966 by W. H. Lewis and Executors of C. S. Lewis). Reprinted by permission of Harcourt Brace Jovanovich, Inc., and Curtis Brown Ltd., on behalf of the Estate of C. S. Lewis.

Sir Compton Mackenzie: from *Sinister Street*. Reprinted by permission of Macdonald and Jane's Publishers Ltd.

Harold Macmillan: from an article in *The Times*, 7 September 1977. Reproduced from *The Times* by permission.

Louis MacNeice: from 'Autumn Journal', *The Collected Poems of Louis MacNeice*, edited by E. R. Dodds (Copyright © The Estate of Louis MacNeice, 1966). Reprinted by permission of Faber & Faber Ltd., and Oxford University Press, Inc.

ACKNOWLEDGEMENTS

J. C. Masterman: from *To Teach the Senators Wisdom*. Reprinted by permission of Curtis Brown Ltd., on behalf of the author.

F. Meyrick: from *Memories*. Reprinted by permission of John Murray (Publishers) Ltd.

James Morris: from *Oxford*. Reprinted by permission of Faber & Faber Ltd.

Dorothy Newcomen: from an article in *The Ship*, December 1937. Reprinted by permission of the St. Anne's College Association of Senior Members.

Harold Nicolson: extract from *Diaries and Letters: 1939–1945*. Reprinted by permission of Collins Publishers.

Alfred Noyes: from *Forty Singing Seamen*. Reprinted by permission of the Estate of the late Dr. Alfred Noyes and William Blackwood & Sons Ltd.

Sir Charles Oman: from *Memories of Victorian Oxford* (Methuen & Co. Ltd.). Reprinted by permission of Associated Book Publishers Ltd. Extracts from the *All Souls Betting Book 1873–1919*, edited by Charles Oman in 1938, are reprinted by permission of All Souls College, Oxford, and the executors of Sir Charles's Estate.

G. R. Parkin: from *The Rhodes Scholarships*. Reprinted by permission of Houghton Mifflin Company, and Constable & Co. Ltd.

Anthony Powell: from *Maurice Bowra: A Celebration*, edited by Hugh Lloyd-Jones. Reprinted by permission of Gerald Duckworth & Co. Ltd.

Dilys Powell: from an article in *Isis*, 28 May 1924. Reprinted by permission of *Isis, The Oxford University Magazine*.

W. H. and W. J. C. Quarrell: from *Oxford in 1710* (Basil Blackwell, Oxford), translated from the original by Z. C. von Uffenbach.

Sir Arthur Quiller-Couch: From 'Alma Mater', *The Vigil of Venus and Other Poems*. Reprinted by permission of J. M. Dent & Sons Ltd., and Miss Foy Quiller-Couch.

The Viscountess Rhondda: from *This Was My World*. Reprinted by permission of Curtis Brown Ltd., on behalf of the author.

L. Rice-Oxley: from *Oxford Renowned* (Methuen & Co. Ltd.). Reprinted by permission of Associated Book Publishers Ltd.

M. R. Ridley: from 'Oxford' (Newdigate Prize 1913, published by Blackwells). Reprinted by permission of Mrs. J. E. L. Ridley.

Annie Rogers: from *Degrees by Degrees*. Reprinted by permission of Oxford University Press.

George Santayana: from *Soliloquies in England*. Reprinted by permission of Constable & Co. Ltd.

ACKNOWLEDGEMENTS

Dorothy L. Sayers: from *Gaudy Night* (Victor Gollancz Ltd.). Reprinted by permission of David Higham Associates Ltd., on behalf of the author.

Thomas Seccombe: from *In Praise of Oxford*. Reprinted by permission of Constable & Co. Ltd., as publishers.

George Bernard Shaw: from an article in *The Saturday Review*, 1898. Reprinted by permission of The Society of Authors on behalf of the Bernard Shaw Estate. Quotation from a letter to Dennis Hird, Principal of Ruskin College, from *The Story of Ruskin College*. Reprinted by permission of the Principal of Ruskin College.

J. G. Sinclair: from *Portrait of Oxford* (Veracity Press).

J. A. Spender: from an article entitled 'Balliol in the 80s' from the magazine *Oxford*, Spring 1936. Reprinted by permission of *Oxford*.

Stephen Spender: from *World Within Worlds*. Reprinted by permission of Hamish Hamilton Ltd., as publishers, and A. D. Peters & Co. Ltd.

J. I. M. Stewart: from J. I. M. Stewart's contribution to *My Oxford*, edited by Ann Thwaite. Reprinted by permission of Robson Books Ltd., Publishers.

Charles Tennyson: from *Cambridge From Within*. Reprinted by permission of Chatto & Windus Ltd., and the author.

W. Tuckwell: from *Reminiscences of Oxford*. Reprinted by permission of John Murray (Publishers) Ltd.

Arthur Waugh: from *One Man's Road* (Chapman & Hall Ltd.). Reprinted by permission of Associated Book Publishers Ltd., and Auberon Waugh.

Evelyn Waugh: notices from *Isis*, reprinted by permission of *Isis*, *The Oxford University Magazine*. Extracts from *A Little Learning*, and *Brideshead Revisited* (Chapman & Hall Ltd.). Reprinted by permission of A. D. Peters & Co. Ltd., on behalf of the Estate of Laura Waugh and the Beneficiaries of the E. Waugh Settlement.

L. Woolley: from *Dead Towns and Living Men*. Reprinted by permission of Oxford University Press.

W. B. Yeats: from 'All Souls' Night', *The Collected Poems of W. B. Yeats* (Macmillan London and Basingstoke). Reprinted by permission of A. P. Watt & Son on behalf of M. B. Yeats, and Miss Anne Yeats. Reprinted also by permission of Macmillan Publishing Co., Inc., from the American Edition of the *Collected Poems* (Copyright 1928 by Macmillan Publishing Co., Inc., renewed 1956 by Georgie Yeats).

While every effort has been made to secure permission, it has in a few cases proved impossible to trace the author or his executor.

Index

1. AUTHORS

2. SUBJECTS